Edited by Akira Harada

Supramolecular Polymer Chemistry

Related Titles

Schalley, C. A. (ed.)
Analytical Methods in Supramolecular Chemistry
Second Edition
2012
ISBN: 978-3-527-31505-5

Urban, M. W. (Ed.)
Handbook of Stimuli-Responsive Materials
2011
ISBN: 978-3-527-32700-3

Samori, P., Cacialli, F. (Eds.)
Functional Supramolecular Architectures
for Organic Electronics and Nanotechnology
2011
ISBN: 978-3-527-32611-2

Sauvage, J.-P., Gaspard, P. (Eds.)
From Non-Covalent Assemblies to Molecular Machines
2011
ISBN: 978-3-527-32277-0

Atwood, J. L., Steed, J. W. (Eds.)
Organic Nanostructures
2008
ISBN: 978-3-527-31836-0

van Leeuwen, P. W. N. M. (Ed.)
Supramolecular Catalysis
2008
ISBN: 978-3-527-32191-9

Diederich, F., Stang, P. J., Tykwinski, R. R. (Eds.)
Modern Supramolecular Chemistry
Strategies for Macrocycle Synthesis
2008
ISBN: 978-3-527-31826-1

Edited by Akira Harada

Supramolecular Polymer Chemistry

WILEY-VCH Verlag GmbH & Co. KGaA

The Editor

Prof. Akira Harada
Osaka University
Department of Macromolecular Science
Graduate School of Science
1-1 Machikaneyama-cho, Toyonaka
Osaka 560-0043
Japan

Cover
The graphic material used in the cover illustration was kindly provided by the editor Prof. Akira Harada.

■ All books published by **Wiley-VCH** are carefully produced. Nevertheless, authors, editors, and publisher do not warrant the information contained in these books, including this book, to be free of errors. Readers are advised to keep in mind that statements, data, illustrations, procedural details or other items may inadvertently be inaccurate.

Library of Congress Card No.: applied for

British Library Cataloguing-in-Publication Data
A catalogue record for this book is available from the British Library.

Bibliographic information published by the Deutsche Nationalbibliothek
The Deutsche Nationalbibliothek lists this publication in the Deutsche Nationalbibliografie; detailed bibliographic data are available on the Internet at http://dnb.d-nb.de.

© 2012 Wiley-VCH Verlag & Co. KGaA, Boschstr. 12, 69469 Weinheim, Germany

All rights reserved (including those of translation into other languages). No part of this book may be reproduced in any form – by photoprinting, microfilm, or any other means – nor transmitted or translated into a machine language without written permission from the publishers. Registered names, trademarks, etc. used in this book, even when not specifically marked as such, are not to be considered unprotected by law.

Cover Design   Adam Design, Weinheim
Typesetting   Thomson Digital, Noida, India
Printing and Binding   Fabulous Printers Pte Ltd, Singapore

Printed in Singapore
Printed on acid-free paper

Print ISBN: 978-3-527-32321-0
ePDF ISBN: 978-3-527-63980-9
ePub ISBN: 978-3-527-63979-3
Mobi ISBN: 978-3-527-63981-6
oBook ISBN: 978-3-527-63978-6

Contents

Preface *XIII*
List of Contributors *XV*

Part One Formation of Supramolecular Polymers *1*

1 Multiple Hydrogen-Bonded Supramolecular Polymers *3*
Wilco P.J. Appel, Marko M.L. Nieuwenhuizen, and E.W. Meijer
1.1   Introduction *3*
1.1.1 Historical Background *3*
1.1.2 Supramolecular Chemistry *4*
1.1.3 Supramolecular Polymerization Mechanisms *4*
1.2   General Concepts of Hydrogen-Bonding Motifs *6*
1.2.1 Arrays of Multiple Hydrogen Bonds *6*
1.2.2 Preorganization through Intramolecular Hydrogen Bonding *8*
1.2.3 Tautomeric Equilibria *9*
1.3   Hydrogen-Bonded Main-Chain Supramolecular Polymers *10*
1.3.1 The Establishment of Supramolecular Polymers *10*
1.3.2 Supramolecular Polymerizations *13*
1.3.3 Hydrophobic Compartmentalization *14*
1.4   From Supramolecular Polymers to Supramolecular Materials *16*
1.4.1 Thermoplastic Elastomers *16*
1.4.2 Phase Separation and Additional Lateral Interactions in Supramolecular Polymers in the Solid State *18*
1.4.3 Supramolecular Thermoplastic Elastomers Based on Additional Lateral Interactions and Phase Separation *19*
1.5   Future Perspectives *23*
      References *25*

2 Cyclodextrin-Based Supramolecular Polymers *29*
Akira Harada and Yoshinori Takashima
2.1   Introduction *29*
2.2   Supramolecular Polymers in the Solid State *29*
2.2.1 Crystal Structures of CD Aliphatic Tethers *30*
2.2.2 Crystal Structures of β-CDs Aromatic Tethers *31*

2.3 Formation of Homo-Intramolecular and Intermolecular Complexes by CDs–Guest Conjugates 33
2.3.1 Supramolecular Structures Formed by 6-Modified α-CDs 33
2.3.2 Supramolecular Structures Formed by 6-Modified β-CDs 39
2.3.3 Supramolecular Structures Formed by 3-Modified α-CDs 40
2.3.4 Hetero-Supramolecular Structures Formed by Modified CDs 42
2.4 Formation of Intermolecular Complexes by CD and Guest Dimers 44
2.5 Artificial Molecular Muscle Based on c2-Daisy Chain 45
2.6 Conclusion and Outlook 48
References 48

3 Supra-Macromolecular Chemistry: Toward Design of New Organic Materials from Supramolecular Standpoints 51
Kazunori Sugiyasu and Seiji Shinkai
3.1 Introduction 51
3.2 Small Molecules, Macromolecules, and Supramolecules: Design of their Composite Materials 53
3.2.1 Interactions between Small Molecules and Macromolecules 53
3.2.2 Interactions between Small Molecules and Molecular Assemblies 56
3.2.3 Interactions between Molecular Assemblies 58
3.2.4 Interactions between Macromolecules 60
3.2.5 Interactions between Macromolecular Assemblies 63
3.2.6 Interactions between Macromolecules and Molecular Assemblies 65
3.3 Conclusion and Outlook 67
References 68

4 Polymerization with Ditopic Cavitand Monomers 71
Francesca Tancini and Enrico Dalcanale
4.1 Introduction 71
4.2 Cavitands 72
4.3 Self-Assembly of Ditopic Cavitand Monomers 75
4.3.1 Structural Monomer Classification of Supramolecular Polymerization 75
4.3.2 Homoditopic Cavitands Self-Assembled via Solvophobic π-π Stacking Interactions 77
4.3.3 Heteroditopic Cavitands Combining Solvophobic Interactions and Metal–Ligand Coordination 78
4.3.4 Heteroditopic Cavitands Combining Solvophobic Interactions and Hydrogen Bonding 82
4.3.5 Heteroditopic Cavitands Self-assembled via Host–Guest Interactions 84
4.3.6 Homoditopic Cavitands Self-assembled via Host–Guest Interactions 88
4.4 Conclusions and Outlook 91
References 92

Part Two	**Supramolecular Polymers with Unique Structures** 95

5	**Polymers Containing Covalently Bonded and Supramolecularly Attached Cyclodextrins as Side Groups** 97
	Helmut Ritter, Monir Tabatabai, and Bernd-Kristof Müller
5.1	Polymers with Covalently Bonded Cyclodextrins as Side Groups 97
5.1.1	Synthesis and Polymerization of Monofunctional Cyclodextrin Monomers 98
5.1.2	Polymer-Analogous Reaction with Monofunctional Cyclodextrin 100
5.1.3	Structure–Property Relationship of Polymers Containing Cyclodextrins as Side Group 102
5.2	Side Chain Polyrotaxanes and Poly*pseudo*rotaxanes 105
5.2.1	Side Chain Polyrotaxanes 106
5.2.2	Side Chain Polypseudorotaxane (Polymer (Polyaxis)/Cyclodextrin (Rotor)) 111
	References 120

6	**Antibody Dendrimers and DNA Catenanes** 127
	Hiroyasu Yamaguchi and Akira Harada
6.1	Molecular Recognition in Biological Systems 127
6.1.1	Supramolecular Complex Formation of Antibodies 127
6.1.2	Supramolecular Complexes Prepared by DNAs 129
6.1.3	Observation of Topological Structures of Supramolecular Complexes by Atomic Force Microscopy (AFM) 129
6.2	Antibody Supramolecules 130
6.2.1	Structural Properties of Individual Antibody Molecules 130
6.2.2	Supramolecular Formation of Antibodies with Multivalent Antigens 130
6.2.2.1	Supramolecular Formation of Antibodies with Divalent Antigens 131
6.2.2.2	Direct Observation of Supramolecular Complexes of Antibodies with Porphyrin Dimers 133
6.2.2.3	Applications for the Highly Sensitive Detection Method of Small Molecules by the Supramolecular Complexes between Antibodies and Multivalent Antigens 134
6.2.3	Supramolecular Dendrimers Constructed by IgM and Chemically Modified IgG 136
6.2.3.1	Preparation of Antibody Dendrimers and their Topological Structures 136
6.2.3.2	Binding Properties of Antibody Dendrimers for Antigens 136
6.3	DNA Supramolecules 139
6.3.1	Imaging of Individual Plasmid DNA Molecules 139
6.3.2	Preparation of Nicked DNA by the Addition of DNase I to Plasmid DNA 140
6.3.3	Catenation Reaction with Topoisomerase I 141

6.3.4		AFM Images of DNA Catenanes *143*	
6.3.5		DNA [n]Catenanes Prepared by Irreversible Reaction with DNA Ligase *144*	
6.4		Conclusions *145*	
		References *146*	

7 Crown Ether-Based Polymeric Rotaxanes *151*
Terry L. Price Jr. and Harry W. Gibson

- 7.1 Introduction *151*
- 7.2 Daisy Chains *153*
- 7.3 Supramolecular Polymers *156*
- 7.4 Dendritic Rotaxanes *157*
- 7.5 Dendronized Polymers *158*
- 7.6 Main chain Rotaxanes Based on Polymeric Crowns (Including Crosslinked Systems) *161*
- 7.7 Side Chain Rotaxanes Based on Pendent Crowns *166*
- 7.8 Poly[2]rotaxanes *170*
- 7.9 Poly[3]rotaxanes *173*
- 7.10 Polymeric End Group Pseudorotaxanes *176*
- 7.11 Chain Extension and Block Copolymers from End Groups *176*
- 7.12 Star Polymers from Crown Functionalized Polymers *179*
  References *181*

Part Three Properties and Functions *183*

8 Processive Rotaxane Catalysts *185*
Johannes A.A.W. Elemans, Alan E. Rowan, and Roeland J.M. Nolte

- 8.1 Introduction *185*
- 8.2 Results and Discussion *185*
- 8.2.1 Catalysis *185*
- 8.2.2 Threading *187*
- 8.3 Conclusion *192*
  References *192*

9 Emerging Biomedical Functions through 'Mobile' Polyrotaxanes *195*
Nobuhiko Yui

- 9.1 Introduction *195*
- 9.2 Multivalent Interaction using Ligand-Conjugated Polyrotaxanes *196*
- 9.3 The Formation of Polyrotaxane Loops as a Dynamic Interface *197*
- 9.4 Cytocleavable Polyrotaxanes for Gene Delivery *199*
- 9.5 Conclusion *201*
- 9.6 Appendix *203*
  References *204*

10	**Slide-Ring Materials Using Polyrotaxane** *205*	
	Kazuaki Kato and Kohzo Ito	
10.1	Introduction *205*	
10.2	Pulley Effect of Slide-Ring Materials *208*	
10.3	Synthesis of Slide-Ring Materials *209*	
10.4	Scattering Studies of Slide-Ring Gels *211*	
10.5	Mechanical Properties of Slide-Ring Gels *213*	
10.6	Sliding Graft Copolymers *215*	
10.7	Recent Trends of Slide-Ring Materials *216*	
10.7.1	Introduction: Diversification of the Main Chain Polymer *216*	
10.7.2	Organic–Inorganic Hybrid Slide-Ring Materials *219*	
10.7.3	Design of Materials from Intramolecular Dynamics of Polyrotaxanes *224*	
10.8	Concluding Remarks *226*	
	References *227*	
11	**Stimuli-Responsive Systems** *231*	
	Akihito Hashidzume and Akira Harada	
11.1	Introduction *231*	
11.2	Stimuli and Responses *231*	
11.2.1	Stimuli *231*	
11.2.1.1	Temperature *231*	
11.2.1.2	Pressure, Force, Stress, and Ultrasound *232*	
11.2.1.3	pH *233*	
11.2.1.4	Chemicals *233*	
11.2.1.5	Electromagnetic Waves or Light *233*	
11.2.1.6	Redox *234*	
11.2.2	Responses *234*	
11.2.2.1	Movement *235*	
11.2.2.2	Capture and Release of Chemicals *235*	
11.2.2.3	Chemical Reactions *235*	
11.2.2.4	Change in Viscoelastic Properties, or Gel-to-Sol and Sol-to-Gel Transitions *236*	
11.2.2.5	Change in Color *236*	
11.3	Examples of Stimuli-Responsive Supramolecular Polymer Systems *236*	
11.3.1	Temperature-Responsive Systems *236*	
11.3.2	Pressure-, Force-, and Sonication-Responsive Systems *239*	
11.3.3	pH-Responsive Systems *241*	
11.3.4	Chemical-Responsive Systems *246*	
11.3.5	Photo-Responsive Systems *249*	
11.3.6	Redox-Responsive Systems *255*	
11.3.7	Multi-Stimuli-Responsive Systems *259*	
11.4	Concluding Remarks *261*	
	References *261*	

12	**Physical Organic Chemistry of Supramolecular Polymers** *269*
	Stephen L. Craig and Donghua Xu
12.1	Introduction and Background *269*
12.2	Linear Supramolecular Polymers *270*
12.2.1	N,C,N-Pincer Metal Complexes *270*
12.2.2	Linear SPs *272*
12.2.3	Theory Related to the Properties of Linear SPs *274*
12.2.4	Linear SPs in the Solid State *275*
12.3	Cross-Linked SPs Networks *276*
12.3.1	Reversibility in Semidilute Unentangled SPs Networks *276*
12.3.2	Properties of Semidilute Entangled SPs Networks *283*
12.3.3	The Sticky Reptation Model *285*
12.4	Hybrid Polymer Gels *286*
12.5	Conclusion *288*
	References *288*

13	**Topological Polymer Chemistry: A Quest for Strange Polymer Rings** *293*
	Yasuyuki Tezuka
13.1	Introduction *293*
13.2	Systematic Classification of Nonlinear Polymer Topologies *293*
13.3	Topological Isomerism *296*
13.4	Designing Unusual Polymer Rings by Electrostatic Self-Assembly and Covalent Fixation *298*
13.5	Conclusion and Future Perspectives *302*
	References *303*

14	**Structure and Dynamic Behavior of Organometallic Rotaxanes** *305*
	Yuji Suzaki, Tomoko Abe, Eriko Chihara, Shintaro Murata, Masaki Horie, and Kohtaro Osakada
14.1	Introduction *305*
14.1.1	Crystals of Pseudorotaxanes *307*
14.1.2	Synthesis of Ferrocene-Containing [2]Rotaxanes by the Threading-Followed-by-End-Capping Strategy *312*
14.1.3	Dethreacting Reaction of Rotaxane-Like Complex *316*
14.1.4	Photochemical Properties of Ferrocene-Containing Rotaxanes *318*
14.1.5	Ferrocene-Containing [3]Rotaxane and Side-Chain Polyrotaxane *320*
14.1.5.1	Strategies and Synthesis of [3]Rotaxanes *320*
14.1.5.2	Strategies and Synthesis of Side-Chain Type Polyrotaxane *321*

14.2	Conclusion *324*	
14.3	Appendix: Experimental Section *324*	
	References *326*	

15 Polyrotaxane Network as a Topologically Cross-Linked Polymer: Synthesis and Properties *331*
Toshikazu Takata, Takayuki Arai, Yasuhiro Kohsaka, Masahiro Shioya, and Yasuhito Koyama

15.1	Introduction *331*
15.2	Linking of Wheels of Main-Chain-Type Polyrotaxane – Structurally Defined Polyrotaxane Network *331*
15.3	Linking of Macrocyclic Units of Polymacrocycle with Axle Unit to Directly Yield a Polyrotaxane Network *336*
15.3.1	Polyrotaxane Networks Having Crown Ethers as the Wheel at the Cross-link Points (I) *336*
15.3.2	Polyrotaxane Network Having Crown Ethers as the Wheel at the Cross-link Points (II) *337*
15.3.3	Polyrotaxane Network Having Cyclodextrins as Cross-link Points: Effective Use of Oligocyclodextrin *339*
15.4	Linking of Wheels of Polyrotaxane Cross-linker to Afford Polyrotaxane Network: Design of the Cross-linker *342*
15.5	Conclusion *344*
	References *345*

16 From Chemical Topology to Molecular Machines *347*
Jean-Pierre Sauvage

16.1	Introduction *347*
16.2	Copper(I)-Templated Synthesis of Catenanes: the 'Entwining' Approach and the 'Gathering and Threading' Strategy *347*
16.3	Molecular Knots *349*
16.4	Molecular Machines Based on Catenanes and Rotaxanes *353*
16.5	Two-Dimensional Interlocking Arrays *354*
16.6	A [3]rotaxane Acting as an Adjustable Receptor: Toward a Molecular 'Press' *355*
16.7	Conclusion *356*
	References *356*

Index *361*

Preface

The chemistry of molecular recognition began more than 50 years ago with the discovery of crown ethers as selective host molecules for alkali metal ions by Dr. Pedersen. In the last 30 years, the chemistry of molecular recognition has greatly expanded. For example, Cram *et al.* incorporated host–guest chemistry and Lehn created supramolecular chemistry. To date, numerous studies have been published on supramolecular complexes.

Moreover, in biological systems, macromolecular recognition by other macromolecules plays an important role in maintaining life (e.g., DNA duplication as well as enzyme–substrate and antigen–antibody interactions). Supramolecular polymer complexes are crucial for the construction of biological structures such as microtubules, microfilaments, and cell–cell interactions.

Synthetic supramolecular polymers have great potential in the construction of new materials with unique structures and functions, because polymers contain vast amounts of information on their main-chains and side-chains. For example, in 1990, supramolecular polymers consisting of cyclodextrins and synthetic polymers were reported. Prof. Lehn's textbook, *Supramolecular Chemistry*, which was published in 1995, mentions supramolecular polymers. Prof. Meijer and Prof. Zimmerman reported supramolecular polymers linked by multiple hydrogen bonds. Since then numerous other reports on supramolecular polymers have been published.

This book is geared toward current supramolecular polymer researchers as well as other interested individuals, including young researchers and students. Each chapter is written by experts who are actively engaged in supramolecular polymer research and have published important papers in the field.

I am honored to be a part of this project, and have eagerly anticipated receiving each chapter. They have all exceeded my expectations, and together they form a book that will become a cornerstone in the field of supramolecular polymer research and, I believe, will help to shape research in the future.

Finally, I would like to express my sincere appreciation to the authors and to all who have assisted in the preparation of this book.

Osaka *Akira Harada*
May 2011

List of Contributors

Tomoko Abe
Tokyo Institute of Technology
Chemical Resources Laboratory
R1-3, 4259 Nagatsuta, Midori-ku
Yokohama 226-8503
Japan

Wilco P.J. Appel
Eindhoven University of Technology
Institute for Complex Molecular
Systems, Laboratory of Macromolecular
and Organic Chemistry
Den Dolech 2
5612 AZ Eindhoven
The Netherlands

Takayuki Arai
Tokyo Institute of Technology
Department of Organic and Polymeric
Materials
2-12-1 (H-126), Ookayama, Meguro-ku
Tokyo 152-8552
Japan

Eriko Chihara
Tokyo Institute of Technology
Chemical Resources Laboratory
R1-3, 4259 Nagatsuta, Midori-ku
Yokohama 226-8503
Japan

Stephen L. Craig
Duke University
Center for Biologically Inspired
Materials and Material Systems
Department of Chemistry
3221 FFSC
124 Science Drive
Durham, NC 27708-0346
USA

Enrico Dalcanale
University of Parma
Department of Organic and Industrial
Chemistry
Viale G. P. Usberti 17/A
43124 Parma
Italy

Johannes A.A.W. Elemans
Radboud University Nijmegen
Cluster for Molecular Chemistry
Heyendaalseweg 135
6525 AJ Nijmegen
The Netherlands

Harry W. Gibson
Virginia Polytechnic Institute & State
University
Department of Chemistry
2105 Hahn Hall
Blacksburg, VA 24061-0001
USA

List of Contributors

Akira Harada
Osaka University
Graduate School of Science
Department of Macromolecular Science
1-1 Machikaneyama-cho, Toyonaka
Osaka 560-0043
Japan

Akihito Hashidzume
Osaka University
Graduate School of Science
Department of Macromolecular Science
1-1 Machikaneyama-cho, Toyonaka
Osaka 560-0043
Japan

Masaki Horie
National Tsing Hua University
Department of Chemical Engineering
Hsinchu, 30013
Taiwan

Kohzo Ito
The University of Tokyo
Graduate School of Frontier Sciences
Department of Advanced Materials
Science, Group of New Materials and
Interfaces
5-1-5 Kashiwanoha, Kashiwa
Chiba 277-8561
Japan

Kazuaki Kato
The University of Tokyo
Graduate School of Frontier Sciences
Department of Advanced Materials
Science, Group of New Materials and
Interfaces
5-1-5 Kashiwanoha, Kashiwa
Chiba 277-8561
Japan

Yasuhiro Kohsaka
Tokyo Institute of Technology
Department of Organic and Polymeric
Materials
2-12-1 (H-126), Ookayama, Meguro-ku
Tokyo 152-8552
Japan

Yasuhito Koyama
Tokyo Institute of Technology
Department of Organic and Polymeric
Materials
2-12-1 (H-126), Ookayama, Meguro-ku
Tokyo 152-8552
Japan

E.W. Bert Meijer
Eindhoven University of Technology
Institute for Complex Molecular
Systems, Laboratory of Macromolecular
and Organic Chemistry
Den Dolech 2
5612 AZ Eindhoven
The Netherlands

Bernd-Kristof Müller
Pharmpur GmbH
Messerschmittring 33
86343 Königsbrunn
Germany

Shintaro Murata
Tokyo Institute of Technology
Chemical Resources Laboratory
R1-3, 4259 Nagatsuta, Midori-ku
Yokohama 226-8503
Japan

List of Contributors

Marko M.L. Nieuwenhuizen
Eindhoven University of Technology
Institute for Complex Molecular
Systems, Laboratory of Macromolecular
and Organic Chemistry
Den Dolech 2
5612 AZ Eindhoven
The Netherlands

Roeland J.M. Nolte
Radboud University Nijmegen
Cluster for Molecular Chemistry
Heyendaalseweg 135
6525 AJ Nijmegen
The Netherlands

Kohtaro Osakada
Tokyo Institute of Technology
Chemical Resources Laboratory
R1-3, 4259 Nagatsuta, Midori-ku
Yokohama 226-8503
Japan

Terry L. Price Jr.
Virginia Polytechnic Institute & State
University
Department of Chemistry
2105 Hahn Hall
Blacksburg, VA 24061-0001
USA

Helmut Ritter
Heinrich Heine University
Institute of Organic and
Macromolecular Chemistry
Universitätsstr. 1
40225 Düsseldorf
Germany

Alan E. Rowan
Radboud University Nijmegen
Institute for Molecules and Materials
Heyendaalseweg 135
6525 AJ Nijmegen
The Netherlands

Jean-Pierre Sauvage
University Louis Pasteur/CNRS
Institute of Chemistry, Laboratory of
Organic – Inorganic Chemistry
Institut Le Bel, U.M.R. 7177
67070 Strasbourg-Cedex
France

Seiji Shinkai
Fukuoka, Japan and Sojo University
Institute of Systems, Information
Technologies and Nanotechnologies
(ISIT)
Kumamoto
Japan

Masahiro Shioya
Tokyo Institute of Technology
Department of Organic and Polymeric
Materials
2-12-1 (H-126), Ookayama, Meguro-ku
Tokyo 152-8552
Japan

Kazunori Sugiyasu
National Institute for Materials Science
(NIMS)
Organic Nanomaterials Center
Macromolecules Group
1-2-1 Sengen
Tsukuba 305-0047
Japan

Yuji Suzaki
Tokyo Institute of Technology
Chemical Resources Laboratory
R1-3, 4259 Nagatsuta, Midori-ku
Yokohama 226-8503
Japan

Monir Tabatabai
Heinrich Heine University
Institute of Organic and
Macromolecular Chemistry
Universitätsstr. 1
40225 Düsseldorf
Germany

Yoshinori Takashima
Osaka University
Graduate School of Science
Department of Macromolecular Science
1-1 Machikaneyama-cho, Toyonaka
Osaka 560-0043
Japan

Toshikazu Takata
Tokyo Institute of Technology
Department of Organic and
Polymeric Materials
2-12-1 (H-126), Ookayama, Meguro-ku
Tokyo 152-8552
Japan

Francesca Tancini
University of Parma
Department of Organic and Industrial
Chemistry
Viale G. P. Usberti 17/A
43124 Parma
Italy

Yasuyuki Tezuka
Tokyo Institute of Technology
Department of Organic and
Polymeric Materials
2-12-1-S8-41 Ookayama, Meguro-ku
Tokyo 152-8552
Japan

Donghua Xu
Duke University
Center for Biologically Inspired
Materials and Material Systems
Department of Chemistry
3221 FFSC
124 Science Drive
Durham, NC 27708-0346
USA

Hiroyasu Yamaguchi
Osaka University
Graduate School of Science
Department of Macromolecular Science
1-1 Machikaneyama-cho, Toyonaka
Osaka 560-0043
Japan

Nobuhiko Yui
Tokyo Medical and Dental University
Institute of Biomaterials and
Bioengineering
2-3-10, Kanda-Surugadai, Chiyoda
Tokyo 101-0062
Japan

and

JST CREST
Tokyo 102-0075
Japan

Part One
Formation of Supramolecular Polymers

1
Multiple Hydrogen-Bonded Supramolecular Polymers

Wilco P.J. Appel, Marko M.L. Nieuwenhuizen, and E.W. Meijer

1.1
Introduction

1.1.1
Historical Background

Since the introduction of the first synthetic polymer more than a hundred years ago by Leo Hendrik Baekeland, covalent polymers have become indispensable in everyday life. The term 'polymeric' was first introduced in 1832 by Jöns Jacob Berzelius to describe a compound with a higher molecular weight than that of the normal compound but with an identical empirical formula as a result of the repetition of equal units [1]. In 1920, Hermann Staudinger defined polymers, which he called macromolecules, to be multiple covalently bound monomers. For this work he was awarded with the Nobel Prize in 1953 [2]. Today, our knowledge of organic synthesis and polymer chemistry allows the preparation of virtually any monomer and its associated polymer. In addition, an in-depth understanding of 'living' types of polymerization facilitates tuning of the molecular weight and molecular weight distribution, at the same time creating the possibility to synthesize a wide variety of copolymers [3].

The macroscopic properties of polymers are directly linked to their molecular structure. As a result, polymer chemists devised synthetic approaches to control the sequence architecture. More recently, the importance of introducing supramolecular interactions between macromolecular chains has become evident, and many new options have been introduced. The final step in this development would be to develop polymers entirely based on reversible, noncovalent interactions. Rather than linking the monomers in the desired arrangement via a series of polymerization reactions, the monomers are designed in such a way that they autonomously self-assemble into the desired structure. As with covalent polymers, a variety of structures of these so-called supramolecular polymers are possible. Block or graft copolymers, as well as polymer networks, can be created in this way.

The first reports on supramolecular polymers date back to the time when many scientists studied the mechanism by which aggregates of small molecules gave rise to increased viscosities. To the best of our knowledge it was Louise Henry who proposed the idea of molecular polymerization by associative interactions in 1878, approximately at the same time that van der Waals proposed his famous equation of state, which took intermolecular interactions in liquids into account, and was only 50 years after Berzelius coined the term polymers. Stadler and coworkers were the first to recognize that hydrogen bonds can be used to bring polymers together [4]. Lehn and coworkers synthesized the first main-chain supramolecular polymer based on hydrogen bonding [5]. In our group, we introduced the self-complementary ureido-pyrimidinone (UPy) quadruple hydrogen-bonding motif that shows a high dimerization constant and a long lifetime. In this chapter, we review the field of supramolecular polymers based on multiple hydrogen bonds and discuss some general approaches to the creation of supramolecular materials based on multiple hydrogen-bonded supramolecular polymers.

1.1.2
Supramolecular Chemistry

Jean-Marie Lehn defined supramolecular chemistry as '... *a highly interdisciplinary field of science covering the chemical, physical, and biological features of chemical species of higher complexity, which are held together and organized by means of intermolecular (noncovalent) binding interactions* [5].' This exciting new field introduced the possibility of self-sorting of subunits during the self-assembly process. At the same time large, complex structures can be created by the assembly of small supramolecular building blocks, thereby allowing the elimination of elaborate synthetic procedures. Complex self-assembly processes are widely recognized to have played an important part in different elements of the origin of life. As a result, many researchers explored different aspects of the field of supramolecular chemistry, using noncovalent interactions to self-assemble molecules into well-defined structures. Noncovalent interactions can vary in type and strength, ranging from very weak dipole-dipole interactions to very strong metal-ligand or ion-ion interactions with binding energies that can approach that of covalent bonds [6]. The most obvious benefits of noncovalent interactions are their reversible nature and their response to external factors such as temperature, concentration, and the polarity of the medium. A subtle interplay between these external factors allows precise control of the self-assembly process. Due to their directionality and the possibility to tune the dynamics and lifetime, hydrogen bonds are among the most interesting assembly units for supramolecular polymers. Before focusing on hydrogen bonding, we shall first address the different mechanisms for the formation of supramolecular polymers.

1.1.3
Supramolecular Polymerization Mechanisms

The mechanism of noncovalent polymerization in supramolecular chemistry is highly dependent on the interactions that play their part in the self-assembly process.

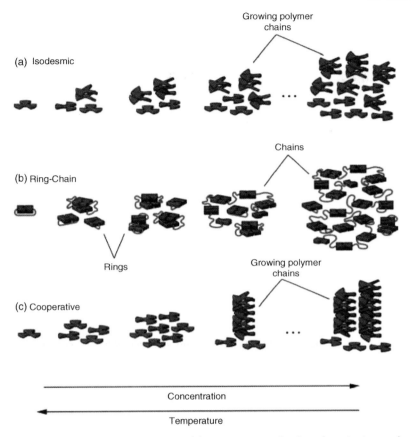

Figure 1.1 Schematic representation of the major supramolecular polymerization mechanisms. Reprinted with permission from Nature Publishing Group [7].

In contrast to covalent bonds, noncovalent interactions depend on temperature and concentration, thereby affecting the degree of polymerization. The mechanisms of supramolecular polymerizations can be divided in three major classes, these being isodesmic, cooperative, or ring-chain equilibria (Figure 1.1) [7].

Isodesmic polymerizations occur when the strength of noncovalent interactions between monomers is unaffected by the length of the chain. Because each addition is equivalent, no critical temperature or concentration of monomers is required for the polymerization to occur. Instead, the length of the polymer chains rises as the concentration of monomers in the solution is increased, or as the temperature decreases.

The ring-chain mechanism is characterized by an equilibrium between closed rings and linear polymer chains. In this mechanism, below a certain monomer concentration the ends of any small polymer chain react with each other to generate closed rings. Above this critical concentration, linear chain formation becomes more favored, and polymer growth is initiated. The degree of polymerization changes abruptly once the critical conditions are reached. The critical polymerization concentration is largely dependent on the length and rigidity of the monomers. Especially at low concentrations, the presence of cyclic oligomers can drastically influence the macroscopic properties.

Cooperative polymerizations occur in the growth of ordered supramolecular polymers in which there are additional interactions present besides the formation of linear polymers, such as those that form helices. This involves two distinct phases of self-assembly: a less favored nucleation phase followed by a favored polymerization phase. In this mechanism, the noncovalent bonds between monomers are weak, hindering the initial polymerization. After the formation of a nucleus of a certain size, the association constant is increased, and further monomer addition becomes more favored, at which point the polymer growth is initiated. Long polymer chains will form only above a minimum concentration of monomer and below a certain temperature, resulting in a sharp transition from a regime dominated by free monomers and small aggregates to a regime where almost all of the material is present as large polymers. For further details about supramolecular polymerization mechanisms we would refer the reader to a recent review by our group [7].

1.2
General Concepts of Hydrogen-Bonding Motifs

The existence of the hydrogen bond was first suggested by Moore and Winmill in 1912 [8], and it was defined in 1920 by Latimer and Rodebush as *'a hydrogen nucleus held between 2 octets, constituting a weak bond'* [9]. In that time the concept of hydrogen bonding was used to explain physical properties and chemical reactivities due to intramolecular and intermolecular hydrogen bonding. Nowadays, we interpret hydrogen bonds as highly directional electrostatic attractions between positive dipoles or charges on hydrogen and other electronegative atoms. In the field of supramolecular chemistry, hydrogen bonding is currently one of the most widely applied noncovalent interactions.

1.2.1
Arrays of Multiple Hydrogen Bonds

Hydrogen bonding is especially suitable as a noncovalent interaction because of the high directionality of the hydrogen bonds. In general, the strength of a single hydrogen bond depends on the strength of the hydrogen bond donor (D) and acceptor (A) involved, and can range from weak $CH-\pi$ interactions to very strong $FH-F^-$ interactions. When multiple hydrogen bonds are arrayed to create linear hydrogen-bonding motifs, both their strength and directionality are increased. However, the binding strength of the motif is dependent not only on the type and number of hydrogen bonds, but also on the order of the hydrogen bonds in the motif.

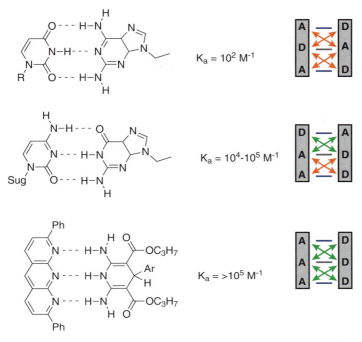

Figure 1.2 Influence of attractive and repulsive secondary interactions on the association constant of threefold hydrogen-bonding motifs [10, 11]. Reprinted with permission from The Royal Society of Chemistry [13].

This important aspect of linear hydrogen-bonding motifs was pointed out by Jorgensen *et al.*, who found a large variation in the association constants of threefold hydrogen-bonding motifs. Although the ADA – DAD and DAA – ADD arrays exhibit an equal amount of hydrogen bonds, the association constants of these motifs were significantly different. This was attributed to the different order of the hydrogen bonds [10]. Since the hydrogen bonds in the motifs are in close proximity, the distance of a hydrogen-bonding donor or acceptor to the neighbor of its counterpart is also relatively small, creating attractive or repulsive electrostatic secondary cross-interactions (Figure 1.2). This theory was later confirmed by Zimmerman *et al.*, who completed the series with the AAA – DDD array and indeed found a significantly higher dimerization constant due to the presence of solely attractive secondary interactions [11].

These so-called secondary interactions have a significant influence on the association constant of the corresponding motif, changing the association constant of the triple hydrogen-bonding motif by at least three orders of magnitude. Based on these results, Schneider *et al.* developed a method to calculate the free association energy for linear hydrogen-bonding motifs taking into account the secondary interactions, each contributing 2.9 kJ mol^{-1} to the binding energy, and expanded it to quadruple hydrogen-bonding motifs [12].

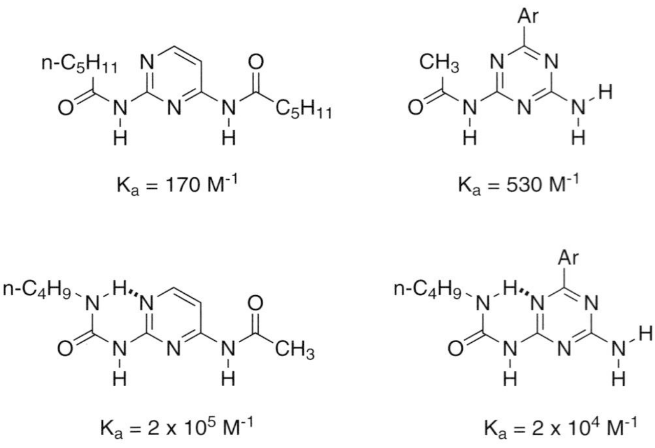

Figure 1.3 Quadruple hydrogen-bonding motifs with their corresponding dimerization constants, revealing the effect of the intramolecular hydrogen bond on the dimerization constant [14].

1.2.2
Preorganization through Intramolecular Hydrogen Bonding

Throughout the development of supramolecular chemistry, our knowledge of hydrogen-bonding motifs expanded rapidly. To attain high association constants, multiple hydrogen-bonding motifs were developed. Our group developed quadruple hydrogen-bonding motifs based on diaminotriazines and diaminopyrimidines in which a remarkably high dimerization constant was achieved when an amide moiety was replaced by a ureido moiety (Figure 1.3) [14]. A large deviation in the values of the experimentally determined dimerization constants of the ureido molecules was observed when compared to the calculations as proposed by Scheider *et al.* However, the experimental values for the amide molecules were in agreement with the calculated values. The large difference between the experimental and the predicted dimerization constants was attributed to the presence of an intramolecular hydrogen bond between the ureido NH and the nitrogen in the ring. This intramolecular hydrogen bond stabilizes the cis conformation of the ureido moiety and forces the carbonyl in plane with the aromatic ring. This causes prearrangement of the DADA hydrogen-bonding motif and results in an increase in the association constant by two or three orders of magnitude.

To reduce the number of repulsive secondary interactions, thereby increasing the association constant, our group introduced the self-complementary 2-ureido-4[1H]-pyrimidinone (UPy) quadruple hydrogen-bonding DDAA motif [15]. The intramolecular hydrogen bond prearranges the motif, resulting in a nearly planar DDAA motif (Figure 1.4) [16]. Due to the reduced number of repulsive secondary interactions and the intramolecular hydrogen bond, the dimerization constant was found to be $6 \times 10^7 \, \text{M}^{-1}$ in chloroform, with a long lifetime of 0.1 s [17].

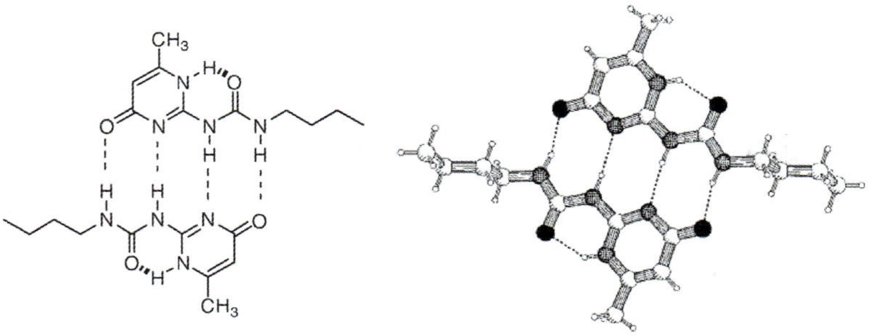

Figure 1.4 2-Ureido-4[1H]-pyrimidinone dimer and its corresponding single-crystal structure. Reprinted with permission from the American Chemical Society [16].

1.2.3
Tautomeric Equilibria

Although the UPy motif exhibits a high dimerization constant, the type of aggregate that is obtained during self-assembly is highly dependent on the substituent on the 6-position of the pyrimidinone ring, since different tautomeric forms can be present [16]. With electron-withdrawing or -donating substituents, the tautomeric equilibrium is shifted to the pyrimidin-4-ol tautomer, which is self-complementary as a DADA hydrogen-bonding motif (Figure 1.5). Due to more repulsive secondary

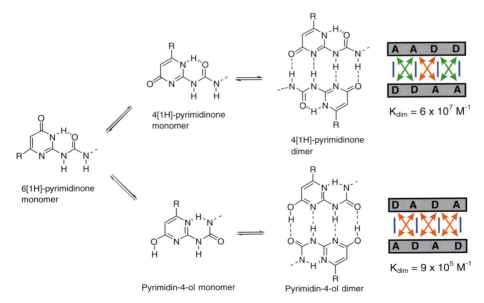

Figure 1.5 Tautomeric equilibria in the 2-ureido-pyrimidinone motif. Reprinted with permission from The Royal Society of Chemistry [13].

interactions, the dimerization constant of this DADA motif is lowered to $9 \times 10^5 \, \text{M}^{-1}$ in chloroform [18]. The tautomeric equilibrium showed a high dependence on the solvent, and also showed concentration dependence. This illustrates that understanding the tautomeric equilibria is crucial for predicting the properties of hydrogen-bonding motifs.

Nowadays, the synthesis of new hydrogen-bonding motifs is almost unrestricted. Current hydrogen-bonding motifs used in supramolecular chemistry are not only purely derived from organic chemistry, but are also derived from hydrogen bonding as found in nature, for example by using the hydrogen-bonding motifs found in DNA base pairs [19] or using peptide mimics (Figure 1.6) [20, 21]. Since the start of supramolecular chemistry, many different hydrogen-bonding motifs have been reported, ranging from monovalent up to dodecavalent hydrogen bonds [21], with dimerization constants up to $7 \times 10^9 \, \text{M}^{-1}$ [22]. However, it has to be noted that some of the reported hydrogen-bonding motifs require a multistep synthetic pathway, which lowers the overall yield tremendously, thereby making them less attractive to use.

1.3
Hydrogen-Bonded Main-Chain Supramolecular Polymers

1.3.1
The Establishment of Supramolecular Polymers

In macromolecular chemistry, the monomeric units are held together by covalent bonds. In 1990, Jean-Marie Lehn introduced a new area within the field of polymer chemistry by creating a polymer in which the monomeric units were held together by hydrogen bonds, resulting in a liquid crystalline supramolecular polymer (Figure 1.7) [23]. This initiated the field of supramolecular polymer chemistry, generating materials with reversible interactions, and thereby introducing the opportunity to produce materials with properties that otherwise would have been impossible or difficult to obtain.

Inspired by this work, Griffin *et al.* developed main-chain supramolecular polymers based on pyridine/benzoic acid hydrogen bonding, also obtaining liquid crystalline supramolecular polymers [24]. Our group introduced supramolecular polymers based on the ureido-pyrimidinone motif. Due to the high dimerization constant present in the UPy motif, supramolecular polymers were formed with a high degree of polymerization even *in semi-dilute solution* [15].

We have defined supramolecular polymers as '...*polymeric arrays of monomeric units that are brought together by reversible and highly directional secondary interactions, resulting in polymeric properties in dilute and concentrated solutions, as well as in the bulk. The monomeric units of the supramolecular polymers themselves do not possess a repetition of chemical fragments. The directionality and strength of the supramolecular bonding are important features of these systems, that can be regarded as polymers and behave according to well-established theories of polymer physics.* In the past the term "living polymers" has been used for this type of polymer. However, to exclude confusion with the important field of

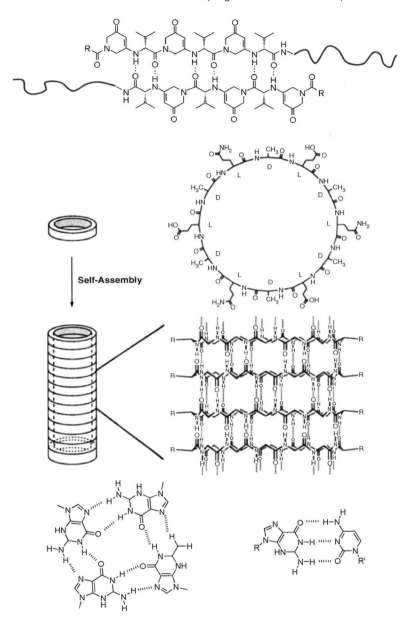

Figure 1.6 Hydrogen-bonding motifs inspired on self-assembly as found in nature. Reprinted with permission from the American Chemical Society [19–21].

living polymerizations, we prefer to use the term supramolecular polymers.' [25] The irony is that in the field of polymer science, Hermann Staudinger fought many scientific battles to prove that polymer molecules consist of covalently bonded monomers rather than noncovalent aggregates of small molecules. Almost a hundred years later,

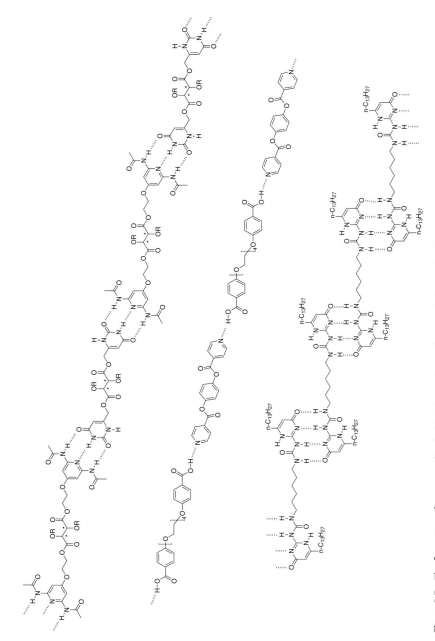

Figure 1.7 The formation of a supramolecular liquid crystalline polymer by hydrogen bonding as introduced by Lehn *et al.* (top), Griffith *et al.* (middle) and our group. (bottom).

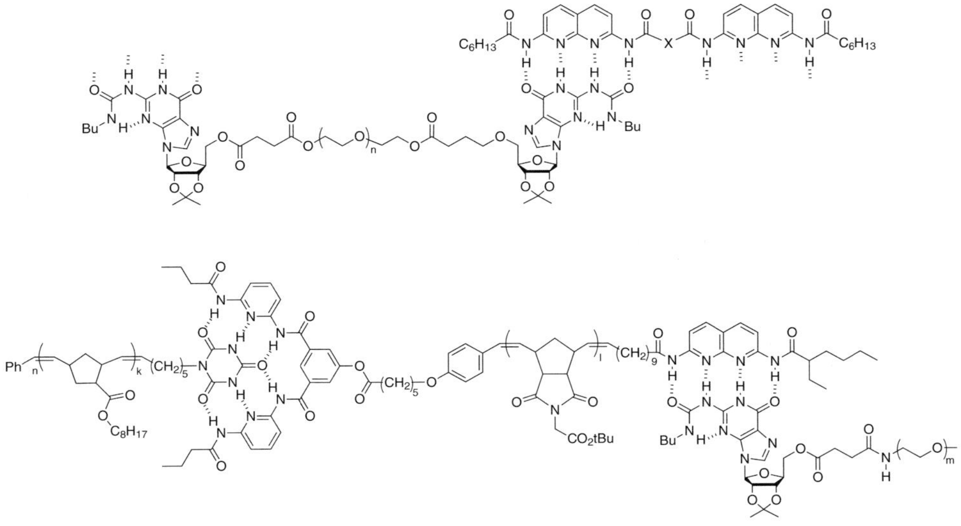

Figure 1.8 Alternating (top) and triblock (bottom) supramolecular copolymers created in solution by using the directionality of complementary hydrogen-bonding motifs [28b,29].

material properties typical of macromolecules can also be obtained by the noncovalent aggregation of small molecules.

In macromolecular chemistry, different types of polymers are distinguished, ranging from linear polymers and graft copolymers to networks. Soon after the introduction of supramolecular polymers, it was recognized that by replacing the covalent bonds between the monomeric units by hydrogen bonds, these polymers can be made in a supramolecular fashion, and one year *before* the introduction of the linear supramolecular polymer by Lehn the group of Fréchet introduced supramolecular graft copolymers [26]. Using multiple hydrogen-bonding moieties attached to one molecule, one can also generate supramolecular polymer networks [27]. With the development of new hydrogen-bonding motifs and a better understanding of the concept of supramolecular polymers, nowadays even alternating [28] or triblock [29] supramolecular hydrogen-bonding copolymers can be created using the high directionality of different hydrogen-bonding motifs (Figure 1.8).

1.3.2
Supramolecular Polymerizations

The polymerization of multivalent linear supramolecular polymers based solely on hydrogen bonding without any additional interactions will in general result in an isodesmic polymerization mechanism. As a consequence, the degree of polymerization (DP) that is obtained will be highly dependent on the dimerization constant and the concentration (Figure 1.9) [30]. Therefore, the obvious approach to increase

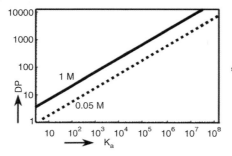 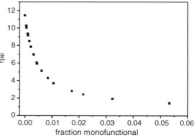

Figure 1.9 Theoretical dependence of the degree of polymerization as a function of association constant and concentration for an isodesmic polymerization mechanism (left) and the influence of monofunctional chainstopper on the polymerization (right). Reprinted with permission from the American Association for the Advancement of Science [15] and the American Chemical Society [25].

the degree of polymerization is to create hydrogen-bonding motifs with high dimerization constants. However, the synthetic accessibility of these motifs and their attachment to other molecules is highly important since incomplete functionalization or other monofunctional impurities present at less than one percent can act as a chainstopper. This has a huge effect on the degree of polymerization, as was demonstrated by viscosity measurements (Figure 1.9) [15, 31]. When one uses the AA-BB type of supramolecular polymers in which the hydrogen-bonding motifs are not self-complementary but need a complementary counterpart, this results in the need for perfect stoichiometry in order to attain high degrees of polymerization, since even a small excess of either one will act as a chainstopper [32]. To avoid this problem, when creating supramolecular polymers a self-complementary hydrogen-bonding motif is preferred.

An important factor that cannot be neglected when going from small molecules to supramolecular polymers is the influence of modifications of the molecular structure on the association constant of the hydrogen-bonding motif [33]. This can be caused by steric effects when attaching large molecules to the motif [34], and it is observed that the polarity of the attached molecule influences the association constant drastically [35]. This will therefore influence the degree of polymerization significantly.

1.3.3
Hydrophobic Compartmentalization

The isodesmic type of polymerization of main-chain hydrogen bonded supramolecular polymers results in a low degree of polymerization and results in the need for hydrogen-bonding motifs with a high dimerization constant in order to obtain long polymers in solution. To overcome this issue, several different strategies can be applied.

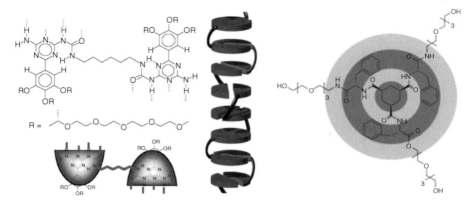

Figure 1.10 Helical supramolecular ureido-triazine polymer (left) and cyclohexane hydrogelator (right) in which the hydrogen-bonding motif is shielded from the solvent by hydrophobic interactions, creating aggregates in water. Reprinted with permission from the National Academy of Sciences [38b] and The Royal Society of Chemistry [39c].

It is widely believed that supramolecular polymers in water based on purely hydrogen bonding are not possible due to the competition of the intermolecular hydrogen bonds with hydrogen-bonding with water molecules [36]. However, hydrophobic compartmentalization is widely found in nature and can shield the hydrogen bonds from the aqueous environment. This decreases the competitive hydrogen bonding of water molecules with the desired intermolecular hydrogen bonds. At the same time this creates a more apolar local environment for the hydrogen-bonding motifs, which strengthens the hydrogen-bonding interactions. Due to their weak interaction energy, the hydrophobic interactions are highly dependent on the temperature and can be induced or eliminated depending on the solvent [37]. However, using hydrophobic compartmentalization to shield the hydrogen-bonding motif from the environment, it is possible to attain supramolecular hydrogen-bonding polymers [38] and hydrogels [39] in water (Figure 1.10).

Additional interactions can be introduced into hydrogen-bonding supramolecular polymers by using hydrophobic compartmentalization. As shown in Figure 1.10, π-π interactions occur between the aromatic cores, creating chiral columnar structures. An important result of these additional interactions is the change of polymerization mechanism from isodesmic to cooperative, creating supramolecular polymers with a high degree of polymerization. This circumvents the requirement for a high dimerization constant in order to obtain supramolecular polymers with a high degree of polymerization. Additional π-π interactions in hydrogen bonded supramolecular polymers are not uncommon and can be applied to obtain higher-order structures [40].

1.4
From Supramolecular Polymers to Supramolecular Materials

1.4.1
Thermoplastic Elastomers

The introduction of polyamides and polyurethanes as polymeric materials created the possibility of having elastomeric materials which are processable at higher temperatures. The intermolecular hydrogen bonding between the polymer chains generates noncovalent crosslinks that crystallize upon hydrogen bonding. The crosslinks are broken upon heating the material, resulting in a dramatic decrease in viscosity, giving it its thermoplastic elastomeric behavior [41]. These polymers could be classified as supramolecular polymers due to their noncovalent crosslinks. However, the entanglements of the high-molecular-weight polymer chain have a significant influence on the macroscopic properties, thereby disqualifying them as true supramolecular polymers.

Inspired by the outstanding mechanical properties and processability of polyamides and polyurethanes, new polymers have been developed in which the amide and urethane moiety were replaced by urea moieties. Ureas can form stronger bifurcated hydrogen bonds than those formed by amides and urethanes. Indeed, when reacting amine-functionalized oligomers with diisocyanates, bis-urea thermoplastic elastomers were obtained which showed a nanofiber morphology, as observed with atomic force microscopy (AFM) (Figure 1.11) [42]. The aggregation of the bis-urea is cooperative due to the synergistic aggregation of the second urea within the bis-urea motif and the less favorable formation of dimers due to alignment of dipole moments. In addition, the bis-urea motif bundles together and crystallizes into long nano-fibers that act as supramolecular crosslinks. This reinforces the material and gives it its good mechanical properties [43]. Using so-called supramolecular self-

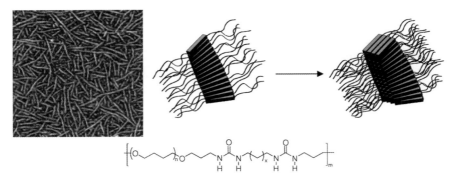

Figure 1.11 Atomic force microscopy phase image (500 × 500 nm) of nano-fibers as observed in thermoplastic elastomers based on the bis-urea motif (left) and the schematic aggregation of bis-urea stacks into the nano-fibers (right). Reprinted with permission from the American Chemical Society [43].

sorting, matching bis-urea molecules were selectively incorporated into the material [42a,d], and these were used to introduce, for example, bioactive molecules to bis-urea supramolecular biomaterials to improve cell adhesion and proliferation for tissue engineering [44]. Moreover, the incorporated bis-urea molecules were used to tune the mechanical properties of the bis-urea polymer [42c].

While the bis-urea crystallization results in favorable material properties, its high melting point severely reduces the mobility of the hydrogen-bonding moieties at room temperature. As a result, these supramolecular materials do not possess self-healing properties.

Leibler *et al.* introduced a system based on dimer fatty acids to synthesize amidoethyl imidazolidone, di(amidoethyl) urea, and diamido tetraethyl triurea oligomers (Figure 1.12) [45]. The system consists of a network of hydrogen bonds, which do not crystallize. At low temperatures the material is crosslinked by hydrogen bonds and behaves as a soft rubber, whereas at high temperatures the hydrogen bonds are broken and the material behaves like a viscoelastic liquid which can be molded, extruded, and reshaped. While the pure oligomer mixture exhibits a glass transition temperature of 28 °C, it can be plasticized with dodecane or water to lower the glass transition temperature. Due to the absence of crystallization and a glass transition temperature below room temperature, this material exhibits remarkable self-healing properties. The material is capable of regaining its mechanical properties after being macroscopically broken by simple mending at room temperature, although the re-establishment of the macroscopic properties and the hydrogen-bonding network takes time.

The examples discussed above show the potential of supramolecular polymers to create novel materials with new and advanced properties. The importance of thermal properties such as glass transition temperatures or melt temperatures dominates the macroscopic properties of the material. When the glass transition temperature is above room temperature, the mobility of the hydrogen-bonding moieties is limited.

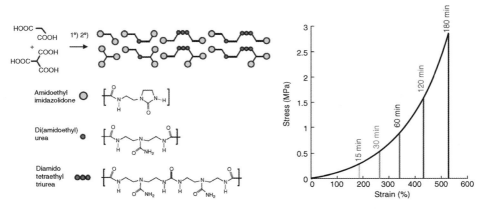

Figure 1.12 A supramolecular rubber based on hydrogen bonding generates a self-healing material at room temperature. The mechanical properties recover in time as the hydrogen-bonding network is restored. Reprinted with permission from Nature Publishing Group [45a].

This prevents the rearrangement of hydrogen bonds and results in a lack of self-healing properties. However, the presence of a glass transition temperature or a melt temperature above room temperature will improve the mechanical properties of the material by acting as crosslinks. The desired macroscopic properties of the material will therefore depend on its application.

1.4.2
Phase Separation and Additional Lateral Interactions in Supramolecular Polymers in the Solid State

Small molecule supramolecular systems as reported by Lehn form supramolecular polymers that show liquid crystalline behavior in bulk. However, these systems are rigid and give brittle materials with inferior mechanical properties at room temperature. To improve the mechanical properties, telechelic amorphous or semi-crystalline oligomers have been functionalized with hydrogen-bonding motifs [46, 47]. Upon functionalization of the oligomer with a hydrogen-bonding motif, materials with properties that resemble the covalent high-molecular-weight counterparts were obtained. However, due to the reversibility of the hydrogen bonds, at high temperatures the noncovalent interactions are broken, resulting in a material exhibiting the properties of the low-molecular-weight oligomers. This could be especially suitable for the synthesis of materials with improved processing properties at elevated temperatures. By using amorphous or semi-crystalline oligomers with multiple functionizable end groups, flexibility is introduced within the molecule and crystallinity is reduced. At the same time, the telechelic oligomer used influences the material properties of the supramolecular polymer.

Phase separation in block copolymers is well known and originates from the immiscibility of one block in the other block and vice versa. By adding hydrogen-bonding motifs to telechelic oligomers, a block copolymer-like molecule is obtained in which the hydrogen-bonding end groups can phase separate from the oligomer in the bulk, depending on their polarity difference and aggregation behavior. Examples illustrate that by using block copolymers with weak hydrogen-bonding blocks on the exterior, quasi-telechelic supramolecular polymers are obtained [48]. Chien *et al.* introduced telechelic supramolecular polymers based on poly(tetrahydrofuran) with benzoic acid end groups [46a]. The supramolecular polymers showed a tendency for micro-phase separation with a high-temperature melting point. This additional endotherm was attributed to the melting of hard segments which originate from the crystallization of benzoic acid end groups driven by benzoic acid dimerization. The hard segments are phase separated, creating physical crosslinks which increased the mechanical properties tremendously [46b]. Similar findings were obtained when using supramolecular polymers with benzoic acid hydrogen-bonding moieties in the side-chain [49]. Whether these self-assembly processes are driven by phase separation of the different blocks or by hydrogen bonding remains uncertain.

Hayes *et al.* investigated the influence of the strength of the hydrogen-bonding motif on the phase separation and mechanical properties of telechelic supramolec-

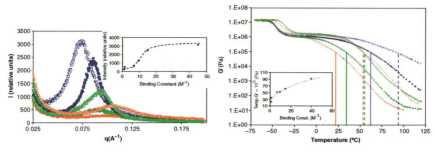

Figure 1.13 Influence of the binding constant of telechelic supramolecular polymers on phase separation as observed with SAXS and corresponding rheological analysis. Reprinted with permission from the American Chemical Society [50].

ular polymers. A clear influence of the dimerization constant on the phase separation was found, which coincides with a change in the mechanical properties as observed by rheological measurements (Figure 1.13) [50]. This clearly shows the influence of hydrogen bonding on the phase separation of telechelic supramolecular polymers and subsequent mechanical properties.

1.4.3
Supramolecular Thermoplastic Elastomers Based on Additional Lateral Interactions and Phase Separation

Phase separation is particularly interesting for supramolecular polymers based on weak hydrogen-bonding motifs, since the phase separation can increase the local concentration. This results in a higher degree of polymerization and a change in supramolecular polymerization mechanism from isodesmic to cooperative. This approach was demonstrated by Rowan et al. who synthesized supramolecular thermoplastic elastomers based on hydrogen-bonding telechelic poly(tetrahydrofuran) [51]. Although weak complementary nucleobase hydrogen-bonded motifs were used ($K_a = 21$ M^{-1} in CDCl$_3$), the supramolecular polymer exhibits good mechanical properties. The formation of such a thermoplastic elastomer is not expected based on solely linear chain extension due to the weak hydrogen bonding and was shown to be related to the phase separation and π-π stacking of the hydrogen-bonding end groups within the soft oligomer matrix (Figure 1.14). The phase separation results in crystallization of the end groups with melting points at 108 °C and 135 °C. Detailed studies revealed that the nucleobase end groups aggregate on top of each other, creating supramolecular cross-links. In a similar system based on a poly(ε-caprolactone) oligomers, the hydrogen-bonding end groups were later visualized with AFM. In combination with X-ray diffraction studies, it was shown that the end groups were arranged in lamellae [52].

Phase separation of the hydrogen-bonding end groups can be induced by introducing additional lateral interactions when the end groups themselves do not exhibit lateral interactions. This was demonstrated by functionalizing telechelic poly(ethyl-

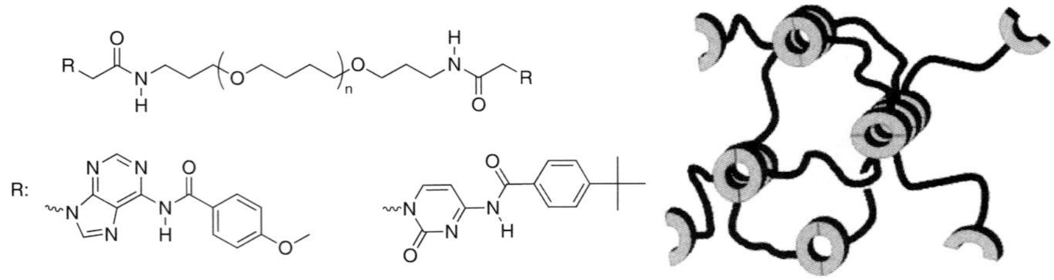

Figure 1.14 Nucleobase hydrogen-bonded supramolecular polymers and their schematic aggregation into phase-separated hard segments. Reprinted with permission from the American Chemical Society [51b].

ene-butylene) oligomers with the ureido-pyrimidinone (UPy) motif. The corresponding supramolecular polymer displays a remarkable increase in macroscopic properties, creating a supramolecular thermoplastic elastomer (Figure 1.15) [47b]. Although the UPy exhibits an extremely high dimerization constant, it was not expected to result in a thermoplastic elastomer upon isodesmic supramolecular polymerization of this molecule, since both the poly(ethylene-butylene) oligomer and its high-molecular-weight counterpart are amorphous, with a glass transition temperature well below room temperature.

The increase in macroscopic properties is a result of the aggregation of the end groups, not only polymerizing in a linear fashion, but also forming stacks of dimers due to the urethane moiety in the end groups that induces lateral aggregation (Figure 1.16) [53, 54]. Due to these lateral interactions, supramolecular crosslinks are obtained that crystallize into nanofibers which could be observed with AFM.

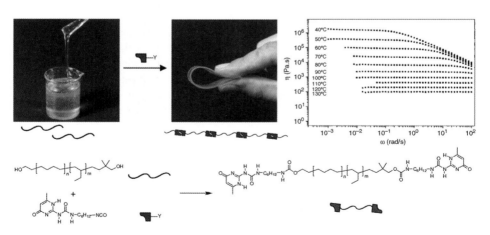

Figure 1.15 A supramolecular thermoplastic elastomer obtained by functionalization of a short telechelic poly(ethylene-butylene) oligomer with an ureido-pyrimidinone hydrogen-bonding moiety and its dynamic melt viscosity as a function of temperature. Reprinted with permission from Wiley-VCH [47b].

1.4 From Supramolecular Polymers to Supramolecular Materials | 21

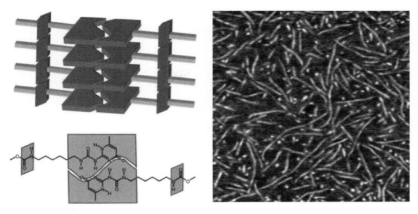

Figure 1.16 Schematic representation of the lateral interactions creating supramolecular crosslinks (left) and the nanofibers as visualized with AFM (500 × 500 nm phase image). Reprinted with permission from Wiley-VCH [54b].

A more detailed study of the influence of the lateral interactions in the end groups by eliminating or reinforcing the lateral interactions confirmed their importance [53]. The supramolecular polymer with no lateral interactions is a sticky gum and shows no distinct phase separation, rheology measurements confirming the presence of UPy-UPy hydrogen bonding (Figure 1.17). Upon introduction and reinforcement of the lateral interactions in the UPy urethane (UPy-T) and UPy urea (UPy-U) motifs respectively, the mechanical properties increase drastically, resulting in thermoplastic elastomers.

The influence of the strength of the lateral interactions is clearly visible, as the UPy-T nanofibers are ill-defined and display a melt at 69 °C, whereas the UPy-U nanofibers are well-defined with a melt at 129 °C (Figure 1.18). An important result of these lateral interactions is the change in polymerization mechanism. In solution,

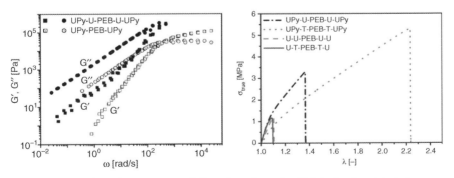

Figure 1.17 Rheological master curves (left) and tensile testing (right) of various telechelic supramolecular poly(ethylene-butylene) polymers. Reprinted with permission from the American Chemical Society [53].

22 1 Multiple Hydrogen-Bonded Supramolecular Polymers

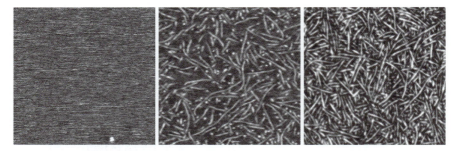

Figure 1.18 AFM phase images (500 × 500 nm) of UPy-PEB-UPy, UPy-T-PEB-T-UPy and UPy-U-PEB-U-UPy respectively.

UPy-urea model compounds reveal an isodesmic polymerization mechanism into stacks, with a lateral association constant of $3 \times 10^2 \, M^{-1}$ in CDCl$_3$ [55]. However, in the bulk the polymerization mechanism becomes cooperative due to phase separation and results in the crystallization of long nanofibers.

The usability of these materials was exemplified by the creation of supramolecular biomaterials, in which telechelic poly(ε-caprolactone) was functionalized with UPy groups to generate a supramolecular biocompatible material. Using the noncovalent nature of the material, UPy-functionalized peptides can be incorporated in the material by simple mixing (Figure 1.19) [56]. The bioactive molecules are anchored into the supramolecular material via the UPy hydrogen-bonding units, establishing the possibility to obtain a dynamic biomaterial that closely resembles the extracellular matrix due to its noncovalent character.

Using this modular approach, materials with different bioactive molecules can easily be made without resynthesizing the whole construct. The incorporation of UPy-functionalized cell adhesion peptides into the supramolecular biomaterial increased cell adhesion, spreading, and proliferation compared to the bare construct, revealing the applicability of this approach. Due to the significant mechanical

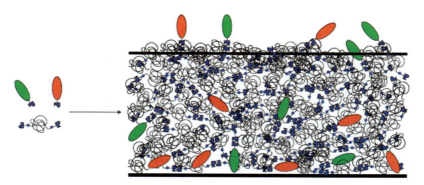

Figure 1.19 Modular approach to supramolecular biomaterials using the noncovalent interactions for the anchoring of bioactive molecules. Reprinted with permission from Nature Publishing Group [56a].

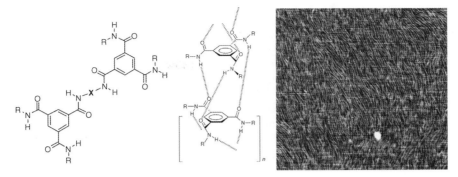

Figure 1.20 Supramolecular polymers based on the benzene-1,3,5-tricarboxamide motif (left) and the nano-fibers as observed with AFM (phase image, 450 × 450 nm, right). Reprinted with permission from the American Chemical Society [59].

properties of these materials, it is possible to electrospin fibrous membranes with diameters less than 1 μm [57].

The need to incorporate lateral interactions in supramolecular polymers could be circumvented by using hydrogen-bonding motifs that comprise the possibility to chain extend and simultaneously act as supramolecular crosslinks [58]. When telechelic poly(ethylene-butylene) was functionalized with the benzene-1,3,5-tricarboxamide (BTA) motif, a supramolecular thermoplastic elastomer was obtained [59]. The BTA motif is capable of chain extending by hydrogen bonding to neighboring BTA molecules. Due to the fact that one BTA motif exhibits two binding sites for other BTA molecules, being above and below the face of the BTA molecule, this results in chain extension as well as supramolecular cross-linking.

The polymerization mechanism is cooperative due to the unfavorable arrangement of the carbonyl groups in the initial aggregation steps and additional dipole-dipole interactions [60]. This results in nano-fibers with a transition to the isotropic phase around 200 °C (Figure 1.20). At room temperature, the material is liquid crystalline, giving it high elastomeric properties but results in a low toughness.

1.5
Future Perspectives

The developments within the fields of polymer chemistry and organic chemistry have enabled the synthesis of complex monomers and polymers. The state-of-the-art knowledge in the field of supramolecular chemistry gives increasing control over self-assembled systems and paves the way for the creation of supramolecular materials based on noncovalent interactions. In the past decades, the hydrogen bond has proven itself to be a most suitable candidate for applications where structuring interaction is required in supramolecular synthesis. With increasing knowledge of hydrogen-bonding motifs and their self-assembly behavior in solution and in the solid state, the exploration of supramolecular synthesis of multi-component systems with different supramolecular motifs has started in recent years. Using purely hydrogen bonding, multi-component systems have shown promising results as, for example, supramolecular block copolymers and bioactive biomaterials.

It is now the time to acquire structures with a well-defined molecular as well as supramolecular structure by combining the vast knowledge on traditional polymer chemistry with the current knowledge on supramolecular chemistry. The next step in the development of hydrogen bonding in supramolecular polymer chemistry is to gain control over the self-assembly, possibly by the use of supramolecular protective groups or by turning 'on' or 'off' the supramolecular polymerizations by switchable hydrogen-bonding motifs. This would be a step toward materials with well-defined properties.

As an example, we have combined our knowledge of polymer chemistry and supramolecular chemistry to obtain supramolecular nano-particles based on hydrogen bonding (Figure 1.21) [61]. Using polymer chemistry, covalent polymers with a narrow molecular weight distribution were synthesized which bear a small fraction of covalently protected UPy groups on their side-chains. Upon deprotection with UV-light, the UPy motif was switched 'on' and supramolecular cross-linking via UPy hydrogen bonding was obtained. Using dilute conditions, single-chain supramolecular nanoparticles were obtained, which mimic the supramolecular folding of proteins. This could be a first step toward artificial proteins and enzymes, suggesting endless possibilities for supramolecular polymers. Moreover, when the temperature of a film of these nanoparticles is raised, intermolecular hydrogen bonds are formed

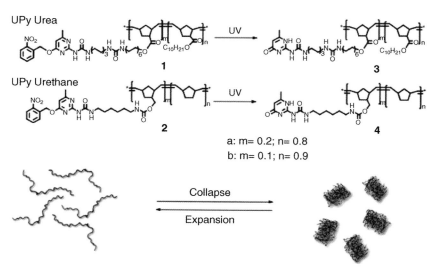

Figure 1.21 The creation of supramolecular nano-particles based on intramolecular cross-linking via hydrogen bonding. Reprinted with permission from the American Chemical Society [61a].

and the system is converted into a three-dimensional hydrogen-bonded network. Thus the full potential of reversible supramolecular interactions is applied in material processing.

References

1 Morawetz, H. (1985) *Polymers: The Origins and Growth of a Science*, John Wiley and Sons.
2 Staudinger, H. (1920) *Ber. Deut. Chem. Ges.*, **53** (6), 1073.
3 Thematic Issue of Chemical Reviews (2009) **109** (11), 4961–6540.
4 (a) Stadler, R. and Burgert, J. (1986) *Makromol. Chem.*, **187**, 1681–1690; (b) Stadler, R. and de Lucca Freitas, L. (1986) *Colloid Polym. Sci.*, **264**, 773; (c) de Lucca Freitas, L. and Stadler, R. (1987) *Macromolecules*, **20**, 2478.
5 (a) Lehn, J.-M. (1990) *Angew. Chem. Int. Ed.*, **26**, 1304–1319; (b) Lehn, J.-M. (1993) *Science*, **260**, 1762–1763.
6 Goshe, A.J., Steele, I.M., Ceccarelli, C., Rheingold, A.L., and Bosnich, B. (2002) *PNAS*, **99**, 4823–4829.
7 De Greef, T.F.A., Smulders, M.M.J., Wolffs, M., Schenning, A.P.H.J., Sijbesma, R.P., and Meijer, E.W. (2009) *Chem. Rev.*, **109**, 5687–5754.
8 Moore, T.S. and Winmill, T.F. (1912) *J. Chem. Soc. Trans.*, **101**, 1635–1676.
9 Latimer, W.M. and Rodebush, W.H. (1920) *J. Am. Chem. Soc.*, 1419–1433.
10 (a) Jorgensen, W.L. and Pranata, J. (1990) *J. Am. Chem. Soc.*, **112**, 2008–2010; (b) Pranata, J., Wierschke, S.G., and Jorgensen, W.L. (1991) *J. Am. Chem. Soc.*, **113**, 2810–2819.
11 Murray, T.J. and Zimmerman, S.C. (1992) *J. Am. Chem. Soc.*, **114**, 4010–4011.
12 Blight, B.A., Hunter, C.A., Leigh, D.A., McNab, H., Thomson, P.I.T. (2011) *Nature Chemistry*, **3**, 244–248.
13 Sijbesma, R.P. and Meijer, E.W. (2003) *Chem. Commun.*, 5–16.
14 Beijer, F.H., Kooijman, H., Spek, A.L., Sijbesma, R.P. and Meijer, E.W. (1998) *Angew. Chem. Int. Ed.*, **37** (1), 75–78.
15 Sijbesma, R.P., Beijer, F.H., Brunsveld, L., Folmer, B.J.B., Ky Hirschberg, J.H.K., Lange, R.F.M., Lowe, J.K.L., and Meijer, E.W. (1997) *Science*, **278**, 1601–1604.
16 Beijer, F.H., Sijbesma, R.P., Kooijman, H., Spek, A.L., and Meijer, E.W. (1998) *J. Am. Chem. Soc.*, **120**, 6761–6769.
17 Söntjens, S.H.M., Sijbesma, R.P., van Genderen, M.H.P., and Meijer, E.W. (2000) *J. Am. Chem. Soc.*, **122**, 7487–7493.
18 de Greef, T.F.A., Ercolani, G., Ligthart, G.B.W.L., Meijer, E.W. and Sijbesma, R.P. (2008) *J. Am. Chem. Soc.*, **130**, 13755–13764.
19 (a) Spada, G.P., Carcuro, A., Colonna, F.P., Garbesi, A. and Gottarelli, G. (1988) *Liquid Crystals*, **3** (5), 651–654; (b) Marlow, A.L., Mezzina, E., Spada, G.P., Masiero, S., Davis, J.T., and Gottarelli, G. (1999) *J. Org. Chem.*, **64**, 5116–5123; (c) Sivakova, S. and Rowan, S.J. (2005) *Chem. Soc. Rev.*, **34**, 9–21.
20 (a) Zeng, H., Yang, X., Flowers, R.A., and Gong, B. (2002) *J. Am. Chem. Soc.*, **124** (12), 2903–2910; (b) Jahnke, E., Lieberwirth, I., Severin, N., Rabe, J.P., and Frauenrath, H. (2006) *Angew. Chem. Int. Ed.*, **45**, 5383–5386; (c) Isimjan, T.T., de Bruyn, J.R., and Gillies, E.R. (2010) *Macromolecules*, **43**, 4453–4459; (d) Skrzeszewska, P.J., de Wolf, F.A., Cohen Stuart, M.A., and van der Gucht, J. (2010) *Soft Matter*, **6**, 416–422.
21 Khazanovich, N., Granja, J.R., McRee, D.E., Milligan, R.A. and Reza Ghadiri, M. (1994) *J. Am. Chem. Soc.*, **116**, 6011–6012.
22 Zeng, H., Yang, X., Brown, A.L., Martinovic, S., Smith, R.D., and Gong, B. (2003) *Chem. Commun.*, 1556–1557.
23 Fouquey, C., Lehn, J.-M., and Levelut, A.-M. (1990) *Adv. Mater.*, **5**, 254–257.
24 (a) Lee, C.-M., Jariwala, C.P. and Griffin, A.C. (1994) *Polymer*, **35** (21), 4550–4554; (b) St. Pourcain, C.B. and Griffin, A.C. (1995) *Macromolecules*, **28**, 4116–4121.
25 Brunsveld, L., Folmer, B.J.B., Meijer, E.W., and Sijbesma, R.P. (2001) *Chem. Rev.*, **101**, 4071–4097.

26 (a) Kato, T. and Fréchet, J.M.J. (1989) *Macromolecules*, **22**, 3819–3821; (b) Shandryuk, G.A., Kuptsov, S.A., Shatalova, A.M., Plate, N.A., and Talroze, R.V. (2003) *Macromolecules*, **36**, 3417–3423; (c) Burd, C. and Weck, M. (2005) *Macromolecules*, **38**, 7225–7230; (d) Pollino, J.M. and Weck, M. (2005) *Chem. Soc. Rev.*, **34**, 193–207; (e) Hammond, M.R. and Mezzenga, R. (2008) *Soft Matter*, **4**, 952–961.

27 (a) Hilger, C. and Stadler, R. (1990) *Macromolecules*, **23**, 2097–2100. (b) Hilger, C. and Stadler, R. (1992) *Macromolecules*, **25**, 6670–6680; (c) St. Pourcain, C.B. and Griffin, A.C. (1995) *Macromolecules*, **28**, 4116–4121; (d) Lange, R.F.M., van Gurp, M. and Meijer, E.W. (1999) *J. Polym. Sci, Part A*, **37** (19), 3657–3670; (e) Chino, K. and Ashiura F M. (2001) *Macromolecules*, **34**, 9201–9204; (f) Rieth, L.R., Eaton, R.F., and Coates, G.W. (2001) *Angew. Chem. Int. Ed.*, **40** (11), 2153–2156; (g) Berl, V., Schmutz, M., Krische, M.J., Khoury, R.G., and Lehn, J.-M. (2002) *Chem. Eur. J.*, **8** (5), 1227–1244.

28 (a) Park, T. and Zimmerman, S.C. (2006) *J. Am. Chem. Soc.*, **128**, 13986–13987; (b) Scherman, O.A., Ligthart, G.B.W.L., Ohkawa, H., Sijbesma, R.P., and Meijer, E.W., *PNAS* (2006) **103** (32), 11850–11855.

29 Yang, S.K., Ambade, A.V., and Weck, M. (2010) *J. Am. Chem. Soc.*, **132**, 1637–1645.

30 Tessa ten Cate, A. and Sijbesma, R.P. (2002) *Macromol. Rapid Commun.*, **23**, 1094–1112.

31 (a) Knoben, W., Besseling, N.A.M., and Cohen Stuart, M.A. (2007) *J. Chem. Phys.*, **126**, 024907; (b) van Beek, D.J.M., Spiering, A.J.H., Peters, G.W.M., te Nijenhuis, K., and Sijbesma, R.P. (2007) *Macromolecules*, **40**, 8464–8475.

32 Burke, K.A., Sivakova, S., McKenzie, B., Mather, P.T., and Rowan, S.J. (2006) *J. Polym. Sci, Part A*, **44**, 5049–5059.

33 Lortie, F., Boileau, S., and Bouteiller, L. (2003) *Chem. Eur. J.*, **9**, 3008–3014.

34 (a) Gillies, E.R. and Fréchet, J.M.J. (2004) *J. Org. Chem.*, **69**, 46–53; (b) Pensec, S., Nouvel, N., Guilleman, A., Creton, C., Boué, F., and Bouteiller, L., *Macromolecules* (2010) **43**, 2529–2534; (c) Feldman, K.E., Kade, M.J., Meijer, E.W., Hawker, C.J., and Kramer, E.J. (2010) *Macromolecules*, **43**, 5121–5127.

35 (a) de Greef, T.F.A., Nieuwenhuizen, M.M.L., Stals, P.J.M., Fitié, C.F.C., Palmans, A.R.A., Sijbesma, R.P., and Meijer, E.W. (2008) *Chem. Commun.*, 4306–4308; (b) Mes, T., Smulders, M.M.J., Palmans, A.R.A., and Meijer, E.W., *Macromolecules* (2010) **43**, 1981–1991; (c) de Greef, T.F.A., Nieuwenhuizen, M.M.L., Sijbesma, R.P., and Meijer, E.W., *J. Org. Chem.* (2010) **75**, 598–610.

36 Rehm, T. and Schmuck, C. (2008) *Chem. Commun.*, 801–813.

37 Obert, E., Bellot, M., Bouteiller, L., Andrioletti, F. Lehen-Ferrenbach, C., and Boué, F. (2007) *J. Am. Chem. Soc.*, **129**, 15601–15605.

38 (a) Ky Hirschberg, J.H.K., Brunsveld, L., Ramzi, A., Vekemans, J.A.J.M., Sijbesma, R.P., and Meijer, E.W. (2000) *Nature*, **407**, 167–170; (b) Brunsveld, L., Vekemans, J.A.J.M., Ky Hirschberg, J.H.K., Sijbesma, R.P., and Meijer, E.W. (2002) *PNAS*, **99** (8), 4977–4982; (c) Boekhoven, J., van Rijn, P., Brizard, A.M., Stuart, M.C.A., and van Esch, J.H. (2010) *Chem. Commun.*, **46**, 3490–3492.

39 (a) van Bommel, K.J.C., van der Pol, C., Muizebelt, I., Friggeri, A., Heeres, A., Meetsma, A., Feringa, B.L., and van Esch, J. (2004) *Angew. Chem. Int. Ed.*, **43**, 1663–1667; (b) Friggeri, A., van der Pol, C., van Bommel, K.J.C., Heeres, A., Stuart, M.C.A., Feringa, B.L., and van Esch, J. (2005) *Chem. Eur. J.*, **11**, 5353–5361; (c) Boekhoven, J., van Rijn, P., Brizard, A.M., Stuart, M.C.A., and van Esch, J.H. (2010) *Chem. Commun.*, **46**, 3490–3492.

40 (a) Barberá, J., Puig, L., Romero, P., Serrano, J.L., and Sierra, T. (2006) *J. Am. Chem. Soc.*, **128**, 4487–4492; (b) Kolomiets, E., Buhler, E., Candau, S.J., Lehn, J.-M. (2006) *Macromolecules*, **39**, 1173–1181; (c) Seki, T., Yagai, S., Karatsu, T., and Kitamura, A. (2008) *J. Org. Chem.*, **73**, 3328–3335.

41 Korshak, V.V. and Frunze, T.M. (1955) Bulletin of the Academy of Sciences of the USSR. *Div. Chem. Sci. (English Translation)*, **1**, 143–149.

42 (a) Koevoets, R.A., Versteegen, R.M., Kooijman, H., Spek, A.L., Sijbesma, R.P., and Meijer, E.W. (2005) *J. Am. Chem. Soc.*, **127**, 2999–3003; (b) Versteegen, R.M., Sijbesma, R.P., and Meijer, E.W. (2005) *Macromolecules*, **38**, 3176–3184; (c) Wisse, E., Govaert, L.E., Meijer, H.E.H., and Meijer, E.W. (2006) *Macromolecules*, **39**, 7425–7432; (d) Botterhuis, N.E., Karthikeyan, S., Veldman, D., Meskers, S.C.J., and Sijbesma, R.P. (2008) *Chem. Commun.*, 3915–3917; (e) Botterhuis, N.E., Karthikeyan, S., Spiering, A.J.H., and Sijbesma, R.P. (2010) *Macromolecules*, **43**, 745–751.

43 Wisse, E., Spiering, A.J.H., Pfeifer, F., Portale, G., Siesler, H.W., and Meijer, E.W. (2009) *Macromolecules*, **42**, 524–530.

44 Wisse, E., Spiering, A.J.H., van Leeuwen, E.N.M., Renken, R.A.E., Dankers, P.Y.W., Brouwer, L.A., van Luyn, M.J.A., Harmsen, M.C., Sommerdijk, N.A.J.M., and Meijer, E.W. (2006) *Biomacromolecules*, **7**, 3385–3395.

45 (a) Cordier, P., Tournilhac, F., Soulié-Ziakovic, C., and Leibler, L. (2008) *Nature*, **451**, 977–980; (b) Montarnal, D., Tournilhac, F., Hidalgo, M., Couturier, J.-L., and Leibler, L., *J. Am. Chem. Soc.* (2009) **131**, 7966–7967.

46 (a) Peter Lillya, C., Baker, R.J., Hütte, S., Henning Winter, H., Lin, Y.-G., Shi, J., Charles Dickinson, L., and Chien, J.C.W. (1992) *Macromolecules*, **25**, 2076–2080; (b) Duweltz, D., Lauprêtre, F., Abed, S., Bouteiller, L., and Boileau, S. (2003) *Polymer*, **44**, 2295–2302.

47 (a) Ky Hirschberg, J.H.K., Beijer, F.H., van Aert, H.A., Magusin, P.C.M.M., Sijbesma, R.P., and Meijer, E.W. (1999) *Macromolecules*, **32**, 2696–2705; (b) Folmer, B.J.B., Sijbesma, R.P., Versteegen, R.M., van der Rijt, J.A.J., and Meijer, E.W. (2000) *Adv. Mater.*, **12** (12), 874–878; (c) Rowan, S.J., Suwanmala, P., and Sivakova, S. (2003) *J. Polym. Sci, Part A*, **41** (22), 3589–3596; (d) Sivakova, S., Bohnsack, D.A., Mackay, M.E., Suwanmala, P., and Rowan, S.J. (2005) *J. Am. Chem. Soc.*, **127**, 18202–18211.

48 (a) Bohle, A., Brunklaus, G., Hansen, M.R., Schleuss, T.W., Kilbinger, A.F.M., Seltmann, J., and Spiess, H.W. (2010) *Macromolecules*, **43**, 4978–4985; (b) Ibarboure, E. and Rodríguez-Hernández, J. (2010) *Eur. Polym. J.*, **46**, 891–899.

49 (a) Hilger, C. and Stadler, R. (1990) *Makromol. Chem.*, **191**, 1347–1361; (b) Hilger, C. and Stadler, R. (1991) *Makromol. Chem.*, **192**, 805–817; (c) Schirle, M., Beckmann, J., and Stadler, R. (1992) *Angew. Makromol. Chem.*, 261–282.

50 Woodward, P.J., Merino, D.H., Greenland, B.W., Hamley, I.W., Light, Z., Slark, A.T., and Hayes, W. (2010) *Macromolecules*, **43**, 2512–2517.

51 (a) Rowan, S.J., Suwanmala, P., and Sivakova, S. (2003) *J. Polym. Sci, Part A*, **41**, 3589–3596; (b) Sivakova, S., Bohnsack, D.A., Mackay, M.E., Suwanmala, P., and Rowan, S.J. (2005) *J. Am. Chem. Soc.*, **127**, 18202–18211.

52 Lin, I.-H., Cheng, C.-C., Yen, Y.-C., and Chang, F.-C. (2010) *Macromolecules*, **43**, 1245–1252.

53 Kautz, H., van Beek, D.J.M., Sijbesma, R.P., and Meijer, E.W. (2006) *Macromolecules*, **39** (13), 4265–4267.

54 (a) van Beek, D.J.M., Spiering, A.J.H., Peters, G.W.M., te Nijenhuis, K., and Sijbesma, R.P. (2007) *Macromolecules*, **40**, 8464–8475; (b) Botterhuis, N.E., van Beek, D.J.M., van Gemert, G.M.L., Bosman, A.W., and Sijbesma, R.P. (2008) *J. Polym. Sci, Part A*, **46**, 3877–3885.

55 Nieuwenhuizen, M.M.L., de Greef, T.F.A., van der Bruggen, R.L.J., Paulusse, J.M.J., Appel, W.P.J., Smulders, M.M.J., Sijbesma, R.P., and Meijer, E.W. (2010) *Chem. Eur. J.*, **16**, 1601–1612.

56 (a) Dankers, P.Y.W., Harmsen, M.C., Brouwer, L.A., van Luyn, M.J.A., and Meijer, E.W. (2005) *Nature Mater.*, **4**, 568–574; (b) Dankers, P.Y.W., van Leeuwen, E.N.M., van Gemert, G.M.L., Spiering, A.J.H., Harmsen, M.C., Brouwer, L.A., Janssen, H.M., Bosman, A.W., van Luyn, M.J.A., and Meijer, E.W. (2006) *Biomaterials*, **27**, 5490–5501.

57 Dankers, P.Y.W., Boomker, J.M., Huizinga-van der Vlag, A., Smedts, F.M.M., Harmsen, M.C., and van Luyn, M.J.A. (2010) *Macromol. Biosci.*, **10**, 1345–1354.

58 (a) Lightfoot, M.P., Mair, F.S., Pritchard, R.G., and Warren, J.E. (1999) *Chem. Commun.*, 1945; (b) Smulders, M.M.J., Schenning, A.P.H.J., and Meijer, E.W. (2008) *J. Am. Chem. Soc.*, **130**, 606–611; (c) Fan, E., Yang, J., Geib, S.J., Stoner, T.C., Hopkins, M.D., and Hamilton, A.D. (1995) *J. Chem. Soc., Chem. Commun.*, 1251–1252; (d) Hanabusa, K., Kawakami, A., Kimura, M., and Shirai, H. (1997) *Chem. Lett.*, **5**, 429–430; (e) van Bommel, K.J.C., van der Pol, C., Muizebelt, I., Friggerri, A., Heeres, A., Meetsma, A., Feringa, B.L., and van Esch, J.H. (2004) *Angew. Chem., Int. Ed.*, **43**, 1663–1667.

59 Roosma, J., Mes, T., Leclère, P., Palmans, A.R.A., and Meijer, E.W. (2008) *J. Am. Chem. Soc.*, **130**, 1120–1121.

60 (a) Stals, P.J.M., Everts, J.C., de Bruijn, R., Filot, I.A.W., Smulders, M.M.J., Martín-Rapún, R., Pidko, E.A., de Greef, T.F.A., Palmans, A.R.A., and Meijer, E.W. (2010) *Chem. Eur. J.*, **16**, 810–821; (b) Filot, I.A.W., Palmans, A.R.A., Hilbers, P.A.J., van Santen, R.A., Pidko, E.A., and de Greef, T.F.A. (2010) *J. Phys. Chem. B*, **114**, 13667–13674.

61 (a) Johan Foster, E., Berda, E.B., and Meijer, E.W. (2009) *J. Am. Chem. Soc.*, **131**, 6964–6966; (b) Berda, E.B., Johan Foster, E., and Meijer, E.W. (2010) *Macromolecules*, **43** (3), 1430–1437.

2
Cyclodextrin-Based Supramolecular Polymers
Akira Harada and Yoshinori Takashima

2.1
Introduction

The supramolecular polymer is a polymer based on the association of monomers through noncovalent interactions that spontaneously form by aggregation in solution or in a solid. One characteristic of supramolecular polymers is the reversibility of the connecting events and annealing or self-healing of the defects. Since Lehn introduced the concept of these polymers in his book, *Supramolecular Chemistry* [1], several research groups have rapidly developed supramolecular polymers [1–5]. Systems specifically based on hydrogen bonding as well as other types of noncovalent interactions have been actively reported following advances. These developments that have occurred in the synthesis of complex structures have arisen from an understanding of chemical design principles. To date, several reviews have reported the development of supramolecular polymers [1,3d].

Supramolecular polymers can be roughly classified as main-chain or side-chain types. These supramolecular polymers have been constructed by specific noncovalent interactions between the end groups of classical covalent polymers and between repeating units on side chains of classical covalent polymers, respectively. However, few reports have examined the formation of supramolecular polymers consisting of cyclodextrins. This paper focuses on daisy-chain type supramolecular polymers (oligomers) consisting of cyclodextrins.

2.2
Supramolecular Polymers in the Solid State

X-ray crystallography has provided much information about the three-dimensional structure of the host – guest complexes. French and Rundle were the first to apply single-crystal X-ray diffraction to α-cyclodextrin (α-CD) and β-cyclodextrin (β-CD) [6]. Although James *et al.* [7] have reported crystallographic data for several α-CD complexes, their resolution for structural analysis of the iodine complex was very low. Several years later, Hybl *et al.* [8] reported the full X-ray structure of an α-CD complex

Supramolecular Polymer Chemistry, First Edition. Edited by Akira Harada.
© 2012 Wiley-VCH Verlag GmbH & Co. KGaA. Published 2012 by Wiley-VCH Verlag GmbH & Co. KGaA.

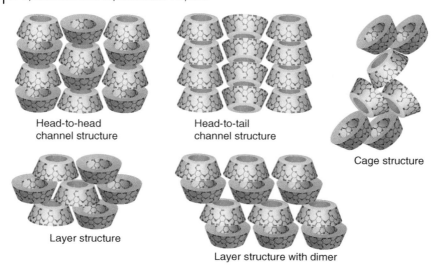

Figure 2.1 The classification of crystal packing structures of CD inclusion complexes.

with potassium acetate. Since then, several reviews have summarized research on the crystal structures of cyclodextrins and their inclusion complexes [9–13].

The crystal structures of CDs and their inclusion complexes can be classified as cage-, channel-, and layer-type structures (Figure 2.1). These classes depend on the relationship between CDs and the guest molecules. Recently, some monosubstituted CDs have exhibited catalytic activity as well as forming supramolecular complexes. Modified CDs have been thoroughly reviewed [14–16]. For example, the crystal structure of 6-O-(*tert*-butylthio)-β-CD molecules is arranged along a screw axis where the *tert*-butylmercaptan group is inserted into the next β-CD cavity to form a helically extended polymeric structure [17]. This type of modified CD behaves as both a donor of the guest group and as an acceptor; these types of structures are called 'supramolecular polymers' in the solid states. This section focuses on the crystal structures of monosubstituted CDs.

2.2.1
Crystal Structures of CD Aliphatic Tethers

The crystal structure of 6-O-(*tert*-butylthio)-β-CD has molecules arranged along a twofold axis with the *tert*-butylmercaptan group inserted into the next β-CD cavity to form an extended polymeric structure [17]. In addition to providing the first example of a monosubstituted CD derivative, this is the first evidence of a supramolecular polymer of an inclusion complex of a monosubstituted CD (Figure 2.2). Since then the crystal structures of 6-monosubstituted β-CDs have been reported, including 6-O-[(R)-2-hydroxypropyl]-β-CD, 6-O-[(S)-2-hydroxypropyl]-β-CD [19], 6-O-(6-cyclo(L-histidyl-L-leucyl))-β-CD [20], 6-O-azido-α-CD, 6-O-allyl-α-CD [21], 6-O-[(6-aminohexyl)

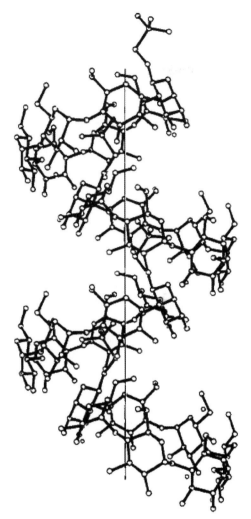

Figure 2.2 The crystal structure of 6-O-(*tert*-butylthio)-β-CD drawing of a helical polymer related by 2_1 screw axis. Reprinted with permission from American Chemical Society 1982, [17].

amino]-β-CD [22], 6-O-(1-propyl)amino-β-CD, and 6-O-[(R)-1-cyclohexylethyl] amino-β-CD [23].

2.2.2
Crystal Structures of β-CDs Aromatic Tethers

Tabushi and Higuchi *et al.* have conducted an X-ray crystallographic study on 6-O-phenylthio-β-CD and 6-O-phenylsulfinyl-β-CD [18]. The guest phenylsulfinyl group

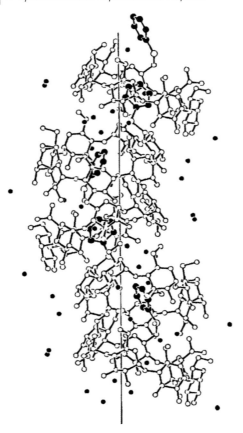

Figure 2.3 The crystal structure of 6-O-phenylthio-β-CD drawing the helical polymer penetrated by 4_1 screw axis. Reprinted with permission from Royal Society of Chemistry 1987, [18].

in 6-O-phenylsulfinyl-β-CD is more deeply included in the host CD cavity than that of 6-O-phenylthio-β-CD. However, both CDs have packing structures where the molecules are arranged around a screw axis to give unique polymeric inclusion column structures from a single species. The macrocycles in helical columns are related by a fourfold screw axis for 6-O-phenylthio-β-CD (Figure 2.3) and a twofold screw axis for 6-O-phenylsulfinyl-β-CD.

Harata et al. have reported the crystal structures of 6-O-[(R)-1-phenylethyl]amino-β-CD and 6-O-[(1R,2S)-2-hydroxyindan-1-yl]amino-β-CD [23]. They discussed the relationship between crystal packing and the inclusion of the substituent group. In each crystal, the substituent groups are inserted into the adjacent β-CD cavity from the secondary hydroxyl side. This type of host–guest self-association through intermolecular inclusion generates a one-dimensional polymeric chain. Moreover, the crystal structures of 6-O-anilino-β-CD [24], 6-O-(4-formylphenyl)-β-CD [25], 6-O-phenylselenyl-β-CD, 6-O-(4-bromophenyl)-β-CD [26], 6-O-(4-nitrophenyl)-β-CD [26],

2.3
Formation of Homo-Intramolecular and Intermolecular Complexes by CDs–Guest Conjugates

6-O-(4-formylphenyl)-β-CD [26], and 6-O-(p-carboxyphenylamino)-β-CD [27] have been reported.

2.3
Formation of Homo-Intramolecular and Intermolecular Complexes by CDs–Guest Conjugates

Numerous modified CDs have been prepared and characterized as their supramolecular structures in aqueous solutions. Most of these form intramolecular complexes. Although chromophore-modified CDs act as chemosensors by changing the location of the substitution part from inside to outside of the CD cavity with the inclusion of other guest molecules (Figure 2.4) [28–30], 6-benzoyl-β-CD, a simple CD–guest conjugate, does not show an appreciable interaction between the benzoyl and β-CD moieties [31]. This result suggests that spacer groups are required to efficiently form intermolecular complexes. Tables 2.1–2.4 summarize the chemical structures of CDs with guest molecules.

2.3.1
Supramolecular Structures Formed by 6-Modified α-CDs

As a typical guest group for CDs, cinnamate compounds (hydroxycinnamoyl, cinnamoyl, and cinnamide groups) have been introduced to α-CD at the C6 position because the flexibility of the guest groups is easily adjusted by bonds. Thus, 6-hydrocinnamoyl-α-CD, 6-cinnamoyl-α-CD, and 6-cinnamide-α-CD (6-HyCiO-α-CD, 6-CiO-α-CD and 6-CiNH-α-CD) have been synthesized, and their complexation has been studied [32]. The phenyl proton peaks in the ^{1}H NMR spectra of 6-Ci-α-CDs (6-HyCiO-α-CD, 6-CiO-α-CD, and 6-CiNH-α-CD) in D_2O exhibit similar shifts to those of ethyl hydrocinnamate upon the addition of CDs, indicating that the

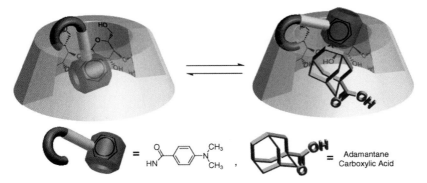

Figure 2.4 Schematic representation of fluorescent chemosensor by intramolecular complexes having a chromophore. Parts reprinted with permission from American Chemical Society 2009, [59].

Table 2.1 Chemical Structures of 6-modified α-CDs with guest molecules.

α-CDs		R
	6-HyCiO-α-CD	
	6-CiO-α-CD	
	6-CiNH-α-CD	
	6-NH₂CiO-α-CD	
	(6-TNBCiO-α-CD)₂ (6-TNBCiO-α-CD)₃	
	6-NH₂Sti-α-CD	
	(6-TNBSti-α-CD)₂	
	6-NH₂StiO-α-CD	
	(6-TAzStiO-α-CD)₂	
	[α-CD(OMe)₁₇(6-NH₂Azo)]₂	
	[α-CD(OMe)₁₇(6-NpAzo)]₂	
	[α-CD(OMe)₁₇(6-PhAzo)]₂	R = H, OH
	[α-CD(OMe)₁₇(6-OHAzo2)]₄	
	[α-CD(OMe)₁₇(6-NpAzoO)]₅	

2.3 Formation of Homo-Intramolecular and Intermolecular Complexes by CDs–Guest Conjugates

Table 2.2 Chemical Structures of 6-modified β-CDs with guest molecules.

α-CDs		R
	6-CiO-β-CD	(cinnamoyl ester group)
	6-HyCiO-β-CD	(hydroxycinnamoyl ester)
	6-HH₂CiO-β-CD	(4-aminocinnamoyl ester, –NH₂)
	6-CiNH-β-CD	(cinnamoyl amide)
	6-AzoEg-β-CD	(azobenzene ethylene glycol)
	6-OctVP-β-CD	(octyl viologen)
	6-AzaCy-β-CD	(aza-crown)
	6-FcAzi-β-CD	(ferrocenyl azide)

phenyl ring is included in the CD cavity. Although the shifts depend on the concentration in D_2O, they are independent of their concentrations in DMSO-d_6, indicating that 6-Ci-α-CDs form intermolecular complexes in D_2O (Figure 2.5a). The phenyl and CD signals in the ^1H NMR and ROESY spectra of 6-Ci-α-CDs have correlation peaks. This observation suggests that the phenyl ring is included in another CD cavity. The molecular weight of 6-CiO-α-CD measured by vapor pressure osmometry (VPO) becomes saturated around 3×10^3 at 30 mM, implying that 6-CiO-α-CD forms an oligomer. The molecular weight of 6-CiNH-α-CD measured by VPO becomes saturated about 2×10^3, which is twice that of the monomer unit. Therefore, 6-CiNH-α-CD forms a supramolecular dimer (Figure 2.5b). 6-(4-Aminocinnamoyl)-α-CD (6-NH₂CiO-α-CD) forms supramolecular oligomers in aqueous solutions. The molecular size of the supramolecular oligomers has been determined by blocking each end of the 4-aminocinnamoyl group with 2,4,6-trinitrobenezene sulfonate sodium salt (TNBS Na). Supramolecular cyclic dimers and trimers (i.e., cyclic daisy chains) are formed. In these daisy chains, all the monomers are linked by mechanical bonds (Figure 2.6a) [33].

Table 2.3 Chemical Structures of 3-modified α-CDs with guest molecules.

α-CDs		R
	3-CiNH-α-CD	
	3-*p*-NH$_2$CiNH-α-CD	
	3-*p*-^{t}BocCiNH-α-CD	
	3-*trans*-Sti-α-CD	
	3-CiHexNH-α-CD	
	3-(*N*-CiHex-*N*-CiLys)-α-CD	

6-(4-Aminostilbene)-α-CD (6-NH$_2$Sti-α-CD) has been prepared by treatment of tosylate-α-CD with diamino stilbene and potassium iodide in *N*-methylpyrrolidin-2-one. At a concentration of 34 mM in an aqueous carbonate buffer, 6-NH$_2$Sti-α-CD reacts with TNBS Na to give rotaxane dimer (6-TNBSti-α-CD)$_2$ (Figure 2.6b) [34]. TNBS Na is a typical stopper for α-CD as it inhibits the decomposition of the supramolecular complex in organic solvents, but TNBS Na is also an electric acceptor as it inhibits photoisomerization of photochromic molecules such as stilbene compounds. To resolve this issue, Easton and his coworkers introduced a methoxytriazine group as a novel bulky stopper α-CD. Treatment of hermaphroditic stilbenyl-α-CD (6-NH$_2$StiO-α-CD) with dichlorotriazine affords the [*c*2]-daisy chain ((6-TAzStiO-α-CD)$_2$) in 13% yield [35].

Kaneda and his coworkers have prepared Janus dimers, cyclic tetramer, and [5] supercyclodextrin with permethylated α-CD. The [5]supercyclodextrin [α-CD (OMe)$_{17}$(6-NpAzoO)]$_5$, which forms a nanosized cyclopentameric array held by only mechanical bonds, is synthesized by the pentakis-azo coupling of a new hermaphrodite monomer with 2-naphthol as a stopper (Figure 2.6c) [36].

Table 2.4 Chemical structures of monomer units of modified CDs to form alternating supramolecular polymers.

Run	CD Monomers
1	6-p-tBocClNH-β-CD; 3-AdHexNH-α-CD
2	3-ClO-α-CD; 2-ClO-α-CD

2 Cyclodextrin-Based Supramolecular Polymers

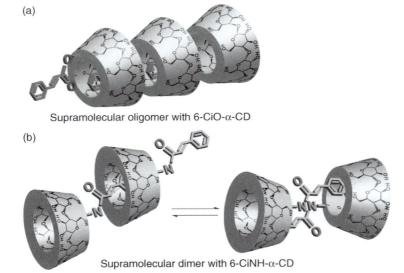

Figure 2.5 Proposed structures of 6-CiO-α-CD (a) and 6-CiNH-α-CD (b).

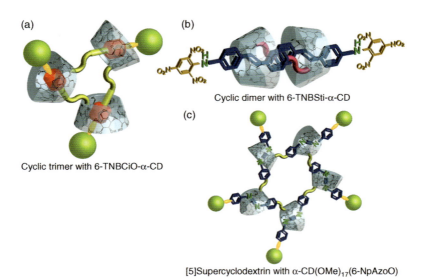

Figure 2.6 Daisy chain oligomer with (a) (6-TNBCiO-α-CD)$_3$, (b) (6-TNBSti-α-CD)$_2$, and (c) [α-CD(OMe)$_{17}$(6-NpAzO)]$_5$. Parts reprinted with permission from American Chemical Society 2009, [59].

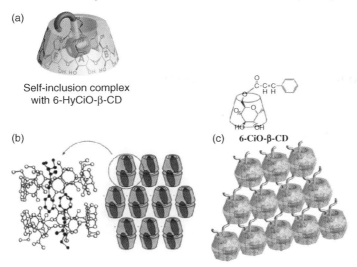

Figure 2.7 Self-inclusion complex of 6-HyCiO-β-CD (a), crystal structure of the 6-CiO-β-CD in the solid state (b), and crystal structure of 6-CiO-β-CD (c). Reprinted with permission from American Chemical Society 2009, [59].

2.3.2
Supramolecular Structures Formed by 6-Modified β-CDs

Table 2.2 summarizes the chemical structures of β-CDs with guest molecules that form supramolecular complexes. The peak shifts of 6-HyCiO-β-CD are independent of the concentration, but the shifts of 6-CiO-β-CD are slightly influenced by the concentrations in D_2O. The ^1H NMR and ROESY spectra of 6-HyCiO-CDs exhibit correlation peaks between the phenyl and CD signals in which the benzene ring is sandwiched between glucose A and glucose D (Figure 2.7a). These results indicate that 6-HyCiO-β-CD forms intramolecular complexes in D_2O, whereas 6-HyCiO-α-CD forms weak intermolecular complexes in D_2O [32]. 6-CiO-β-CD is only slightly soluble in water. X-ray analysis demonstrated that 6-CiO-β-CD yields multilayered supramolecular cyclic dimers (Figure 2.7b). The crystal structure of 6-aminocinnamoyl-β-CD (6-NH_2CiO-β-CD) indicates intermolecular complexes are formed in a tail-to-tail fashion in the solid state (Figure 2.8a) [37]. The substituent group is inserted into an adjacent β-CD cavity from the primary hydroxyl group to form a columnar channel structure.

Trans-azobenzene -modified β-CD (6-AzoEg-β-CD) forms an association dimer, which limits competitive guest complexation. However, exposure to UV and visible lights achieves *trans-cis* and *cis-trans* isomerization, respectively. After photoirradiation of *trans*-6-AzoEg-β-CD, *cis*-6-AzoEg-β-CD decomposes into the intermolecular supramolecular complex [38]. Later researchers have reported supramolecular complexes of 6-OctVP-β-CD, [39] 6^A-deoxy-6^A-(6-(2-(1,4,7,10-tetraoxa-13-azacyclo-

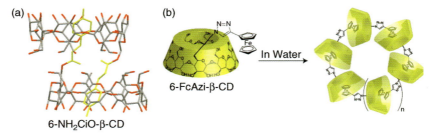

Figure 2.8 Crystal structure of 6-NH$_2$CiO-β-CD drawing dimeric columnar structure by dimeric columnar structure (a) and cyclic supramolecular polymer from 6-FcAzi-β-CD (b) Parts reprinted with permission from American Chemical Society 2005, [37].

pentadecan-13-yl)acetamido)hexylamino)-β-CD, and 6^A-deoxy-6^A-(6-(2-(1,4,7,10,13-pentaoxa-16-azacyclooctadecan-16-yl)acetamido)hexylamino)-β-CD form Janus assemblies and daisy chains [40].

Ferrocene and its derivatives are very attractive guest molecules for β-CD due to their high affinity for β-CD and their redox properties. Previously, ferrocene-CD conjugates (Fc-β-CDs) have been studied as potential artificial enzymes [41] and redox-switching systems [42]. Ritter and his coworkers have reported a novel CD derivative bearing Fc via click chemistry [43]. DLS and cryo-TEM measurements indicate that 6-FcAzi-β-CD forms 'ring-like' self-assembly properties (Figure 2.8b). This host–guest induced effect is due to the incorporation of the Fc moiety into the CD cavity.

2.3.3
Supramolecular Structures Formed by 3-Modified α-CDs

As the previous two sections demonstrate, supramolecular polymers do not easily form from 6-modified CDs in aqueous solutions. Although some aromatic groups of CDs are included from the secondary hydroxyl side of CDs (i.e., the wider side), others are included from the primary hydroxyl side. If the guest group is attached to one of the secondary hydroxyl groups, supramolecular polymers can be realized in aqueous solutions. This section focuses on supramolecular complexes from 3-modified CDs where the guest group at the C3 position is on the secondary hydroxyl side.

Table 2.3 summarizes the chemical structures of 3-modified CDs with guest molecules that form supramolecular complexes. 3-CiNH-α-CD and 3-p-NH$_2$CiNH-α-CD form long supramolecular polymers with molecular weights greater than 1.6×10^4 [44, 45]. Poly[2]rotaxanes (daisy chain) are formed by 3-p-NH$_2$CiNH-α-CD stabilized by bulky stoppers (TNBS and adamantanamide) at each end of the guest groups (Figure 2.9a). MALDI-TOF indicates that poly[2]rotaxane is about 13 units long [46]. 3-p-NH$_2$CiNH-α-CD has been further modified with a *tert*-butoxycarbonyl (*t*-Boc) group at the end of the guest moiety to form more stable supramolecular polymers. VPO indicates that M_n for supramolecular polymers of 3-p-^{t}BocCiNH-α-C exceeds 1.6×10^4 at concentrations above 15 mM.

2.3 Formation of Homo-Intramolecular and Intermolecular Complexes by CDs–Guest Conjugates | 41

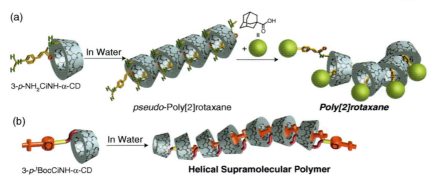

Figure 2.9 Schematic representation of supramolecular polymers constructed by 3-CiNH-α-CD in aqueous solution and Helical supramolecular polymer formed by 3-p-tBocCiNH-α-CD. Figure 2.9 (a) reprinted with permission from Royal Society of Chemistry 2008, [46].

The negative–positive splitting of the Cotton bands in the circular dichroism spectra demonstrates the substitution/substitution interactions among adjacent monomers in the supramolecular polymer in aqueous solutions. The splitting pattern suggests that the supramolecular polymer has a left-handed helical conformation, which has been confirmed by atomic force microscopy (Figure 2.9b) [47]. Moreover, the CD spectra confirm some cooperativity in the formation of helical supramolecular polymers. The intensity of the split Cotton bands increases exponentially as the concentration increases, which further supports cooperativity.

We have examined structural control of supramolecular complexes via photoirradiation. Stilbene derivatives show photoinduced isomerization, which occurs from the trans to the cis form and from the cis to the trans form under irradiation with UV and visible light, respectively. Stilbene compounds are favorable photoswitchable guest compounds because the association constant of α-CD for *trans*-stilbene is larger than that for *cis*-stilbene. X-ray crystallographic analysis of 3-*trans*-Sti-α-CD indicates that a double-threaded dimer is formed. After photoirradiation with visible light ($\lambda = 340$ nm), the diffusion coefficient of 3-*cis*-Sti-α-CD significantly decreases as the concentration increases, suggesting that 3-*trans*-Sti-α-CD forms a double-threaded dimer, but 3-*cis*-Sti-α-CD forms a nonthreaded supramolecular self-assembly, approximately a 15-mer, in a 1 mM aqueous solution (Figure 2.10) [48].

Linear supramolecular polymers formed by 6-CiO-α-CD or 3-CiNH-α-CD demonstrate that the viscosity of supramolecular polymers is lower than that expected from the molecular weight of the supramolecular polymers. The introduction of branched units likely plays an important role in obtaining viscous supramolecular systems. 3-Cinnamoylaminohexanamide-α-CD (3-CiHexNH-α-CD) and 3-N^{α}-cinnamolyaminohexancarbonyl-N^{ε}-cinnamoyl-lysinamide-α-CD (3-(N-CiHexNH-N-CiLys)-α-CD) bearing two guest parts have been employed to improve the degree of polymerization as well as the physical properties [49]. The reduced viscosity of 3-CiNH-α-CD and 3-CiHexNH-α-CD moderately increases as the concentration increases. These results indicate that 3-CiNH-α-CD and 3-CiHexNH-α-CD form linear supramolecular polymers at high concentrations. Although the reduced

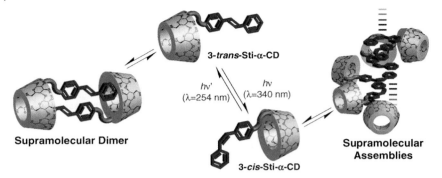

Figure 2.10 Schematic illustration of switching between supramolecular dimer and nonthreaded supramolecular self-assembly consisting of 3-Sti-α-CD with photoirradiation. Reprinted with permission from American Chemical Society 2008, [48].

viscosity of compound 3-(N-CiHexNH-N-CiLys)-α-CD is too low to be measured due to its low solubility in aqueous solutions, mixtures of the 3-(N-CiHexNH-N-CiLys)-α-CD derivative and 3-CiHexNH-α-CD show a higher reduced viscosity (4.0 cm^3 g^{-1}) at 35 mM, implying the formation of supramolecular polymer networks (Figure 2.11).

2.3.4
Hetero-Supramolecular Structures Formed by Modified CDs

An isomeric mixture should form homo- or random-supramolecular polymers during the process of narcissistic self-sorting; molecules with a high affinity for themselves operate in these systems. Social self-sorting is relatively rare in supramolecular polymers because two hetero units often form a thermodynamically unfavorable structure and assemble into a less compact structure. A specific molecular design requires alternating supramolecular polymers to be prepared. Table 2.4 summarizes the chemical structures of the monomeric units of modified CDs that form alternating supramolecular polymers.

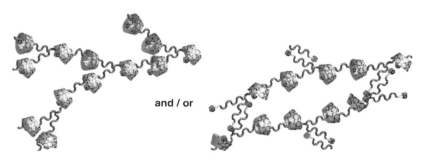

Figure 2.11 Schematic illustration of branched supramolecular polymers formed by 3-CiHexNH-α-CD and 3-(N-CiHexNH-N-CiLys)-α-CD complexes in aqueous solutions.

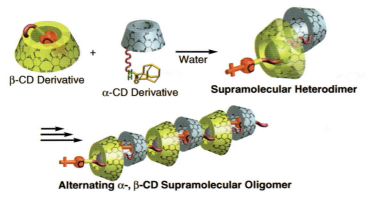

Figure 2.12 Proposed structures of supramolecular polymer incorporating 3-AdHexNH-α-CD and 6-p-^{t}BocCiNH-β-CD. Reprinted with permission from American Chemical Society 2004, [50].

Similarly to 6-HyCiO-β-CD, modified β-CDs with relatively larger cavities tend to form intramolecular self-inclusion complexes in aqueous solutions. Although 3-p-^{t}BocCiNH-α-CD effectively forms a helical supramolecular polymer, 6-p-^{t}BocCiNH-β-CD forms a self-inclusion complex. On the other hand, adding an adamantane guest compound, which is an excellent guest molecule for β-CD, enables 6-p-^{t}BocCiNH-β-CD to decompose into the self-inclusion complex, indicating that the cinnamoyl group is ejected from the β-CD cavity.

To examine these properties, supramolecular copolymers formed from α-CD and β-CD derivatives have been designed to extend our supramolecular polymer research to include more sophisticated structures. Because α- and β-CDs interact strongly with ^{t}Boc-cinnamoyl and adamantyl groups, respectively, an α-CD derivative modified with the adamantyl group via a hexa(methylene) linker (3-AdHexNH-α-CD) and a β-CD derivative modified with a Boc-modified cinnamoyl group (6-p-^{t}BocCiNH-β-CD) have been prepared. Mixing 3-AdHexNH-α-CD in a one-to-one ratio with 6-p-^{t}BocCiNH-β-CD in D_2O forms a heterodimer, which aligns end-to-end in longitudinal rows to form supramolecular oligomers in an alternating manner (Figure 2.12) [50]. M_n for the supramolecular polymer is $<1.0 \times 10^4$ at 10 mM by VPO, and M_n appears to increase as the concentration is further increased.

More recently, isomers of cinnamoyl α-CDs (2-CiO-α-CD and 3-CiO-α-CD) have self-organized to give different types of supramolecular complexes in aqueous solutions. 2-CiO-α-CD forms a double-threaded dimer, which has been characterized by single-crystal X-ray analysis. In contrast, 3-CiO-α-CD forms a supramolecular oligomer, which has been characterized by pulsed field gradient spin-echo NMR. The addition of 2-CiO-α-CD to an aqueous solution of 3-CiO-α-CD triggers the formation of an alternating supramolecular oligomer via a social self-sorting system, but not self- or random- supramolecular complexes (Figure 2.13) because CiO-α-CDs recognizes the difference in the substitution position on the glucopyranose unit [51].

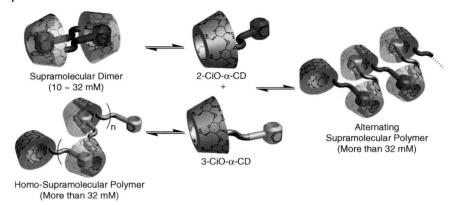

Figure 2.13 Schematic illustration of the formation of supramolecular complexes using CiO-α-CDs [51]. Parts reprinted with permission from American Chemical Society 2009.

2.4
Formation of Intermolecular Complexes by CD and Guest Dimers

Numerous studies have examined the cooperative binding of CD dimers and their guests, but few have investigated the formation of intermolecular complexes of CD dimers. To obtain supramolecular polymers, small cyclic complexes should not be formed. Thus, the conformation of supramolecular complexes from β-CD dimers with terephthalic acid linkers has been investigated using ditopic guest dimers with varying flexibility and adamantyl moieties [52, 53]. ROESY NMR spectra of the β-CD dimer with ditopic adamantane guest dimers show NOE between the protons of the adamantyl substituents and the inner protons of CD. The β-CD dimer forms high-molecular-weight supramolecular polymers with stiff guest molecules as well as cyclic supramolecular oligomers with flexible ditopic guest dimers (Figure 2.14a).

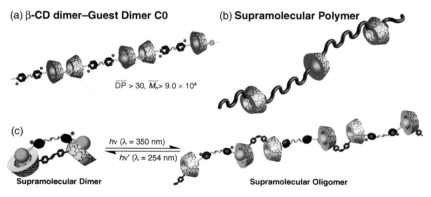

Figure 2.14 Proposed structures of supramolecular polymer incorporating β-CD dimer with adamanane dimer (a), the β-CD-PEG dimer with the adamantane-PEG dimer (b), and external stimulus-responsive supramolecular polymers constructed by a stilbene CD dimer (c). Figure 2.14 (a) reprinted with permission from American Chemical Society 2005, [52]. (b) reprinted with permission from American Chemical Society 2005, [53]. (c) reprinted with permission from American Chemical Society 2007, [54].

A 1:1 mixture of the β-CD-PEG dimer and the adamantane-PEG dimer forms self-organized intermolecular supramolecular polymers (Figure 2.14b).

The conformation of the stilbene bis(β-CD) dimer in an aqueous solution can be photochemically controlled. First, it was speculated that *trans*-stilbene bis(β-CD) dimer would form a supramolecular polymer and photoirradiation would decompose the *cis*-stilbene bis(β-CD) dimer, but these hypotheses contradict the experimental observations. The ROESY spectra of the β-CD dimer with the adamantyl dimer exhibit NOE between the protons of the adamantyl moieties and the inner protons of CD. When the stilbene bis(β-CD) dimer is in the trans conformation, supramolecular dimers or small assemblies form in solution, whereas the cis conformation gives linear supramolecular polymers with high molecular weights. Thus, an external stimulus can control the structure of the supramolecular polymer (Figure 2.14c) [54].

2.5
Artificial Molecular Muscle Based on c2-Daisy Chain

Recently, the subject of nanoscale artificial molecular muscles and motors has attracted much interest. Designs have been developed employing a double-threaded dimer (c2-daisy chain) in which actuation is inspired by natural muscle such as that of the myosin–actin complex. The double-threaded dimer is a good example of a controlled molecular movement in an artificial molecular muscle system.

N,N'-p-xylylene-linked oligo-Janus [2]rotaxanes, which are based on a permethylated α-CD, have been prepared. Their contractible and extendable nature coupled with the photochromism properties has been investigated by NMR studies (Figure 2.15) [55]. Each Janus unit is isomerized by photoirradiation and heating. However, the experimental evidence for the size change of oligo-Janus[2]rotaxanes through photoisomerization has yet to be elucidated.

As described above, 6-CiNH-α-CD and 6-NH$_2$CiNH-α-CD form double-threaded dimers. However, the substituent groups on α-CDs are too short to mimic the

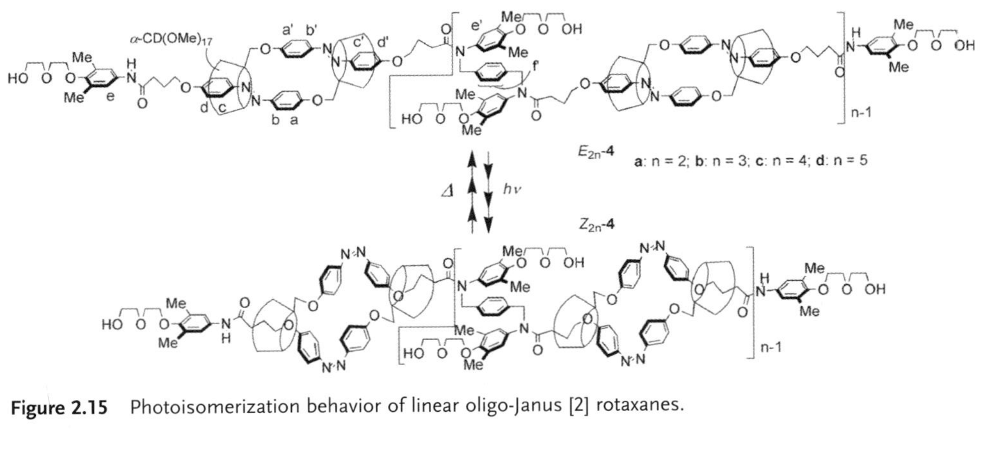

Figure 2.15 Photoisomerization behavior of linear oligo-Janus [2] rotaxanes.

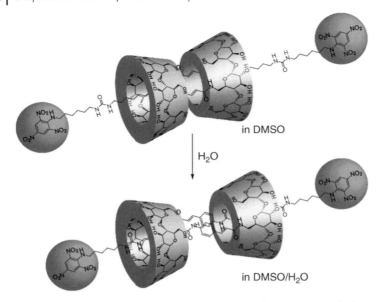

Figure 2.16 A double-threaded dimer bearing a long substituent part and a large stopper group depending on the solvent polarity. Reprinted with permission from American Chemical Society 2007, [56].

contraction and extension of a skeletal muscle. A modified 6-NH$_2$CiNH-α-CD with a long alkyl chain has been prepared to improve the contraction behavior. The double-threaded dimer shows a conformational change as the solvent polarity is increased. To estimate the size of the double-threaded dimer in DMSO-d_6 and H$_2$O/DMSO-d_6 (1:1), the hydrodynamic radii (R_H) of the supramolecular complexes have been determined by the pulse field gradient spin-echo (PFGSE) NMR technique. The apparent volume of double-threaded dimer **3** in DMSO-d_6 and H$_2$O/DMSO-d_6 (1:1) is 8.1 nm^3 and 8.6 nm^3 in a diluted solution (5 mM), respectively. This difference in volume suggests the double-threaded dimer forms a complex with a larger size or is in the stretched state in H$_2$O/DMSO-d_6 (Figure 2.16) [56].

The irradiation wavelength can control the structure of stilbene compounds. A c2-daisy chain with stilbene groups precisely controls the contraction and extraction behavior. Easton has used α-CD and stilbene to construct a reversible molecular mimic of a muscle fiber, which relies on hydrophobic forces and size constraints as well as the photochemical isomerization of the stilbene units to function. *Trans*-stilbene moieties, but not the *cis*-stilbene groups, are included in α-CD cavities. Irradiating at 350 nm causes photoisomerization of the *trans,trans*-daisy chain (*E,E*) to the *cis,trans*- and *cis,cis*-isomer daisy chains (*E,Z*) and (*Z,Z*); the CDs move from the *cis*-stilbene moieties toward the propyl and blocking groups. However, irradiating at 254 nm recovers the *trans,trans*-daisy chain (*E,E*), indicating that the daisy chains (*E,E*), (*E,Z*), and 3(*Z,Z*) represent the extended, intermediate, and contracted states, respectively, and these states depend on the photoirradiation wavelength (Figure 2.17) [57].

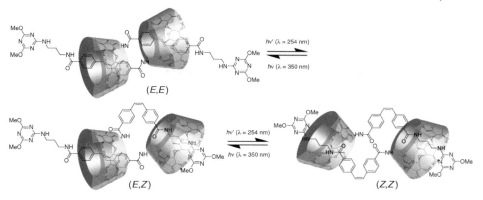

Figure 2.17 Proposed mechanism of external stimulus-responsive supramolecular polymers constructed by a stilbene cyclodextrin dimer.

To observe a large change, a longer linker should be introduced between the two recognition sites. We have synthesized an α- CD derivative substituted on its 3-position with azobenzene (Azo) and heptamethylene (C7) moieties as photoresponsive and nonresponsive recognition sites, respectively, which are connected by a longer linker, oligo(ethylene glycol) (OEG; degree of polymerization (DP) = ∼21) [58]. 2D NMR spectroscopy, CD spectra, and PGSE NMR spectroscopy confirm that the recognition site of the α-CD moiety is switched by photoisomerization of the Azo moiety, which causes a size change in R_H due to the different linker lengths between the two recognition sites (Figure 2.18).

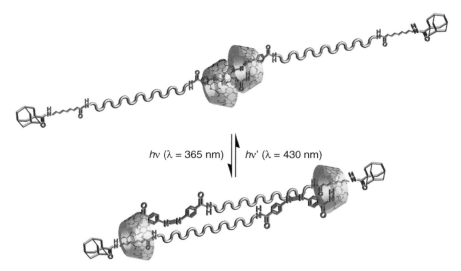

Figure 2.18 Schematic illustration of Janus[2]rotaxane with azobenzene units and poly(ethylene glycol) units photoswitching between the contraction and the extension. Reprinted with permission from Wiley-VCH 2010, [58].

2.6
Conclusion and Outlook

The supramolecular polymers have developed over the past decade. In the future, the self-healing and self-repairing materials are expected to be used for not only architectural materials but also biomedical material. We have focused on the formation of supramolecular polymers consisting of cyclodextrins. CDs have low toxicity for living organisms and high selectivity for guest molecules. Modified cyclodextrin organizes to give a double-threaded dimer or a supramolecular oligomer at high concentrations. Especially, the structure of double-threaded dimers is reminiscent of sarcomere in the smooth muscle fiber. Modified CDs recognized the difference in the substitution position on a glucopyranose unit. Therefore, modified CDs are suitable supramolecular units for the preparation of the artificial molecular muscle. It is important that not only hydrophobic interactions but also π-π stacking and hydrogen-bonding interactions function cooperatively to create supramolecular complexes. However, the external stimuli responsible for CD supramolecular material have yet to be elucidated. Artificial molecular muscle, supramolecular glues, and self-healing materials consisting of CDs are within reach and will be created within the next decade.

References

1. Lehn, J.-M. (1995) *Supramolecular Chemistry: Concepts and Perspectives*, VCH, Weinheim.
2. Ciferri, A. (2000) *Supramolecular Polymers*, Marcel Dekker, NewYork, NY.
3. (a) Brunsveld, L., Folmer, B.J.B., Meijer, E.W., and Sijbesma, R.P. (2001) *Chem. Rev.*, **101**, 4071–4097; (b) Ciferri, A. (2002) *Macromol. Rapid Commun.*, **23**, 511–529; (c) Sijbesma, R.P. and Meijer, E.W. (2003) *Chem. Commun* (1), 5–16; (d) Ciferri, A. (ed.) (2005) *Supramolecular Polymers*, 2nd edn, CRC Press: Taylor and Francis, Boca Raton, FL.
4. (a) Sijbesma, R.P., Beijer, F.H., Brunsveld, L., Folmer, B.J.B., Hirschberg, J.H.K.K., Lange, R.F.M., Lowe, J.K.L., and Meijer, E.W. (1997) *Science*, **278**, 1601–1604; (b) Söntjens, S.H.M., Sijbesma, R.P., van Genderen, M.H.P., and Meijer, E.W. (2000) *J. Am. Chem. Soc.*, **122**, 7487–7493; (c) Ligthart, G.B.W.L., Ohkawa, H., Sijbesma, R.P., and Meijer, E.W. (2005) *J. Am. Chem. Soc.*, **127**, 810–811; (d) Scherman, O.A., Ligthart, G.B.W.L., Sijbesma, R.P., and Meijer, E.W. (2006) *Angew. Chem., Int. Ed.*, **45**, 2072–2076; (e) Greef, T.F.A. and Meijer, E.W. (2008) *Nature*, **453**, 171–173.
5. (c) Castellano, R.K., Rudkevich, D.M., and Rebek, J.Jr. (1997) *Proc. Natl. Acad. Sci. U. S. A.*, **94**, 7132–7137; (b) Vollmer, M.S., Clark, T.D., Steinem, C., and Ghadiri, M.R. (1999) *Angew. Chem., Int. Ed.*, **38**, 1598–1601; (c) Zimmerman, S.C., Zeng, F.W., Reichert, D.E.C., and Kolotuchin, S.V. (1996) *Science*, **271**, 1095–1098; (d) Yang, X., Hua, F., Yamato, K., Ruckenstein, E., Gong, B., Kim, W., and Ryu, C.Y. (2004) *Angew. Chem., Int. Ed.*, **43**, 6471–6474; (e) Hua, F., Yang, X., Gong, B., and Ruckenstein, E. (2005) *J. Polym. Sci., Part A: Polym. Chem.*, **43**, 1119–1128; (f) Park, T., Zimmerman, S.C. and Nakashima, S. (2005) *J. Am. Chem. Soc.*, **127**, 6520–6521.
6. French, D. and Rundle, R.E. (1942) *J. Am. Chem. Soc.*, **64**, 1651–1653.
7. James, W.J., French, D., and Rundle, R.E. (1959) *Acta Crystallogr.*, **12**, 385–389.

8 Hybl, A., Rundle, R.E., and Williams, D.E. (1965) *J. Am. Chem. Soc.*, **87**, 2779–2788.

9 Saenger, W. (1984) Chapter 8, in *Inclusion Compounds*, vol. 2 (eds J.L. Atwood, J.E.D. Davies, and D.D. MacNicol), Academic Press, London.

10 Le Bas, G. and Rysanek, N. (1987) Chapter 3, in *Cyclodextrins and their Industrial Uses* (ed. D. Duchëne), Editions de Santë, Paris.

11 Harata, K. (1991) Chapter 9, in *Inclusion Compounds*, vol. 5 (eds J.L. Atwood, J.E.D. Davies, and D.D. MacNicol), Oxford University Press, New York.

12 Harata, K. (ed.) (1996) Chapter 9, in *Cyclodextrins. Comprehensive Supramolecular Chemistry*, vol. 3 (eds J. Szejtli and T. Osa), Pergamon, Oxford, U.K.

13 Harata, K. (1998) *Chem. Rev.*, **98**, 1803–1827.

14 Easton, C.J. and Lincoln, S.F. (1999) In *Modified Cyclodextrins: Scaffolds and Templates for Supramolecular Chemistry*, Imperial College Press, London.

15 Khan, A.R., Forgo, P., Stine, K.J., and D'Souza, V.T. (1998) *Chem. Rev.*, **98**, 1977–1996.

16 Engeldinger, E., Armspach, D., and Matt, D. (2003) *Chem. Rev.*, **103**, 4147–4173.

17 Hirotsu, K., Higuchi, T., Fujita, K., Ueda, T., Shinoda, A., Imoto, T., and Tabushi, I. (1982) *J. Org. Chem.*, **47**, 1143–1144.

18 Kamitori, S., Hirotsu, K., Higuchi, T., Fujita, K., Yamamura, H., Imoto, T., and Tabushi, I. (1987) *J. Chem. Soc. Perkin Trans. 2*, **1**, 7–14.

19 Harata, K., Rao, C.T., Pitha, J., Fukunaga, K., and Uekama, K. (1991) *Carbohydr. Res.*, **222**, 37–45.

20 Di Blasion, B., Pavone, V., Nastri, F., Isernia, C., Saviano, M., Pedone, C., Cucinotta, V., Impellizzeri, G., Rizzarelli, E., and Vecchio, G. (1992) *Proc. Natl. Acad. Sci. U. S. A.*, **89**, 7218–7221.

21 Hanessian, S., Benalil, A., Simard, M., and Bëlanger-Gariepy, F. (1995) *Tetrahedron*, **51**, 10149–10158.

22 Mentzafos, D., Terzis, A., Coleman, A.W., and de Rango, C. (1996) *Carbohydr. Res.*, **282**, 125–135.

23 Harata, K., Takenaka, Y., and Yoshida, N. (2001) *J. Chem. Soc., Perkin Trans. 2*, **9**, 1667–1673.

24 Liu, Y., You, C.-C., Zhang, M., Weng, L.-H., Wada, T., and Inoue, Y. (2000) *Org. Lett.*, **2**, 2761–2763.

25 Liu, Y., Fan, Z., Zhang, H.-Y., and Diao, C.-H. (2003) *Org. Lett.*, **5**, 251–254.

26 Liu, Y., Fan, Z., Zhang, H.-Y., Yang, Y.-W., Ding, F., Liu, S.-X., Wu, X., Wada, T., and Inoue, Y. (2003) *J. Org. Chem.*, **68**, 8345–8352.

27 Eliadou, K., Giastas, P., Yannakopoulou, K., and Mavridis, I.M. (2003) *J. Org. Chem.*, **68**, 8550–8557.

28 Hamasaki, K., Ikeda, H., Nakamura, A., Ueno, A., Toda, F., Suzuki, I. and Osa, T. (1993) *J. Am. Chem. Soc.*, **115**, 5035–5040.

29 Ueno, A., Minato, S., Suzuki, I., Fukushima, M., Ohkubo, M., Osa, T., Hamada, F., and Murai, K. (1990) *Chem. Lett.*, **4**, 605–608.

30 Wang, Y., Ikeda, T., Ikeda, H., Ueno, A. and Toda, F. (1994) *Bull. Chem. Soc. Jpn.*, **67**, 1598–1607.

31 Tong, L.-H., Hou, Z.-J., Inoue, Y., and Tai, A. (1992) *J. Chem. Soc., Perkin Trans. 2*, **8**, 1253–1257.

32 Harada, A., Kawaguchi, Y., and Hoshino, T. (2001) *J. Inclusion Phenom. Macrocycl. Chem.*, **41**, 115–121.

33 Hoshino, T., Miyauchi, M., Kawaguchi, Y., Yamaguchi, H., and Harada, A. (2000) *J. Am. Chem. Soc.*, **122**, 9876–9877.

34 Onagi, H., Easton, C.J., and Lincoln, S.F. (2001) *Org. Lett.*, **3**, 1041–1044.

35 Dawson, R.E., Maniam, S., Lincoln, S.F., and Easton, C. (2008) *J. Org. Biomol. Chem.*, **6**, 1814–1821.

36 (a) Fujimoto, T., Uejima, Y., Imaki, H., Jung, J.H., Sakata, Y., and Kaneda, T. (2000) *Chem. Lett* (5), 564–565; (b) Fujimoto, T., Sakata, Y., and Kaneda, T. (2000) *Chem. Lett* (7), 764–765; (c) Fujimoto, T., Sakata, Y., and Kaneda, T. (2000) *Chem. Commun.* (21), 2143–2144; (d) Kaneda, T., Yamada, T., Fujimoto, T., and Saka, Y. (2001) *Chem. Lett.* (12), 1264–1265.

37 Miyauchi, M., Hoshino, T., Yamaguchi, H., Kamitori, S., and Harada, A. (2005) *J. Am. Chem. Soc.*, **127**, 2034–2035.
38 Fukushima, M., Osa, T., and Ueno, A. (1991) *J. Chem. Soc., Chem. Commun.*, **1**, 15–17.
39 Park, J.W., Choi, N.H., and Kim, J.H. (1996) *J. Phys. Chem.*, **100**, 769–774.
40 Lock, J.S., May, B.L., Clements, P., Lincoln, S.F., and Easton, C. (2004) *J. Org. Biomol. Chem.*, **2**, 1381–1386.
41 (a) Schreyer, S. and Mikkelsen, S. (1999) *Bioconjug. Chem.*, **10**, 464–469; (b) Suzuki, I., Chen, Q., Kashiwagi, Y., Osa, T., and Ueno, A. (1993) *Chem. Lett.* (10), 1719–1722.
42 (a) Groom, C.A. and Luong, J.H.T. (1994) *Biosens. Bioelectron.*, **9**, 305–313; (b) Luong, J., Brown, S., and Schmid, P. (1995) *J. Mol. Recognit.*, **8**, 132–138.
43 Munteanu, M., Kolb, U., and Ritter, H. (2010) *Macromol. Rapid Commun.*, **31**, 616–618.
44 Miyauchi, M. and Harada, A. (2005) *Chem. Lett.*, **34**, 104–105.
45 Harada, A. (2006) *J. Polym. Sci., Part A: Polym. Chem.*, **44**, 5113–5119.
46 Miyawaki, A., Miyauchi, M., Takashima, Y., Yamaguchi, H., and Harada, A. (2008) *Chem. Commun. (Cambridge, U. K.)* (4), 456–458.
47 Miyauchi, M., Takashima, Y., Yamaguchi, H., and Harada, A. (2005) *J. Am. Chem. Soc.*, **127**, 2984–2989.
48 Yamauchi, K., Takashima, Y., Hashidzume, A., Yamaguchi, H., and Harada, A. (2008) *J. Am. Chem. Soc.*, **130**, 5024–5025.
49 (a) Miyawaki, A., Takashima, Y., Yamaguchi, H., and Harada, A. (2007) *Chem. Lett.*, **36**, 828–829; (b) Miyawaki, A., Takashima, Y., Yamaguchi, H., and Harada, A. (2008) *Tetrahedron*, **64**, 8355–8361.
50 Miyauchi, M. and Harada, A. (2004) *J. Am. Chem. Soc.*, **126**, 11418–11419.
51 Tomimasu, N., Kanaya, A., Takashima, Y., Yamaguchi, H., and Harada, A. (2009) *J. Am. Chem. Soc.*, **131**, 12339–12343.
52 Ohga, K., Takashima, Y., Takahashi, H., Kawaguchi, Y., Yamaguchi, H., and Harada, A. (2005) *Macromolecules*, **38**, 5897–5904.
53 Hasegawa, Y., Miyauchi, M., Takashima, Y., Yamaguchi, H., and Harada, A. (2005) *Macromolecules*, **38**, 3724–3730.
54 Kuad, P., Miyawaki, A., Takashima, Y., Yamaguchi, H., and Harada, A. (2007) *J. Am. Chem. Soc.*, **129**, 12630–12631.
55 Tsuda, S., Aso, Y., and Kaneda, T. (2006) *Chem. Commun.*, **29**, 3072–3074.
56 Tsukagoshi, S., Miyawaki, A., Takashima, Y., Yamaguchi, H., and Harada, A. (2007) *Org. Lett.*, **9** (6), 1053–1055.
57 Dawson, R.E., Lincoln, S.F., and Easton, C. (2008) *J. Chem. Commun.*, **34**, 3980–3982.
58 Li, S., Taura, D., Hashidzume, A., and Harada, A. (2010) *Chem. Asian J.*, **5** (10), 2281–2289.
59 Harada, A., Hashidzume, A., Yamaguchi, H., and Takashima, Y. (2009) *Chem. Rev.*, **109** (11), 5974–6023.

3
Supra-Macromolecular Chemistry: Toward Design of New Organic Materials from Supramolecular Standpoints
Kazunori Sugiyasu and Seiji Shinkai

3.1
Introduction

The quest to understand the principle of weak noncovalent interactions such as hydrogen bonding, electrostatic interaction, van der Waals force, and so on has evoked intense research in supramolecular chemistry. Molecular contrivances with noncovalent interaction sites (molecular recognition sites) have straightforwardly led to chemosensors, artificial enzymes, topological molecules, and so forth [1]. In these systems, the noncovalent interaction sites are arranged so that they can converge on the guest molecule in consideration of the additivity and the complementarity of noncovalent interactions ('convergent' type recognition). In contrast, 'divergent' type molecular recognition, when properly designed, can link the corresponding molecules together, and, particularly when complementary to themselves, they self-assemble to large objects [2]. The programmed direction and the number of noncovalent interactions define not only the size and shape but also the functions of these molecular assemblies; in fact, this adaptability permits vast applications in nanotechnology and materials science. Equally attractive are supramolecular polymers which have structures analogous to those of synthetic polymers, although the repeating units are linked via divergent-type noncovalent interactions [2c].

Molecular assemblies and supramolecular polymers have led to a new horizon in advanced materials. The designed unit molecules and the resultant programmed assemblies of the supramolecular systems continue to offer new and exciting alternatives to conventional covalent bond-based synthetic polymers. The attractive advantages of the supramolecular systems are as follows: (1) reversible, recyclable and/or stimuli responsiveness, (2) flexible design of functions actualized by the program of direction and/or angle of noncovalent interactions (specifically, π–π stacking is an attractive characteristic for organic electronics), and (3) accessibility of nano-sized superstructures such as spheres, tubes, fibers, helices, and so on. From these viewpoints, the definition of 'polymers' as opposed to 'small molecules' could be becoming less meaningful, and molecular assemblies (i.e., supramolecular polymers) nowadays have a great influence on the design of functional materials.

Indeed, these three entities (i.e., small molecules, macromolecules, and supramolecular assemblies) have frequently been associated with each other to create one systematic device.

In combinations of *small molecules and macromolecules*, for instance, macromolecules are known as attractive scaffolds on which molecular recognition events toward small molecules can create amplified signals. Fluorescence signal amplification for ultra-sensitive chemosensors [3] and dynamic helix inversions actualized in poly(phenylacetylene)s backbones [4] are the representative examples. Developments of such polymer chains, which are often described as one-dimensional molecular wires, into two- or three-dimensional condensed matter (e.g., thin films [3a] and liquid crystals [4a]) can further amplify the subtle perturbation; however, reports on the concrete methodologies are still very limited. Inevitably, consideration of multi-step assemblies with a combination of molecular recognition is of great significance for scaling up the different hierarchies in size. From an aspect of functional materials, *small molecules in macromolecules*, termed as 'dopants' or 'additives', are known to be able to change polymers' entire electronic characteristics as well as their mechanical and rheological properties. These small molecules do interact and/or react with polymeric materials. Although the performance of the polymeric materials has been improved and optimized, even enough for practical production, the systematic understanding of their mechanistic details and molecular design based on supramolecular chemistry has not been pursued so much. For example, application of more complex small molecular 'additives' such as 'molecular machines' could realize hitherto unknown structures and functions (Section 3.2.1).

Small molecules with molecular assemblies have so far been well studied. For instance, catalytic reactions on the micellar surface, molecular recognition at the interface, and 'command surfaces' for liquid crystal alignment are successful examples [5]. The 'sergeant-and-soldiers' principle, the effect that a small amount of chiral additives influences overall helix sense of the supramolecular assemblies, can also be included in this category (this effect was originally observed in polymer systems) [2d]. We can quote more advanced systems composed of well-programmed 'host' molecules yielding organogels. This research field has attracted a large amount of attention as one of the new supramolecular materials in the last decade [6]. The new concept of host-guest organogels would be useful for the design of new organic optoelectronics and sensory materials (Section 3.2.2). The systems where *two or more molecular assemblies coexist* are interesting, especially when reflecting on the complexity and the functionality of natural systems (e.g., biological membranes). Feringa, van Esch, and coworkers reported an orthogonal self-assembly composed of surfactant and hydrogelator in a single medium [7a]. Meijer, Schenning, and coworkers have succeeded in the fabrication of photovoltaics from photo- and electroactive supramolecular assemblies [7b]. In relation to these recent examples, 'self-sorting organogels' composed of p-n heterojunction points are discussed below (Section 3.2.3).

As a combination of *macromolecules and macromolecules*, DNA is the best example in which precise molecular recognition between the macromolecules constructs three-dimensionally programmed structures. Creation of such systems by synthetic polymers is quite difficult to realize. As a stereo-complex with a double-stranded helical structure composed of conventional synthetic polymers, it-PMMA/st-PMMA [8], is a well-known but rather exceptional example, and modification and/or functionalization of these polymers can cause the loss of the complexation ability. In supramolecular chemistry, foldermers (mostly oligomers) [9] and DNA analogs [10] have been reported. As the design of these molecules requires skillful organic syntheses, extension toward nanodevices and structural materials still seems difficult. From this point of view, we are focusing on helical polysaccharides (schizophyllan and curdlan) that have a one-dimensional hydrophobic hollow structure [11]. Our research was first motivated by our unexpected finding of triple helix formation by two polysaccharide chains and one polynucleotide chain [11b]. In the latter part of this manuscript, we describe the aspects of macromolecular host-guest chemistry utilizing these polysaccharides as one-dimensional hosts (Section 3.2.4). In *macromolecular assemblies*, simple blending of polymers and microphase separation chemistry in block copylmers have been studied comprehensively, and there is a notable accumulation of physics established therein [12]. We would like to introduce macromolecular assemblies triggered by molecular recognition such as protein-substrate interactions. High selectivity due to the molecular recognition is useful not only for many biochemical purposes but also for novel supramolecular network structures (Section 3.2.5). A combination of *molecular assemblies with macromolecules* would be a powerful approach toward the creation of practical supramolecular materials, since they can compensate each other by stability, processiblity, and functionality. A successful example of such a processable supramolecular/macromolecular composite, with which the anisotropic supramolecular structure was realized up to the cm scale by a simple shearing operation, is described in the last section (Section 3.2.6).

Given the above brief introduction to materials science from supramolecular aspects, there should be a lot of areas to which supramolecular chemists can contribute. We hope that as one of the approaches in these vast studies, we can step forward to new materials design based on molecular and macromolecular recognition, focusing on the orientation and the hierarchy of supramolecular systems. Although examples are still limited, we here outline our recent progress in 'supra-macromolecular chemistry' toward new organic materials from supramolecular standpoints (Figure 3.1).

3.2
Small Molecules, Macromolecules, and Supramolecules: Design of their Composite Materials

3.2.1
Interactions between Small Molecules and Macromolecules

Oriented polymers and/or polymer nanostructures have attracted a large amount of attention. Of particular interest are structures that consist of conjugated polymers

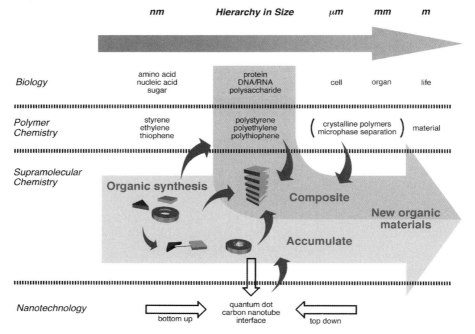

Figure 3.1 Materials design with consideration of the hierarchy in size: black curved arrows are the supramolecular viewpoints, some of which are outlined in this manuscript in relation to the concept of 'supra-macromolecular chemistry'.

(CPs) because of their potential applications as, for example, anisotropic emitters and conductors [13]. From a supramolecular standpoint, Takeuchi and coworkers recently reported a supramolecular bundling approach toward the alignment of CPs [14]. The concept they have proposed resembles that found in animal cells where one-dimensional actin filaments with high affinity elicit the formation of actin bundles. The bundling proteins therein possess two interactive modules for cross-linking actin filaments, and their distinct structural properties determine the type of assembly. Learning from such natural systems, they have designed 'aligner' molecules (Figure 3.2). **Aligner**, a porphyrinatozinc oligomer having two pairs of interactive binding tweezers, elicits homotropic allosterism during the binding with the amino-functionalized conjugated polymers (**PPE**). Thus, the first binding via dative bonds between a pair of porphyrinatozinc tweezers of **Aligner** with **PPE** preorganizes the second pair of tweezers such that it has an even higher affinity toward the second **PPE** binding. This allosteric bundling phenomenon enables the formation of defects to be avoided and thus enables juxtaposition of **PPE**s since the growth of the **Aligner/PPE** sheet structure ideally takes place at the reactive edges where the 'second' tweezers exist. To confirm the cooperativity of the binding of **PPE** to **Aligner**, changes in the absorption spectra of **Aligner** with the addition of a monomeric guest molecule were investigated. Plots of the absorbance as a function of the concentration of the monomeric guest indeed showed a sigmoidal curvature. From

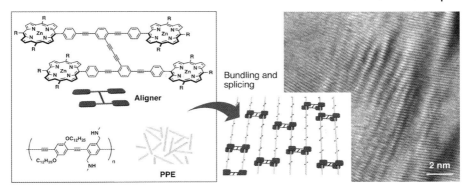

Figure 3.2 Alignment of conjugated polymers via a supramolecular bundling approach.

analyses of the binding isotherm using nonlinear curve fitting and the Hill plot method, the association constants (K_n) and Hill coefficient (n_H) were determined to be $K_1 = 1.6 \times 10^5$, $K_2 = 3.0 \times 10^5$, and $n_H = 1.9$, respectively, clearly indicating the action of highly cooperative binding. AFM images of the **Aligner/PPE** assembly showed sheet structures with an average height of 0.35 nm, which is slightly higher than that of **PPE** itself (0.30 nm). The size of the sheet structure is sensitive to the mixing ratio between **Aligner** and **PPE** and the aging time; for example, with the conditions of [**Aligner**] = 0.08 μM, [**PPE**] = 1.2 μM, and aging for 24 h at r.t., a sheet structure of several hundred square nanometers was obtained. High-resolution TEM images gave direct evidence for aligned **PPE** in the sheet structure. As shown in Figure 3.2, the **Aligner/PPE** assembly showed dark/light alternate layers. The dark regions would be assigned to domains that contain ordered π-stacked layers and/or heaviest atoms (Zn) of **Aligner**; the results of energy-dispersive X-ray spectroscopy support this proposal. The periodicity observed in the low-magnification micrograph (2.0 nm) corresponds to the distance between **PPE** units when they are bundled in a parallel manner by **Aligner** molecules. Thus, the concept based on the cooperative supramolecular bundling provides ordered assemblies of conjugated polymers ('supramolecular textile'), in which interpolymer spacing can be controlled by the molecular design of the aligners (e.g., by the distance between the binding subunits). More recently, Takeuchi and coworkers have succeeded in stabilizing the thermodynamically formed 'temporary' textile into a 'permanent' one by covalently 'stitching' via the ring-closing olefin methathesis reaction; in this case, a new aligner molecule having olefinic groups at the peripheral positions was applied [15].

Sada and coworkers have succeeded in the creation of a double-stranded helix from *achiral* flexible synthetic polymers with an aid of a small molecular 'clip' [16]. The clip they have applied is 2,6-bis(2-oxazolyl)pyridine (**PYBOX**), which was previously found to form an orthogonal hydrogen-bonding pair with secondary dialkylammonium cations (Figure 3.3) [17]. The chiral bis**PYBOX**-porphyrin derivative, **Por(PYBOX)$_2$**, was designed based on porphyrin's self-aggregation ability and detector function for the supramolecular assembly and chirality. Thus, a linear polymer composed of ammonium cations can act as a template to assemble **Por(PYBOX)$_2$**,

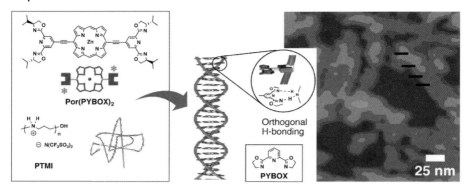

Figure 3.3 Double-stranded helix formation from achiral synthetic polymers.

which yields a supramolecular ladder structure constructed by two polymers as stringers and a porphyrin stack as rungs. Bulky chiral groups at both ends of **PYBOX** are expected to influence the overall sense of the helix. Absorption spectral changes of **Por(PYBOX)$_2$** observed with the addition of **PTMI**, poly(trimethylene iminium), revealed the stacking of the porphyrin molecules. Furthermore, the negative Cotton effect observed for the CD spectrum of **Por(PYBOX)$_2$/PTMI** complex demonstrated that the porphyrin molecules aggregate to form supramolecular assemblies in a helical fashion between the two **PTMI**s' stringers. Conversely, **PTMI** seems to form the double-stranded helix at both edges of the helical stacking of **Por(PYBOX)$_2$** (Figure 3.3). The twisting should be induced in order to relieve steric hindrance between the isopropyl groups and electrostatic repulsion among the ammonium cations. Consequently, the one-dimensional structure observed by AFM seemed relatively rigid and straight. More interestingly, the averaged length of the supramolecular double helix was very consistent with the estimated length of **PTMI** from the degree of polymerization. These results suggest that the double helical structure is programmable by molecular and macromolecular design. Therefore, the obtained nano structure would be useful as a building block for supra-macromolecular materials.

3.2.2
Interactions between Small Molecules and Molecular Assemblies

Supramolecular capsules can be regarded as one of the ultimate hosts since the guest molecules inside are completely shielded from the exterior environment, which enables the stabilization of highly reactive chemical species, avoids deactivation processes in photochemical reactions, realizes stereoselective syntheses, and so on [18]. In addition, the well-preorganized three-dimensional cavity minimizes the entropic penalty for guest complexation; therefore, the large binding constant as well as the high selectivity are to be expected. Octaamide-appended porphyrin (**Por(amide)$_8$**) can form such a supramolecular capsule through the *convergent* hydrogen bonding array sewing two molecules together according to the amide-amide β-sheet

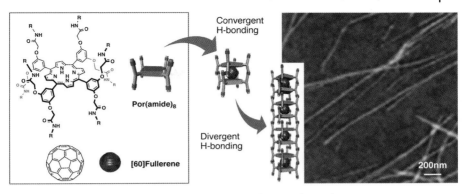

Figure 3.4 Fullerene-binding motivated self-assembly of octaamide-porphyrin.

motif [19]. The dimeric capsule composed of extended porphyrin π-surfaces has a structure analogous to that of the cyclic bisporphyrin host, which is known to bind [60]fullerene with large binding constants [20]. *Divergent* hydrogen bonding sites above and below the capsule are expected to form 'intercapsular' hydrogen bonding, which yields a one-dimensional supramolecular polymer. When [60]fullerene and **Por(amide)$_8$** were mixed, one-dimensional fiber structures were indeed observed using various microscopic techniques (Figure 3.4). Stoichiometry between [60] fullerene and **Por(amide)$_8$** was determined by elemental analysis to be 1 : 2, supporting our expectation as shown in Figure 3.4. In clear contrast, in the absence of [60] fullerene, **Por(amide)$_8$** can form two-dimensional sheet structures via hydrogen bonding developed in a horizontal direction with respect to the porphyrin plane [21]. Thus, the fullerene-motivated morphological changes are due to a strong porphyrin–[60]fullerene interaction as well as skillfully programmed hydrogen bonding sites. It is interesting that the strong fluorescence of **Por(amide)$_8$** is completely quenched in the **Por(amide)$_8$**/[60]fullerene composite. This finding implies that the one-dimensional supramolecular fibers consisting of porphyrins and [60]fullerenes could show unique electron transfer phenomena, which would be applicable to photocurrent generation devices. It should be noted that incorporation of *unsubstituted* fullerene, maintaining its intrinsic properties, is of importance for such photoelectronic devices. Toward this, we have succeeded in the preparation of solution-processable **Por(amide)$_8$**/[60]fullerene composite with unimolecular width; in this case, we used a more 'flexible' porphyrin bearing chiral substituents, **R**, namely α-phenylethyl groups [19b].

Naphthalenediimide-based organogelator (**Naph(imide)$_2$**) forms one-dimensional supramolecular fiber structures via a cooperative action of π-π stacking, hydrogen bonding, and van der Waals force, and yields a very robust organogel [22]. The π-cleft built by naphthalenediimides has been applied to the design of a plethora of novel supramolecular architectures; for example, Stoddart, Sanders, and coworkers have reported the formation of catenanes and rotaxanes utilizing the molecular recognition ability toward alkoxynaphthalenes [23]. The driving force here is a charge transfer interaction, which gives a characteristic absorption band in the visible region. With

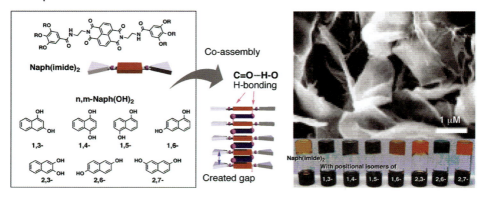

Figure 3.5 Amplified colorimetric sensing by molecular recognition events in molecular assembly.

these intriguing phenomena in mind, we prepared the organogel of **Naph(imide)$_2$** in the presence of several positional isomers of n,m-dihydroxynaphthalenes (Figure 3.5). Interestingly, each isomer induced a color change of organogels from pale yellow to various other colors (Figure 3.5). Noteworthy is the fact that a non-gelling reference compound having solubilizing substituents, **R** (2-ethylhexyl groups), generates drastically less intense and less distinct color changes in the solution phase. Hence, determination of the positional isomers of dihydroxynaphthalene with the naked eye is not possible in the solution state. For example, the bluish-green gel prepared with 2,6-dihydroxynaphthalene shows an impressive 70-fold enhancement in the molar absorption coefficient of the charge transfer band relative to that of the reference solution. From SEM observation, it was found that a one-dimensional fiber structure of **Naph(imide)$_2$** alone was transformed into a two-dimensional sheet structure by co-assembly with dihydroxynaphthalenes. This morphological change is not surprising when it is considered that the intercalation of the donor moiety creates a gap between the gelator molecules; this intercalation triggers the two-dimensional aggregation to maintain the closely packed nature of the alkyl chains. It seems that a combination of various interactions such as hydrogen bonding in addition to charge transfer interaction is of importance for this selective and sensitive sensing scheme in the organogel matrix. Thus, molecular recognition in molecular assemblies leads to hitherto unknown phenomena both in solution and in the solid state as demonstrated by amplified colorimetric sensing in our organogel system.

3.2.3
Interactions between Molecular Assemblies

Continuing efforts have been devoted to the development of organic optoelectronics such as flexible light-emitting diodes, field effect transistors, photovoltaic cells, and so forth. The advantages of the organic substances over their inorganic counterparts, namely (i) preparation by a low-cost solution-process, (ii) well-defined stacks of π-conjugated molecules, and (iii) integration of various molecular functions are of fundamental importance for practical applications. Of these, (iii) will lead to more advanced optoelectronic systems, although concrete examples are still very limited. The supramolecular approach is the most promising methodology to meet these demands. Programmed molecules can recognize the direction and number of noncovalent interactions to form precise supramolecular architectures such that intriguing properties are integrated within them. Particularly interesting is the example of orthogonal molecular assembly (i.e., 'self-sorting') where *different* molecules distinguish between self and nonself [24]. This system will enable us to integrate the supramolecular functions as they are designed, thus enabling us to fulfill the requirement of (iii).

Pery and **Thio**, shown in Figure 3.6, function as opposite semiconducting materials of the n-type and p-type, respectively [25]. They form one-dimensional supramolecular fiber structures and yield organogels in mutual organic solvents (chlorobenzene) because of their structural similarity; the main driving forces for their self-assembly are π-πstacking, van der Waals force, and hydrogen bonding. In the meantime, their molecular lengths are slightly different and they have different numbers of hydrogen bonding sites: two in **Pery** and four in **Thio**. Therefore, self-aggregation constants of **Pery** and **Thio** are different and are determined to be $K_{agg} = 1.3 \times 10^3 \, M^{-1}$ and $>10^5 \, M^{-1}$, respectively. Reflecting the difference in K_{agg}, the dissociation temperature of **Thio** ($T_m = 90\,°C$) is much higher than that of **Pery** ($T_m = 25\,°C$) as monitored by variable-temperature UV-vis spectra. When **Pery** and **Thio** (in chlorobenzene) were put in a septum-capped test tube and the mixture was heated to dissolve these compounds, a transparent and stable organogel was obtained ('self-sorting organogel'). It is interesting that the absorption peaks attributed to **Pery** and **Thio** independently changed upon heating; that is, absorption bands of **Pery** gradually sharpened first ($T_m = 25\,°C$), then those of **Thio** showed a red shift ($T_m > 90\,°C$). These absorption spectral changes in the mixed system are well consistent with those observed for individual gels during the thermal dissociation process. This result indicates that **Pery** and **Thio** self-assemble independently (self-sorting) and do not interfere with each other in the aggregation–dissociation process.

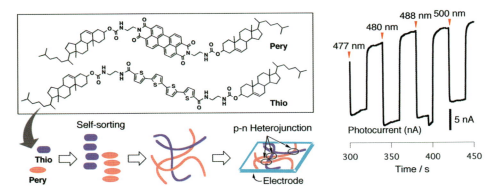

Figure 3.6 Self-sorting supramolecular fiber formation of p-type and n-type semiconducting molecules.

We consider that the kinetic and non-equilibrium nature of the organogel formation are important for self-sorting and segregation of these two gelators [25a].

The advantage of the self-sorting organogel system could be applied to an efficient photovoltaic system in which an orderly π-stacking of the photoactive molecules and a large donor-acceptor interface area are required. This so-called 'bulk heterojunction' facilitates efficient charge separation of photogenerated excitons and produces continuous carrier pathways. With this phenomenon in mind, we prepared a cast film on ITO (Indium Tin Oxide) electrode from the self-sorting organogel (Figure 3.6). Fluorescence measurements of the film revealed that there is a contribution of energy and/or electron transfer between oligothiophene and perylene chromophores. This photochemical process is probably taking place at 'p-n hetero junction points', where supramolecular fibers of **Pery** and **Thio** entangle with each other. A photocurrent generation set-up was prepared with the ITO working electrode and a Pt wire counterelectrode in a solution of electrolyte and electron sacrifice (ascorbic acid). Upon photoirradiation, anodic photocurrent was generated. The photocurrent action spectrum of the film prepared from the self-sorting organogel extends to the visible light region and perfectly overlaps its absorption spectrum, indicating that both **Pery** and **Thio** act as photoactive species for the photocurrent generation. This preliminary experiment demonstrates that self-sorting phenomena of supramolecular functions will offer more potential advantages for complicated and advanced design of organic devices.

3.2.4
Interactions between Macromolecules

Schizophyllan (**SPG**) is a natural polysaccharide consisting of a 1,3-β-linked glucose backbone with a 1,6-β-glucosyl side-chain linked at every third main-chain glucose. **SPG** adopts a triple helix (**t-SPG**) in nature, which can be dissociated into a single chain (**s-SPG**) by dissolving in dimethylsulfoxide (DMSO). The **s-SPG** chain can retrieve the original triple helix by exchanging DMSO for water (renaturation) [11, 26]. Therefore, **s-SPG** is substantially a random flexible polymer but yet prefers one-dimensional helical conformation in certain conditions. Important for our purpose is the fact that the interior surface of the helix is hydrophobic and the exterior glucose side-chain can act as a hydrophilic group for aqueous media (Figure 3.7). This unique amphiphilic core-shell structure resembles that of cyclodextrins. Therefore, **SPG** can be considered as a cylindrical or one-dimensional analog of cyclodextrins, which are well-established host molecules for small hydrophobic compounds. This idea encouraged us to pursue a challenge to determine whether **SPG** can function as a novel one-dimensional host in which various guests, including even larger objects, can be accepted. In the course of our research, we demonstrated that **SPG**-based one-dimensional hosts can accommodate various 'guests' – not only small molecules but also large objects such as synthetic polymers, nanoparticles, and carbon nanotubes [11a]. It seems that flexible glycosyl bonds can find a suitable conformation to wrap up these large guests (that is, according to 'induced fit'). Hence, we can now step forward to macromolecular host-guest chemistry.

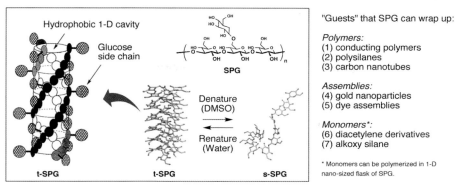

Figure 3.7 Denature/renature process of **SPG** which is applicable to macromolecular host-guest chemistry.

One of the most attractive 'guests' in this era is the family of carbon nanotubes (CNTs), which have already found innumerable applications in materials science and nanotechnology. However, their strong cohesive nature and thus poor solubility have caused researchers trouble for a long time, as these properties seriously hamper efforts to obtain reproducible data and functional materials. One potential breakthrough is a wrapping approach using natural as well as synthetic polymers; amylose, DNA, peptides, poly(vinylpyrrolidone), polythiophenes, and so on have been reported as applicable solubilizers [27]. Of these, polysaccharides are particularly attractive, especially amylose [27a], because (1) it has almost no light absorption in the UV-vis wavelength region, so that the dispersed composites are suitable for photochemical experiments, (2) the sugar-coated surface of the composite should show biocompatiblility, which would make medicinal applications possible, and (3) if the sugar group is recognized by proteins (such as lectins), one may construct novel CNT-based supramolecular network structures. Furthermore, when polysaccharide helically wraps CNTs, the chirality information in the saccharide may be transcribed onto the π-conjugated systems in CNTs. Numata and coworkers took the aforementioned advantages of **SPG** to wrap and solubilize single-walled carbon nanotubes (SWNTs) (Figure 3.8) [28]. A typical procedure is simple and is as follows: a DMSO suspension containing SWNTs and **s-SPG** was first sonicated for several minutes, and then the mixture was gradually diluted with water: the final composition of water/DMSO (v/v) was 94/6 (v/v). This process induces the renaturation of **s-SPG** to **t-SPG**. In the meantime, SWNTs can be entrapped in the hydrophobic one-dimensional cavity of **SPG**. The resultant black dispersion of **SPG**/SWNT composite was clear and stable for over 1 month. The composite was examined using several microscopic techniques such as TEM, SEM, AFM, and CLSM. For example, the AFM image of **SPG**/SWNT showed one-dimensional objects along which a periodical structure with inclined stripes was observed. Furthermore, three-dimensional TEM observation revealed that a thin cylinder of **SPG** is wrapping the SWNT; this cylinder can be viewed from any angle. These results strongly support the fact that **s-SPG** helically wraps one piece of SWNT. Interesting is the characteristic of renature/denature

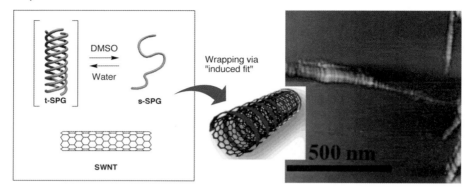

Figure 3.8 Noncovalent wrapping of SWNTs by **SPG**.

reversibility of **SPG**, which enables the wrapping/uncovering processes. When DMSO was added to the aqueous solution of **SPG**/SWNT (up to 50 vol% of DMSO content), SWNTs were immediately precipitated out. This procedure will be useful for the purification of SWNTs.

SPG/conjugated polymer complex can also be prepared by the same procedure via the renaturation process of **SPG** [29]. Thus, when a DMSO solution of **s-SPG** was added to an aqueous solution of **PT** (ionic polythiophene), and then the mixed solution was incubated for 12 h at 60 °C, **SPG/PT** complex was reproducibly obtained (Figure 3.9). In general, a color change of the **PT** solution from yellow to orange in the absence of **SPG** indicates the **PT-PT** interpolymer interaction. In the presence of **SPG**, on the other hand, the absorption maximum of **PT** solution (403 nm) became red-shifted by 51 nm via the complexation with **SPG** (λ_{max} of **SPG/PT** complex is 454 nm). This absorption red shift can be attributed to the increase in the effective conjugation length of the **PT** backbone by **SPGs** support, which changes random-coiled conformation of **PT** to a more planar conformation. Direct evidence of the **SPG-PT** interaction was obtained by CD spectral measurements. Although **PT** itself is CD

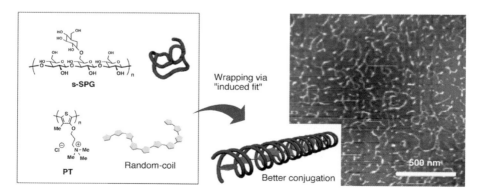

Figure 3.9 Chiral insulated polythiophene by **SPG** wrapping.

inactive, **SPG/PT** complex showed an intense split-type induced CD in the π-π*
transition. The shape and sign of the ICD pattern are characteristic of a right-handed
helix of the **PT** backbone, reflecting the stereochemistry of **SPG**, which also tends to
adopt a right-handed helical structure. Interesting in terms of 'insulated molecular
wire' [30] is the fact that the absorption maximum of **SPG/PT** *film* (456 nm) was
almost identical to that measured in solution (454 nm). In clear contrast, **PT** film
prepared without **SPG** showed a distinct red shift (by 138 nm) in the absorption
maximum in comparison with **PT** solution (λ_{max}: 403 nm in solution → 541 nm in
film). Furthermore, the simple **PT** film showed the vibronic fine structure in the
absorption spectrum, suggesting the strong aggregation of **PT** in the film. These
findings clearly demonstrate that **s-SPG** host can wrap the **PT** backbone and thus
reduce interchain interactions even in the solid phase. Chiral insulated molecular
wire described herein would be of importance for organic electronics, for example, to
fabricate circularly polarized electroluminescence devices.

3.2.5
Interactions between Macromolecular Assemblies

Curdlan (**CUR**) has the same polymer backbone as schizophyllan, that is, 1,3-β-linked
glucose, and thus forms a triple helix. Due to the lack of the solubilizing 1,6-
β-glucosyl side chain, however, **CUR** is scarcely soluble in water, a cheap material.
In order to use this inexpensive material potentially endowed with the function
similar to **SPG**, we have applied regioselective bromination-azidation of **CUR** to
afford 6-azido-6-deoxycurdlan, which can be followed by chemoselective [3 + 2]-
cycloaddition with various functional modules bearing a terminal alkyne group (i.e.,
'click chemistry') [31]. The reaction is very convenient, quantitative, and regioselective, therefore, various functionalized **CUR**s (semi-artificial **CUR**) can be prepared.
According to this method, Numata and coworkers synthesized *water-soluble* cationic
(**CUR-N$^+$**) and anionic curdlan (**CUR-SO$_3^-$**), having ammonium groups and sulfonate groups, respectively [32]. Interestingly, these ionic curdlans cannot form triple
helix structures even in pure water as confirmed by spectroscopic studies, which is
probably due to the ionic repulsion among the introduced ionic side-chains. The
breakthrough suggested by this observation is that we no longer need to go through
the denature process of the triple helix for the composite preparation (Figure 3.7). In
fact, when we mixed *aqueous* solution of **CUR-N$^+$** with those guest molecules used in
our precedent studies (e.g., fluorophores, polycytidylic acid, permethyldecasilane,
and SWNTs), the complexation proceeded spontaneously; noteworthy here is that the
denature/renature process induced with the aid of DMSO (or NaOH aq.) is
indispensable for **SPG** to bind these guests [32a]. Spectroscopic studies as well as
microscopic observation of **CUR-N$^+$**/guest composites are all in line with those data
obtained from **SPG**/guest composites. Accordingly, cationic and anionic composites
with SWNT as a guest could be prepared utilizing **CUR-N$^+$** and **CUR-SO$_3^-$**,
respectively (Figure 3.10) [32b]. The complementary electrostatic interaction between
CUR-N$^+$/SWNT and **CUR-SO$_3^-$**/SWNT would acquire the potential to create a
specific hierarchical architecture. When aqueous solutions of **CUR-N$^+$**/SWNT and

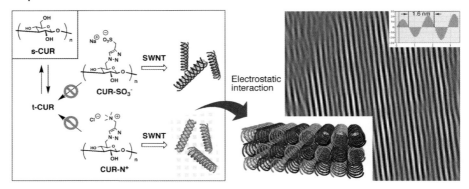

Figure 3.10 Hierarchical architecture via electrostatic interaction between cationic and anionic **CUR**/SWNT composites.

CUR-SO$_3^-$/SWNT were mixed with monitoring zeta-potential, such that the potential charges could be neutralized, a black solution was obtained without any accompanying precipitate formation. The resultant aqueous mixture was cast on a substrate and observed by AFM and TEM. In AFM images, we observed a well-developed sheet structure with micrometer-scale length, which is entirely different from the very fine fibrous structures observed for individual **CUR-N$^+$**/SWNT and **CUR-SO$_3^-$**/SWNT. The scan profile of the sheet structure revealed that several thin layers are piled up to form the sheet-like structure, the surface of which is almost flat. TEM images of the sheet structure showed highly ordered fibrous assemblies. Since the TEM images were obtained without staining, the observed contrast should arise from the presence of SWNT. The periodicity of the dark layer estimated by Fourier-filtered image is 1.6 nm, which is consistent with the diameter of the individual composite obtained by the AFM height profile (1.5 ∼ 1.7 nm, Figure 3.10). Although it is still unclear, at present, why the two-dimensional sheet structure, but not the three-dimensional packed structures, was formed, the results demonstrate that the hierarchical SWNT architecture can be created by just mixing two kinds of aqueous solution in the appropriate ratio.

CUR can be functionalized with various modules via 'click chemistry'. On the other hand, chemical modification of **SPG** is also possible because **SPG** has one distinctive structural feature, namely the presence of the pendent 1,6-β-glucoside having NaIO$_4$-sensitive 1,2-diols that can be converted to aldehyde groups. **SPG-Lac**, functionalized with lactoside-appendages, was thus prepared from aldehyde-functionalized **SPG** via imine formation with aminoethyl-β-lactoside followed by reduction by NaBH$_4$ [33]. The sugar-coated surface of the **SPG**/guest composite should show biocompatiblility, which would make medicinal applications possible. For example, **SPG-Gal** (galactose-functionalized **SPG** synthesized by the same method) was found to enhance cellular ingestion of antisense oligonuculeotides (guests) through endocytosis mediated by a sialoglycoprotein receptor [33c]. Furthermore, binding and/or cross-linking with proteins (such as lectins) on the composite materials would construct novel **SPG**-based supramolecular network structures.

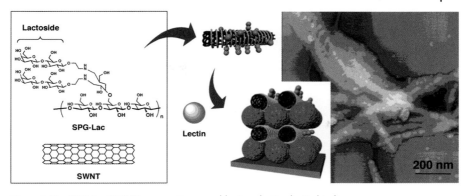

Figure 3.11 SPG-Lac/SWNT composite assembly via selective lectin binding.

SPG-Lac/SWNT composite was prepared according to the usual procedure mentioned above. Molecular recognition of **SPG-Lac**/SWNT composite toward various lectins was assessed by surface plasmon resonance (SPR) using lectin-immobilized Au surfaces [33a]. Among several Au surfaces modified with different lectins, **SPG-Lac**/SWNT composite showed highly specific binding toward that modified with *ricinus communis* agglutinin (RCA$_{120}$, β-Lac-specific); in contrast, the other lectin-immobilized surface showed negligible resonance changes. Furthermore, the specific interactions between **SPG-Lac**/SWNT composite and lectins were visualized by AFM. With RCA$_{120}$ only, **SPG-Lac**/SWNT composite showed dense clustering structures (Figure 3.11). This lectin-affinity of **SPG-Lac**/SWNT composite is advantageous for the fabrication of various SWNT-based sensory systems or superstructures. For example, this system is readily applicable to construct a layer-by-layer structure composed of the **SPG-Lac**/SWNT composite and RCA$_{120}$. Quartz crystal microbalance (QCM) measurements using the RCA$_{120}$-immobilized Au surface revealed a step-by-step decrease in the frequency, proving the construction of the expected layer-by-layer structure. This noncovalent strategy is applicable not only to many biochemical purposes but also to various photo- and electrochemical materials.

3.2.6
Interactions between Macromolecules and Molecular Assemblies

Highly ordered microphase separation is a characteristic of nano-segregated structures obtained from programmed molecular systems such as block copolymers and liquid crystals [12, 34]. These systems provide anisotropic function hierarchically extended from the molecular level even up to the cm scale, which are foreseen in the field of materials science and biotechnology. In contrast, although the supramolecular approach (program of molecular organization) has realized a plethora of desired superstructures and functions, there is currently a paucity of studies considering the different hierarchies of the supramolecular systems over all length scales for device fabrication [2a]. A crucial challenge, in this respect, is to extend these supramolecular architectures to a larger scale (> cm) while retaining their unique functions intact. With this object in mind, we attempted to prepare supramolecule/macromolecule hybrids to take advantage of the processablity of the macromolecular domain [35].

As mentioned in Section 3.2.3, **Pery** forms one-dimensional supramolecular fiber structures in organic solvents such as aromatic solvents [25b], which indicates that aromatic molecules would not interfere with the self-assembling process of **Pery**. Accordingly, when **Pery** and polystyrene (*aromatic* polymer, **PS**) were dissolved in aromatic solvent, supramolecular fiber structures of **Pery** were observed without any influence of **PS**. After drying the solvent, we succeeded in obtaining a self-standing supramolecular/macromolecular hybrid film (Figure 3.12). This hybrid film was stable and processable; for example, a shearing operation can be applied above the glass transition temperature of **PS**. This operation forced supramolecular fibers of **Pery** into the unidirectionally aligned state. Various microscopic techniques such as TEM, SEM, and POM revealed that supramolecular fiber structures are aligned along the shearing direction. Furthermore, this alignment was also confirmed by spectroscopic methods showing anisotropic absorbance at π-π^* transition of **Pery** (400 ∼ 600 nm) and C=O stretching of imide groups (1654 cm^{-1}). These results indicate that the alignment of supramolecular fiber structures is hierarchically achieved from the molecular level to the cm scale. One of the interesting applications of the anisotropic supramolecular function would be photonic and electronic devices. In fact, in flash photolysis time-resolved microwave conductivity measurements, we confirmed a large anisotropic ratio (8.6: parallel/perpendicular to the shearing direction), indicating that intrafiber hopping is predominant over interfiber hopping. The simple and versatile method described herein creates a new foothold which is indispensable for macroscopic organization of supramolecular architectures featuring ordered, anisotropic, and monolithic structures (self-standing films as large as cm scale are available). Diverse examples of functional supramolecular assemblies can readily be employed in this method as well, which means that various combinations of these two components will lead to further development of novel functional 'organic alloys' [35].

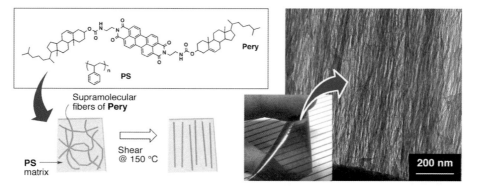

Figure 3.12 Processable supramolecular/macromolecular composite.

3.3
Conclusion and Outlook

In this manuscript, we have outlined strategic aspects of how we can take advantage of molecular recognition to create new supramolecular materials. When we can extend specificity of noncovalent interactions to macromolecular systems and supramolecular assemblies in consideration of hierarchy in size, we shall expect that practical supramolecular materials will be commercialized, probably in the near future. In such approaches, however, multi-step assembly needs to be processed, for which we have not yet established reliable criteria and methodologies.

In practice, any self-assembly is governed by an energetically (i.e. thermodynamically) favorable process which is complex and frequently becomes impracticable, especially when multi-step self-assemblies through different hierarchies need to be programmed. The energetic distribution, barrier, and competition cause the formation of defects during the multi-step processes. As briefly mentioned above when discussing aligners and also reported from other groups [36], self-assembling followed by covalent fixation to obtain the permanently stable object would be one approach. By utilizing the resultant material in the next-step self-assembly as a stable and isolable building block independent of thermodynamics, one may scale up the hierarchy in size toward the supramolecular materials. Another approach would be the control of the kinetic factor of self-assembly. For example, the self-sorting organogel is formed in such a process; one gelator forms a supramolecular assembly instantly while the other is still molecularly dissolved, and after a certain period of time, the second self-assembling takes place. This two-step independent kinetic scheme yields two different molecular assemblies in a single medium. After processing from the solution to the solid phase, we may be able to avoid rearrangement arising from the thermodynamic equilibrium occurring in the solution. Undoubtedly, integration of the various functions and properties of small molecules, macromolecules, and supramolecular systems, rather than their classification, needs to be accomplished in careful and skillful consideration on molecular recognition: Section 3.2.6 gives a good example. As a matter of course, other approaches such as epitaxial or template assemblies on substrates, Langmuir-Blodgett techniques, layer-by-layer methods, and so forth are also of importance in a series and/or parallel processing of supramolecular and macromolecular materials, and there should be many other ways to scale up the hierarchical assemblies from molecular size to the real world size. Furthermore, combinations of these methods are sure to contribute to further development of the concept proposed in this manuscript.

Acknowledgments

The authors acknowledge Prof. K. Sada (Kyushu University), Prof. M. Takeuchi (National Institute for Materials Science), Prof. M. Numata (Kyoto Prefectural University), Prof. N. Fujita (The University of Tokyo), Prof. K. Sakurai (The University of

Kitakyushu), Prof. K. Kaneko (Kyushu University) and other coworkers who performed the research reported in this manuscript as well as related efforts which indirectly influenced this work.

References

1 (a) Lehn, J.-M. (1995) *Supramolecular Chemistry*, VCH; (b) Steed, J.W. and Atwood, J.L. (2005); *Supramolecular Chemistry*, Wiley; (c) deSilva, A.P., Gunaratne, H.Q.N., Gunnlangsson, T., Huxley, A.J.M., McCoy, C.P., Rademacher, J.T., and Rice, T.E. (1997) *Chem. Rev.*, **97**, 1515; (d) James, T.D., Sandanayake, K.R.A.S. and Shinkai, S. (1996) *Angew. Chem., Int. Ed.*, **35**, 1910; (e) Kay, E.R., Leigh, D.A. and Zerbetto, F. (2007) *Angew. Chem., Int. Ed.*, **46**, 72; (f) Balzani, V., Credi, A., Raymo, F.M., and Stoddart, J.F. (2000) *Angew. Chem., Int. Ed.*, **39**, 3349; (g) Shinkai, S., Ikeda, M., Sugasaki, A., and Takeuchi, M. (2001) *Acc. Chem. Res.*, **34**, 494; (h) Ariga, K., Hill, J.P., Lee, M.V., Vinu, A., Charvet, R., and Acharya, S., *Sci. Technol. Adv. Mater.* (2008) **9**, 014109.

2 (a) Hoeben, F.J.M., Jonkheijm, P., Meijer, E.W., and Schenning, A.P.H.J. (2005) *Chem. Rev.*, **105**, 1491; (b) Shimizu, T., Masuda, M., and Minamikawa, H. (2005) *Chem. Rev.*, **105**, 1401; (c) Brunsveld, L., Folmer, B.J.B., Meijer, E.W., and Sijbesma, R.P. (2001) *Chem. Rev.*, **101**, 4071; (d) Palmans, A.R.A. and Meijer, E.W. (2007) *Angew. Chem., Int. Ed.*, **46**, 8948; (e) Service, R.F. (2005) *Science*, **309**, 95.

3 (a) Thomas, S.W., Joly, G.D., and Swager, T.M. (2007) *Chem. Rev.*, **107**, 1339; (b) McQuade, D.T., Pullen, A.E., and Swager, T.M. (2000) *Chem. Rev.*, **100**, 2537.

4 (a) Maeda, K. and Yashima, E. (2006) *Top. Curr. Chem.*, **265**, 47; (b) Yashima, E. and Maeda, K. (2008) *Macromolecules*, **41**, 3.

5 (a) Shinkai, S. (1982) *Prog. Polym. Sci.*, **8**, 1; (b) Cordes, E.H. and Dunlap, R.B. (1969) *Acc. Chem. Res.*, **2**, 329; (c) Ariga, K. and Kunitake, T. (1998) *Acc. Chem. Res.*, **31**, 371; (d) Ichimura, K. (2000) *Chem. Rev.*, **100**, 1847.

6 (a) Terech, P. and Weiss, R.G. (1997) *Chem. Rev.*, **97**, 3133; (b) van Esch, J.H. and Feringa, B.L. (2000) *Angew. Chem., Int. Ed.*, **39**, 2263; (c) van Bommel, K.J.C., Friggeri, A., and Shinkai, S. (2003) *Angew. Chem., Int. Ed.*, **42**, 980; (d) Ishi-i, T. and Shinkai, S. (2005) *Top. Curr. Chem.*, **258**, 119; (e) Sangeetha, N.M. and Maitra, U. (2005) *Chem. Soc. Rev.*, **34**, 821.

7 (a) Heeres, A., van der Pol, C., Stuart, M., Friggeri, A., Feringa, B.L., and van Esch, J. (2003) *J. Am. Chem. Soc.*, **125**, 14252; (b) van Herrikhuyzen, J., Syamakumari, A., Schenning, A.P.H.J., and Meijer, E.W. (2004) *J. Am. Chem. Soc.*, **126**, 10021.

8 (a) Liquori, A.M., Anzuino, G., Coiro, V.M., D'Alagni, M., Santis, P.D., and Savino, M. (1965) *Nature*, **206**, 358; (b) Schomaker, E. and Challa, G. (1989) *Macromolecules*, **22**, 3337; (c) Serizawa, T., Hamada, K.-i. and Akashi, M. (2004) *Nature*, **429**, 52.

9 (a) Huc, I. (2004) *Eur. J. Org. Chem.*, **2004**, 17; (b) Berl, V., Huc, I., Khoury, R.G., Krische, M.J., and Lehn, J.-M. (2000) *Nature*, **407**, 720.

10 (a) Nielsen, P.E. (1999) *Acc. Chem. Res.*, **32**, 624; (b) Tanaka, K., Clever, G.H., Takezawa, Y., Yamada, Y., Kaul, C., Shionoya, M., and Carell, T. (2006) *Nat. Nanotech.*, **1**, 190; (c) Iwaura, R., Hoeben, F.J.M., Masuda, M., Schenning, A.P.H.J., Meier, E.W., and Shimizu, T. (2006) *J. Am. Chem. Soc.*, **128**, 13298.

11 (a) Sakurai, K., Uezu, K., Numata, M., Hasegawa, T., Li, C., Kaneko, K., and Shinkai, S. (2005) *Chem. Commun.*, 4383; (b) Sakurai, K. and Shinkai, S. (2000) *J. Am. Chem. Soc.*, **122**, 4520.

12 Förster, S. and Plantenberg, T. (2002) *Angew. Chem., Int. Ed.*, **41**, 688.

13 Ozin, G.A. and Aresenault, A.C. (2005) *Nanochemistry*, RSC Publishing, Cambridge.

References

14 Kubo, Y., Kitada, Y., Wakabayashi, R., Kishida, T., Ayabe, M., Kaneko, K., Takeuchi, M., and Shinkai, S. (2006) *Angew. Chem., Int. Ed.*, **45**, 1548.

15 (a) Wakabayashi, R., Kubo, Y., Kaneko, K., Takeuchi, M., and Shinkai, S. (2006) *J. Am. Chem. Soc.*, **128**, 8744; (b) Wakabayashi, R., Kubo, Y., Hirata, O., Takeuchi, M., and Shinkai, S. (2005) *Chem. Commun.*, 5742; (c) Wakabayashi, R., Kaneko, K., Takeuchi, M., and Shinkai, S. (2007) *New J. Chem.*, **31**, 790.

16 Sugimoto, T., Suzuki, T., Shinkai, S., and Sada, K. (2007) *J. Am. Chem. Soc.*, **129**, 270.

17 (a) Sada, K., Sugimoto, T., Tani, T., Tateishi, Y., Yi, T., Shinkai, S., Maeda, H., Tohnai, N., and Miyata, M. (2003) *Chem. Lett.*, **32**, 758; (b) Sugimoto, T., Sada, K., Yamaguchi, K., and Shinkai, S. (2004) *Chem. Commun.*, 1226.

18 (b) Rebek Jr., J. (2005) *Angew. Chem., Int. Ed.*, **44**, 2068; (b) Casnati, A., Sansone, F., and Ungaro, R. (2003) *Acc. Chem. Res.*, **36**, 246; (c) Koblenz, T.S., Wassenaar, J., and Reek, J.N.H. (2008) *Chem. Soc. Rev.*, **37**, 247; (d) Ikeda, A. and Shinkai, S. (1997) *Chem. Rev.*, **97**, 1713.

19 (a) Shirakawa, M., Fujita, N., and Shinkai, S. (2003) *J. Am. Chem. Soc.*, **125**, 9902; (b) Shirakawa, M., Fujita, N., Shimakoshi, H., Hisaeda, Y., and Shinkai, S. (2006) *Tetrahedron*, **62**, 2016.

20 (a) Tashiro, K. and Aida, T. (2007) *Chem. Soc. Rev.*, **36**, 189; (b) Tashiro, K., Aida, T., Zheng, J.-Y., Kinbara, K., Saigo, K., Sakamoto, S., and Yamaguchi, K. (1999) *J. Am. Chem. Soc.*, **121**, 9477.

21 Shirakawa, M., Kawano, S.-i., Fujita, N., Sada, K., and Shinkai, S. (2003) *J. Org. Chem.*, **68**, 5037.

22 Mukhopadhyay, P., Iwashita, Y., Shirakawa, M., Kawano, S.-i., Fujita, N., and Shinkai, S. (2006) *Angew. Chem., Int. Ed.*, **45**, 1592.

23 (a) Hamilton, D.G., Feeder, N., Prodi, L., Teat, S.J., Clegg, W., and Sanders, J.K.M. (1998) *J. Am. Chem. Soc.*, **120**, 1096; (b) Vignon, S.A., Jarrosson, T., Iijima, T., Tseng, H.-R., Sanders, J.K.M., and Stoddart, J.F. (2004) *J. Am. Chem. Soc.*, **126**, 9884; (c) Gabriel, G.D., and Iverson, B.L. (2002) *J. Am. Chem. Soc.*, **124**, 15174.

24 (a) Mukhopadhyay, P., Wu, A.X., and Isaacs, L. (2004) *J. Org. Chem.*, **69**, 6157; (b) Mukhopadhyay, P., Zavalij, P.Y., and Isaacs, L. (2006) *J. Am. Chem. Soc.*, **128**, 14093.

25 (a) Sugiyasu, K., Kawano, S.-i., Fujtia, N., and Shinkai, S. (2008) *Chem. Mater.*, **20**, 2863; (b) Sugiyasu, K., Fujita, N., and Shinkai, S. (2004) *Angew. Chem., Int. Ed.*, **43**, 1229; (c) Kawano, S.-i., Fujita, N., and Shinkai, S. (2005) *Chem. Eur. J.*, **11**, 4735.

26 (a) Yanaki, T., Norisuye, T., and Fujita, H. (1980) *Macromolecules*, **13**, 1462; (b) Norisuye, T., Yanaki, T., and Fujita, H. (1980) *J. Polym. Sci., Polym. Phys. Ed.*, **18**, 547; (c) Kashiwagi, Y., Norisuye, T., and Fujita, H. (1981) *Macromolecules*, **14**, 1220.

27 (a) Star, A., Steuerman, D.W., Heath, J.R., and Stoddart, J.F. (2002) *Angew. Chem., Int. Ed.*, **41**, 2508; (b) Zheng, M., Jagota, A., Semke, E.D., Diner, B.A., Mclean, R.S., Lustig, S.R., and Tassi, R.E.R. (2003) *Nat. Mater.*, **2**, 338; (c) Tasis, D., Tagmatarchis, N., Bianco, A., and Prato, M. (2006) *Chem. Rev.*, **106**, 1105.

28 (a) Numata, M., Asai, M., Keneko, K., Bae, A.-h., Hasegawa, T., Sakurai, K., and Shinkai, S. (2005) *J. Am. Chem. Soc.*, **127**, 5875; (b) Numata, M., Asai, M., Kaneko, K., Hasegawa, T., Fujita, N., Kitada, Y., Sakurai, K., and Shinkai, S. (2004) *Chem. Lett.*, **33**, 232.

29 Li, C., Numata, M., Bae, A.-h., Sakurai, K., and Shinkai, S. (2005) *J. Am. Chem. Soc.*, **127**, 4548.

30 (a) Frampton, M.J. and Anderson, H.L. (2007) *Angew. Chem., Int. Ed.*, **46**, 1028; (b) Lee, D. and Swager, T.M. (2004) *Synlett*, 149.

31 (a) Binder, W.H. and Sachsenhofer, R. (2007) *Macromol. Rapid Commun.*, **28**, 14; (b) Hasegawa, T., Umeda, M., Numata, M., Li, C., Bae, A.-H., Fujisawa, T., Haraguchi, S., Sakurai, K., and Shinkai, S. (2006) *Carbohydr. Res.*, **341**, 35.

32 (a) Ikeda, M., Hasegawa, T., Numata, M., Sugikawa, K., Sakurai, K., Fujiki, M., and Shinkai, S. (2007) *J. Am. Chem. Soc.*, **129**, 3979; (b) Numata, M., Sugikawa, K., Kaneko, K., and Shinkai, S. (2008) *Chem. Eur. J.*, **14**, 2398.

33 (a) Hasegawa, T., Fujisawa, T., Numata, M., Umeda, M., Matsumoto, T., Kimura,

T., Okumura, S., Sakurai, K., and Shinkai, S. (2004) *Chem. Commun.*, 2150; (b) Hasegawa, T., Umeda, M., Matsumoto, T., Numata, M., Mizu, M., Koumoto, K., Sakurai, K., and Shinkai, S. (2004) *Chem. Commun.*, 382; (c) Karinaga, R., Anada, T., Minari, J., Mizu, M., Koumoto, K., Fukuda, J., Nakazawa, K., Hasegawa, T., Numata, M., Shinkai, S., and Sakurai, K., *Biomaterials* (2006) **27** 1626.

34 (a) Muthukumar, M., Ober, C.K., and Thomas, E.L. (1997) *Science*, **277**, 1225; (b) Lehn, J.-M. (1993) *Makromol. Chem. Macromol. Symp.*, **69**, 1; (c) Kato, T. (2002) *Science*, **295**, 2414.

35 (a) Sugiyasu, K. (2005) *Doctoral Thesis:* 'Design of Novel Composite Materials by Functional Organogels' *Kyushu University*; (b) Sugiyasu, K., Fujita, N., and Shinkai, S. (2005) *J. Synth. Org. Chem., Jpn.*, **63**, 359; (c) Sugiyasu, K., Fujita, N., Akiba, I., Matsunaga, H., Sakurai, K., Seki, S., Saeki, A., Tagawa, S., and Shinkai, S.,*submitted*.

36 (a) Zhou, W., Li, Y., and Zhu, D. (2007) *Chem. Asian J.*, **2**, 222; (b) Shinkai, S., Takeuchi, M., and Bae, B.-A. (2005) *Supramol. Chem.*, **17**, 181; (c) de Loos, M., van Esch, J., Stokroos, I., Kellogg, R.M., and Feringa, B.L. (1997) *J. Am. Chem. Soc.*, **119**, 12675; (d) Sada, K., Takeuchi, M., Fujita, N., Numata, M., and Shinkai, S. (2007) *Chem. Soc. Rev.*, **36** 415.

4
Polymerization with Ditopic Cavitand Monomers
Francesca Tancini and Enrico Dalcanale

4.1
Introduction

In the last few years the merging of polymer science with supramolecular chemistry has created a new, thriving field of research [1] known under the name of supramolecular polymer chemistry [2]. The driving force behind this methodological breakthrough is the ability to control noncovalent interactions with the same precision as that achieved by synthetic organic chemistry. Some of the most relevant issues associated with the development of supramolecular polymers are: (i) to achieve macroscopic expression of molecular recognition, (ii) to trigger stimuli-specific responses in polymeric materials, and (iii) to move self-assembly from the nano to the meso and macro scale. The positive fallout of this merging is demonstrated by the appearance of supramolecular polymers presenting unique mechanical [3], electronic [4], biological [5], and self-healing properties [6]. Inspired by biological systems in which damage triggers an autonomic healing response, supramolecular self-healing has become a hot topic in polymer science [7].

The characterization of supramolecular polymers is a challenging problem, related to their inherent reversibility. The determination of average molecular weight by direct methods is very often difficult, as in supramolecular polymers the molecular weight increases with concentration. In contrast, in Static Light Scattering (SLS) the measurement of high molecular weights requires dilution of the sample. The concentration conundrum can be solved by turning to indirect methods like Pulsed-Field Gradient Spin-Echo (PGSE) [8], an NMR technique which correlates the hydrodynamic radius of an object with its diffusion time in solution. However, to obtain an average molecular weight estimate from PGSE data, the shape of the object must be defined *a priori*, and this can be a source of error, particularly when the polymer shape in solution is oblate or cylindrical rather than spherical. Dynamic Light Scattering (DLS) is affected by the same problem, since it also determines the hydrodynamic radius of the measured object, whose shape must be defined or correlated to known standards.

The analysis of molecular weight distribution also remains challenging. Size Exclusion Chromatography (SEC), widely used with covalent polymers, is problematic in supramolecular polymers. It can be safely employed only in the case of a negligible dynamic equilibrium between polymeric and monomeric species in solution. Time-resolved photoluminescence (PL) is a viable alternative, provided that the polymer contains chromophores [9].

Surface-imaging techniques like Scanning Tunneling Microscopy (STM) have been used to determine the degree of polymerization and the molecular weight distribution in rigid, rod-like covalent polymers deposited on surfaces [10, 11]. The translation of these techniques to the case of supramolecular polymers is not straightforward, as surface structures often do not match their solution counterparts. Typically, surface adsorption amplifies the self-assembly, leading to larger polymeric structures with respect to solution, and this facilitates lateral interactions among polymeric objects. As a result, like in covalent polymers, the evolution of polymer morphology on surfaces is concentration dependent.

While the self-assembly of supramolecular polymers driven by H-bonding and metal coordination has undergone explosive development, other routes are far less traveled, despite the many different options potentially available. Molecular receptors are ideal candidates as building blocks for supramolecular polymers, thanks to the large pool of structures and interaction modes available. The main obstacle to their employment is the need for relatively high association constants ($>10^5$) to obtain truly polymeric materials. This stringent requirement considerably narrows the number of available options. So far, few classes of molecular receptors have been turned into supramolecular host–guest polymers: cyclodextrins (see Chapter 3), cucurbiturils [12, 13], calixarenes [14–16], crown ethers [17, 18] and cavitands. In the first two cases hydrophobicity is the driving force for polymerization, while in the others specific host–guest interactions are operating. Rebek's polycaps represent a special case, where H-bonding polymerization is driven by guest encapsulation [19].

In this chapter we will show that cavitands occupy a unique position among the above cited monomers, since their exquisite versatility allows them to exploit most of the interaction modes described above, either alone or in combination, to generate supramolecular polymers.

4.2
Cavitands

Cavitands, defined as synthetic organic compounds having enforced cavities of molecular dimensions, are well-known and versatile molecular receptors [20]. In the design of cavitands, the choice of the bridging groups connecting the phenolic hydroxyls of the resorcinarene scaffold is pivotal, since it determines shape, dimensions and complexation properties of the resulting cavity. Over the years, several classes of cavitands have been reported, each of them presenting peculiar molecular recognition properties [20–23].

In this chapter we will limit the discussion to the two classes of cavitands employed so far in the formation of supramolecular polymers, namely the quinoxaline [24] and the phosphonate cavitands [25]. In both cases, a clear understanding of the key interactions

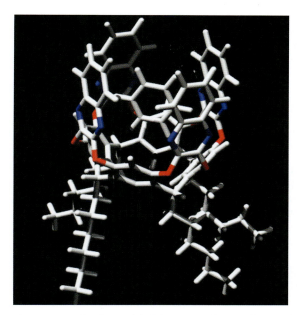

Figure 4.1 Molecular model of the vase conformer of quinoxaline-bridged cavitand including a benzene guest into the cavity.

responsible for their complexation properties was obtained by the comparative analysis of their behavior in the solid state, in solution, and in the gas phase.

The quinoxaline-bridged cavitand (Figure 4.1) in its vase conformation presents a 7.2 Å wide and 8.3 Å deep hydrophobic cavity capable of selectively trapping aromatic guests via CH-π interactions with the quinoxaline walls and the resorcinarene bowl [26, 27]. Besides its preference for aromatic guests, this cavitand has the peculiar property of reversibly switching between the closed vase conformation (C_{4v} symmetry) and an open kite conformation (C_{2v} symmetry) with a flat, extended surface about 19.3 × 15.6 Å in size (Figure 4.2) [28]. Conformational switching can be reversibly induced by temperature [28] or pH [29] changes, with the kite conformation being preferred at low temperatures and low pH values. The kite conformer, called velcrand by Cram [30], is stabilized by the presence of methyl substituents in the apical positions of the resorcinarene skeleton. This kite conformer has the unique property of forming dimers in solution via solvophobic interactions. In the dimer, one of the two cavitand positions itself over the other one rotated by 90° in order to share a large common surface in which the four protruding methyls (two for each velcrand) insert into the four available holes formed by the sloping aryl faces (Figure 4.3). In solution, the velcrand dimerization can be monitored by the chemical shift change of the methyl substituents in apical position. The upward methyls move upfield, while the outward ones move downfield. Being solvophobic in origin, the strength of the interaction can be tuned with solvent polarity. In polar solvents like acetonitrile or nitromethane, the dimerization K_a exceeds 10^5 M^{-1}, while in nonpolar

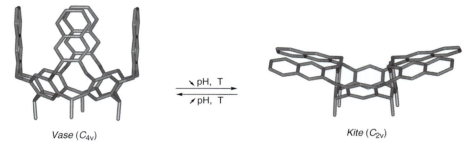

Figure 4.2 Molecular models of the vase and kite conformers of quinoxaline-bridged cavitand. The equilibrium is temperature- and pH-dependent. Hydrogen atoms and 'feet' are omitted for clarity. Reproduced with permission from *Chem. Eur. J.* **2006**, *12*, 4775–4784. Copyright 2006 Wiley-VCH Verlag.

solvents like benzene, the dimerization is suppressed. This solvent-dependent dimerization was chosen as one of the self-assembling modes for cavitand-based supramolecular polymers, in association with either metal coordination or H-bonding. Since these self-assembling modes operate under totally different sets of conditions, they can be implemented jointly without reciprocal interference.

Phosphonate cavitands have established themselves as one of the most promising classes of molecular receptors thanks to their diversified complexation properties, ranging from cationic species such as ammonium, methylpyridinium, and inorganic salts to neutral guests like alcohols. The introduction of four P(V) stereogenic centers as bridging units creates a family of six diastereomeric cavitands, each with a different orientation of the PO groups, that is, inward (i) or outward (o) with respect to the cavity. The key player of the whole class is the tetraphosphonate cavitand Tiiii [25], presenting all four P=O bridging groups oriented inward with respect to the cavity (Figure 4.4). The stereospecific synthesis of this isomer requires bridging a suitable resorcinarenes with four P(III) units and then oxidizing them *in situ* with hydrogen peroxide [31]. Its peculiar complexation ability is the result of three interaction modes, which can be activated either individually or in combination by

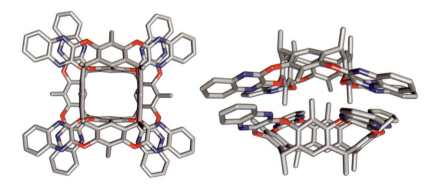

Figure 4.3 Crystal structure of the kite-to-kite velcraplex dimer. Reproduced with permission from Ref. [30].

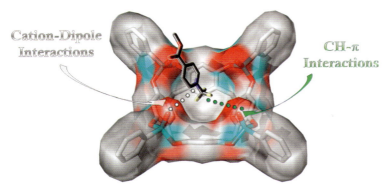

Figure 4.4 The two major interaction modes responsible for the complexation of methylpyridinium guests by tetraphosphonate cavitand Tiiii.

the host according to the guest requirements: (i) multiple ion-dipole interactions between the inward facing P=O groups and the positively charged guests [32], (ii) single or multiple H-bonding involving the P=O groups [33], and (iii) CH_3-π interactions between an acidic methyl group present on the guest and the π-basic cavity of the host [34]. Figure 4.4 highlights the two major interaction modes operating with methylpyridinium guests. Depending on the type and number of interactions activated, the measured K_a in nonpolar solvents can vary between $10^2\,M^{-1}$ for short chain alcohols to $10^7\,M^{-1}$ for methylpyridinium salts [35] and even higher for N,N-methylalkylammonium salts.

The data presented so far clearly indicate that both classes of cavitands satisfy the condition of high association constants, necessary to produce truly polymeric materials.

4.3
Self-Assembly of Ditopic Cavitand Monomers

4.3.1
Structural Monomer Classification of Supramolecular Polymerization

Cavitand-based monomers are particularly appealing building blocks for the realization of supramolecular polymers thanks to their inherent versatility in terms of synthetic modularity and molecular recognition properties. In particular, the opportunity to functionalize the cavitand platform at both the upper and the lower rim allows multiple self-assembly motifs to embed multiple self-assembly motifs on the same molecule, thus leading to complex supramolecular architectures, featuring orthogonal switching modes.

The monomers are required to be ditopic or polytopic in order to ensure polymer growth by the reversible association of their end-groups. All the cavitand monomers reported so far in the literature are ditopic species, bearing two self-assembly-efficient moieties connected through suitable linkers. The two moieties can be identical to each other (for *homoditopic* monomers) or different (for *heteroditopic* monomers).

On the basis of structural considerations, cavitand monomers can be classified into four main classes (Scheme 4.1). In the first class, heteroditopic A-B monomers can be found, in which A:A and B:B interactions are respectively complementary. In this case a single monomer is present, but two interaction modes are active, leading to the formation of A-B:B-A:A-B type polymers. A second class includes heteroditopic A-B monomers, in which A:B interaction is complementary in nature. In this case also a single monomer is involved, but only one interaction mode is active, and A-B:A-B:A-B type polymers are formed. The third class collects the homoditopic A-A monomers, in which the A:A interaction is self-complementary. A single monomer and a single interaction are involved, leading to the formation of A-A:A-A:A-A type polymers. Finally, the fourth class includes the A-A monomers, in which the A:A interaction is not self-complementary. In this case two different homoditopic monomers are required, namely A-A and B-B, and the polymerization process proceeds thanks to the complementarity of the A:B interaction, leading to A-A:B-B:A-A:B-B copolymers.

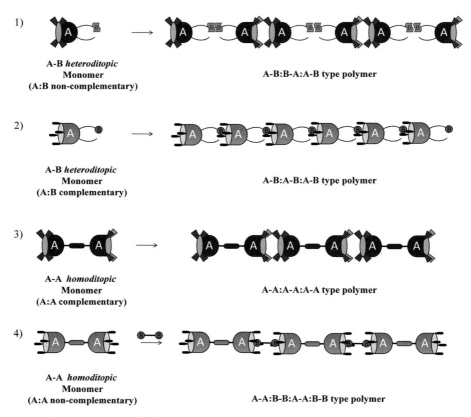

Scheme 4.1 Supramolecular polymerization motifs of cavitand monomers.

Several factors have to be considered in designing cavitand monomers. Having established the number and type of noncovalent interaction modes to exploit, a synthetic strategy has to be envisioned to introduce the effective end-groups on the cavitand platform. To this end, structural considerations are necessary in order to maximize the efficiency of the selected binding motifs.

The preorganization of the monomeric units is pivotal in the determination of the final assembled objects. For example, the equilibrium between cyclic species and high-molecular-weight linear polymers, which is often observed during supramolecular polymerization, can be controlled by acting on the rigidity and on the preorganization of the building-blocks. Also, by manipulating the same parameters, the thermodynamic contributions can be tuned, thus achieving enthalpically and entropically favorable processes. Again, the polymer growth mechanism can be strictly affected by the monomer structure, and suitable structural considerations can sometimes allow us to predict whether an isodesmic, ring–chain, or cooperative mechanism will be followed [36]. In the next paragraphs we will present some examples of cavitand-based supramolecular polymers, which will clarify the above-mentioned principles.

4.3.2
Homoditopic Cavitands Self-Assembled via Solvophobic π-π Stacking Interactions

Paek and coworkers reported in 2004 the only example of homoditopic cavitand-based monomer self-assembling via self-complementary solvophobic π-π stacking interactions [37]. They prepared the bisvelcrand **1** (Scheme 4.2) by a stepwise route

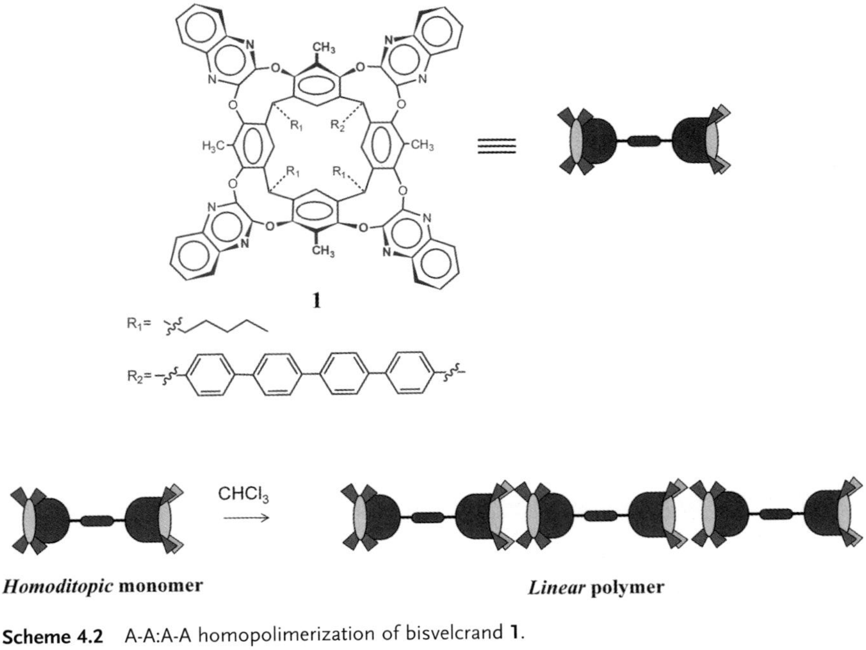

Scheme 4.2 A-A:A-A homopolimerization of bisvelcrand **1**.

that allowed the covalent linkage of two quinoxaline-bridged velcrand units through a tetraphenyl spacer.

The ^{1}H NMR spectrum of **1** in CDCl$_3$ (3.54 mM) at 298 K exhibited the signals for the methyl peaks at 3.17 and 2.33 ppm, as expected for ArCH_3 protons of kite-kite quinoxaline dimers [30], demonstrating the efficiency of the interaction mode chosen to drive the polymerization process. The energy barrier for pseudorotation between the kite conformers of oligobisvelcraplex **1$_n$**, was determined by ^{1}H NMR experiments at variable temperature in C$_6$D$_5$NO$_2$ and estimated equal to 16.7 kcal mol^{-1}. This value, higher than that found for simple quinoxaline dimers (15.7 kcal mol^{-1}), demonstrated that **1** exists in solution in the form of oligomeric species, imposing a larger hindrance to fast equilibrium between kite structures on the ^{1}H NMR time scale. Both pulse-field gradient spin-echo (PGSE) NMR experiments and vapor pressure osmometry (VPO) analysis indicated that bisvelcrand **1** oligomerizes as its concentration increases, reaching a number of linked units ranging between seven and ten in CHCl$_3$ at 298 K. Initially, observations at the optical microscope proved the tendency of **1** to form gels or fibers in high concentration regimes. Scanning electron microscopy (SEM) was used to investigate the microscopic structure of polybisvelcraplex **1$_n$** in the solid state. These studies demonstrated that the formation of polymeric structures depends on the concentration of the solute as well as on the polarity of the solvents. From a chloroform solution of **1** in a concentration range of 0.03–0.40 mM, images of nicely aligned linear strands were observed, while operating at diluted concentration (<0.01 mM) numerous dots were revealed, diagnostic of the presence of isolated monomeric species. Moving to THF, a different organization was achieved characterized by long, twisted bundles with a diameter of about 110–120 nm.

4.3.3
Heteroditopic Cavitands Combining Solvophobic Interactions and Metal–Ligand Coordination

The quest for highly adaptive dynamic materials led our group to engineer a new cavitand monomer operating in multimodal fashion through the implementation of two self-assembly codes, namely solvophobic aggregation and metal coordination [38]. For this purpose we synthesized cavitand **2**, featuring four quinoxaline bridges in kite conformation at the upper rim and two pyridyl terminal groups at the lower rim (Scheme 4.3). By exploiting the dimerization properties of quinoxaline kite velcrand and the capability of the pyridine ligands to coordinate a metal center, we realized a system capable of bimodal, independent self-assembling interactions.

Kite-to-kite dimerization in CD$_2$Cl$_2$ solution was monitored by ^{1}H NMR spectroscopy by following the splitting of several signals, with special attention paid to the ArCH_3 methyl protons, particularly diagnostic for the kite-kite dimer formation. The association constant and the free energy of association for this interaction were determined by ^{1}H NMR experiments at variable temperature ($K_{a\ 273K} = 1.91 \cdot 10^3$ M^{-1}; $\Delta G°_{273K} = -4.10$ kcal mol^{-1}). Comparison between the

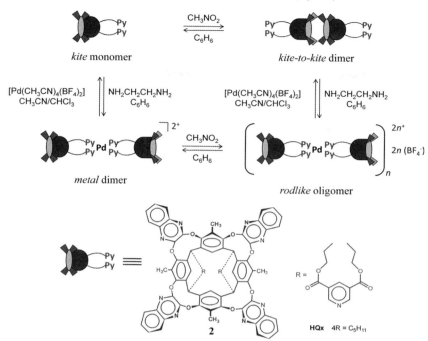

Scheme 4.3 Bimodal self-assembly cycle of the A-B heteroditopic cavitand monomer **2**.

obtained data with those recorded for a related tetrapentyl-footed kite velcrand **HQx**, revealed a reduced tendency of **2** to dimerize ($\Delta G°_2 = -4.10$ kcal mol^{-1} versus $\Delta G°_{HQx} = -6.81$ kcal mol^{-1}). This result was clarified by means of X-ray crystallography, which provided the crystal structure of the self-assembled oligovercraplex 2_n.

X-Ray analysis (Figure 4.5) revealed that the introduction of two pyridyl-based connecting units at the lower rim rigidified the velcrand structure, affecting the spatial orientation of the quinoxaline wings, thus reducing their dimerization capability. The reported crystal structure indicates only 77 short intermolecular atomic distances in the dimer, remarkably less than the 132 short distances observed in the original Cram's velcraplex (see Figure 4.3). Interestingly there is a linear correlation between the number of contacts and the measured K_a in solution: each short VdW contact accounts for a contribution of 0.053 kcal mol^{-1} to the overall binding. The presence of two preorganized pyridine feet at the lower rim jeopardizes the solvophobic interactions, leading to the formation of short oligomers instead of polymers.

The metal-directed self-assembly interaction was previously investigated in a structurally related tetramethylene-bridged cavitand by using [Pd(CH$_3$CN)$_4$](BF$_4$)$_2$ as the metal precursor and ethylenediamine as the bidentate competitive ligand to induce the disassembly of the coordination dimer [39].

The two self-assembly protocols were finally combined by envisioning a bimodal self-assembly cycle involving **2**, and ultimately realized reversible oligomeric structures (Scheme 4.3). By mixing **2** and [Pd(CH$_3$CN)$_4$](BF$_4$)$_2$ in a 2:1 molar ratio

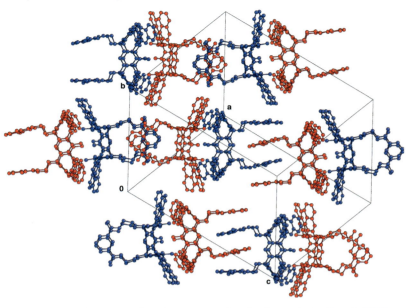

Figure 4.5 Crystal structure of **2·2** dimers. Reproduced with permission from Ref. [38].

in $CHCl_3/CH_3CN$, complex 2_n was formed and precipitated as a yellow powder. Both ^1H NMR experiments and MALDI-TOF measurements corroborated the oligomeric nature of 2_n. While the solvophobic disassembly was achieved by moving to a competitive apolar solvent, such as benzene, the coordinative dissociation was selectively carried out by addition of ethylenediamine.

The bimodal self-assembly protocol was finally transferred and amplified on surfaces. In particular, tapping mode scanning force microscopy (TM-SFM) was exploited to explore the supramolecular architectures formed on highly oriented pyrolytic graphite (HOPG). In order to prevent the lateral aggregation between individual self-assembled objects, a linear alkane ($C_{32}H_{64}$) was initially added to a dilute solution of **2**. After removal of the liquid by spinning off, and after thermal annealing at 313 K for 15 min, two types of nanostructures formed on the graphite surfaces, namely rodlike and grainlike objects. In particular, the rodlike structures, featuring lengths of up to 150 nm and widths of (7 ± 3) nm were consistent with the lateral aggregation of five to seven supramolecular rods.

A different approach to the self-assembly of supramolecular architectures through combined solvophobic and metal–ligand interactions was described by Paek et al. [40]. In this case, the new velcrand **3**, formed by a 2-methylresorcin[4]arene-based quinoxaline kite velcrand unit functionalized at the lower rim with a p-pyridylphenyl foot, was prepared to dimerize by metal coordination. The originality of this approach consists in the different structural possibilities of the self-assembled nanostructure based on **3**. In fact, as shown in Scheme 4.4, after a first dimerization by solvophobic interactions, a subsequent oligomerization occurs upon addition of a metal center, generating linear or cyclic structures, depending on the geometry of the

4.3 Self-Assembly of Ditopic Cavitand Monomers

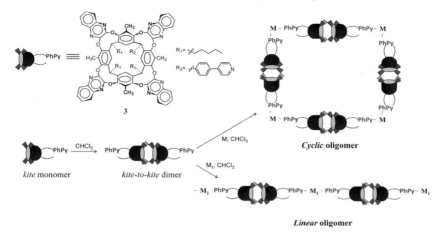

Scheme 4.4 Chain and ring formation upon addition of a metal center to a solution of **3·3** dimers.

metal involved in the coordination event. For example, velcrand **3** was expected to form linear metal dimers with *trans*- Pd(DMSO)$_2$Cl$_2$, but an orthogonal metal dimer with *cis*-Pt(dppp)(OTf)$_2$ (Scheme 4.5), thus leading to the formation of linear or cyclic oligovercraplexes respectively.

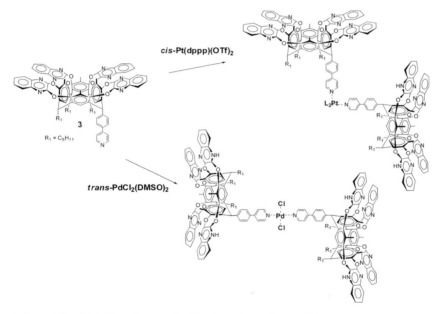

Scheme 4.5 Metal-directed assembly of bent and linear dimers of **3**.

While kite-to-kite dimerization was studied by ^{1}H NMR experiments in CDCl$_3$, the metal-coordination motif was investigated by the combined use of electrospray ionization mass spectroscopy (ESI-MS) and PGSE NMR. The formation of square planar cyclic tetramers by the simple mixing of **3** with Pt(dppp)(OTf)$_2$ in a 2:1 molar ratio was demonstrated by ESI-MS analysis, which provided the specific molecular ion peaks of the tetrameric oligovercraplex **3**$_4$. On the other hand, PGSE-NMR experiments supported the presence of a concentration-dependent decrease of the diffusion coefficient for CDCl$_3$ solutions of **3** coordinated with Pt(dppp)(OTf)$_2$. This result is indicative of the formation of larger self-assembled structures with concentration increase, as expected for supramolecular architectures. In particular, the possibility to shift ring–chain equilibria toward the formation of linear species by acting on the monomer concentration is a well-known topic in the field of supramolecular polymerization [36], and we describe it in more detail the next paragraph.

4.3.4
Heteroditopic Cavitands Combining Solvophobic Interactions and Hydrogen Bonding

An alternative strategy to the one outlined above was recently described by our group [41]. We prepared the new quinoxaline kite velcrand **4**, bearing a single ureidopyrimidone (upy) moiety at the lower rim (Scheme 4.6) and able to combine multiple hydrogen bonding with solvophobic interactions, to generate dual-coded supramolecular structures. The presence of a single H-bonding unit at the lower rim removes the problem of velcrand distortion observed in our previous work [38].

^{1}H NMR experiments in CDCl$_3$ were carried out to track the self-assembly of monomer **4** in solution. In particular, the presence of diagnostic signals for kite-to-kite dimerization (3.10 ppm and 2.13 ppm) and for H-bond formation (region between 13.14 and 9.75 ppm) demonstrated that the two chosen binding motifs were operational and noninteracting between each other. A simple ^{1}H NMR dilution experiment put a lower rim limit on the dimerization constants: $K_a > 10^5$ M^{-1}. ^{1}H NMR experiments in a low concentration regime also revealed the presence of a ring–chain equilibrium. The splitting of the H-bonded N*H* signals into three sets of peaks corroborated the concomitant formation of linear and cyclic oligomers (Figure 4.6).

This behavior was expected, taking into account the high flexibility of the upy alkyl foot in monomer **4**. Cyclic species were destroyed by addition of an equimolar amount of 2,7-acetamido-1,8-naphthyridine. The naphthyridine derivative dissociated the upy-upy homodimers by turning them in robust upy-napy heterodimers [42, 43]. In this way, geometrically equivalent napy·**4**·**4**·napy tetrameric species were formed, responding to the simplification observed in the NMR spectrum (Figure 4.7).

The influence of temperature and monomer concentration on the ring–chain equilibrium was investigated in detail. The predominant formation of linear species was achieved by heating the solution, as well as by increasing the concentration of **4**. The polymerization potential of the cavitand monomer is hampered by the formation of cyclic structures at low concentrations, as shown by NMR experiments and supported by SLS and DLS measurements. The reversible assembly and disassembly

4.3 Self-Assembly of Ditopic Cavitand Monomers | 83

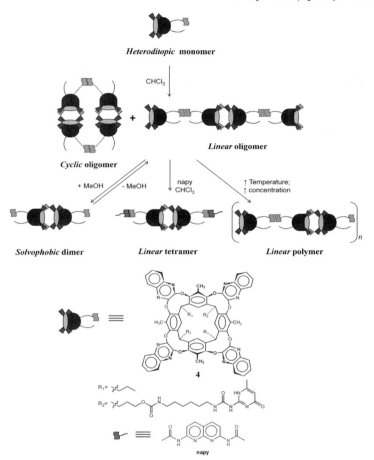

Scheme 4.6 Cyclic versus linear polymerization of velcrand **4** driven by solvent polarity, temperature and concentration.

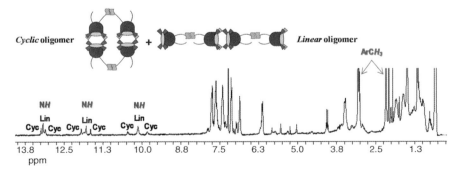

Figure 4.6 Partial ^{1}H NMR spectrum in CDCl$_3$ of [4·4]$_n$ and the proposed equilibrium between linear and cyclic structures.

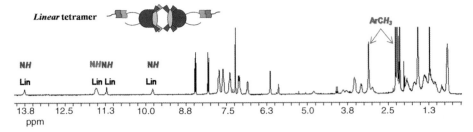

Figure 4.7 Partial ^{1}H NMR spectrum in CDCl$_3$ for **napy·4·4·napy** tetramer formation.

of 4_n was obtained by selective intervention on the H-bond motif via addition of a competing solvent such as methanol. Instead, the solvophobic kite-to-kite dimerization is resilient, being effective even in low polarity solvents. SLS and DLS measurements were performed to study the growth trend of oligomer 4_n. Both techniques revealed that the number of connected monomer units increased with the concentration of **4**.

4.3.5
Heteroditopic Cavitands Self-assembled via Host–Guest Interactions

Let us now turn our attention to cavitand-based supramolecular polymers driven by host–guest interactions. For this purpose phosphonate cavitands have been used, since they fulfill the requirement for high association constants between the host and suitable guests (see Chapter 6.2). By exploiting the outstanding complexation properties of tetraphosphonate cavitands toward methylpyridinium guests, we self-assembled a new class of supramolecular polymers featuring not only a reversible nature but also a remarkable plasticity, allowing structural switches from linear to star-branched architectures [44]. In particular, cavitand **5** was prepared (Scheme 4.7), functionalized at the upper rim with four phosphonate bridges in their all inward-facing configuration, and at the lower rim with a single methylpyridinium unit.

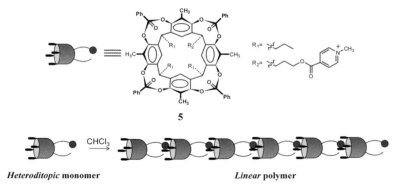

Scheme 4.7 A-B heteroditopic Polymerization mode of cavitand **5**.

In this way both the host cavity and the guest moiety are placed in the same molecule on opposite sides, to exclude the possibility of self-association.

The complexation properties of tetraphosphonate cavitand toward methylpyridinium guest were studied by isothermal titration calorimetry (ITC). In order to have a suitable model for a single polymerization step, we synthesized: (a) a cavitand derivative **6** without self-complementary functionalization at the lower rim, (b) the structurally related but complexation-inefficient tetrathiophosphonate analog **7**, featuring the same methylpyridinium foot. Substitution of the four P=O groups with the P=S moieties completely prevents complexation by eliminating cation–dipole interactions between the cavitand and the charged nitrogen of the pyridinium unit [45]. Its larger size does not preclude the insertion of methyl groups into the cavity, therefore the introduction of four P=S bridges has little influence on CH-π interactions, which, however, are not sufficient for complexation. For cavitands **5** and **7** both the iodide and triflate salts were prepared to evaluate the counterion influence on polymer formation (Figure 4.8). The association constants related to the formation of **6·7a** and **6·7b** dimers were determined by ITC, obtaining in both cases values exceeding 10^7 M^{-1} in dichloromethane (Table 4.1).

The complexation being driven not only by enthalpy but also by entropy, the effect of the solvent was demonstrated to be pivotal in the process. In particular an increase of two orders of magnitude was observed moving from methanol to methylene chloride (Table 4.1).

Additional information about the major interactions responsible for the complexation came from X-ray analysis. The crystal structures of the **6·7b** dimer (Figure 4.9) shows the two major interactions responsible for the complexation, already described in Section 4.2. Moreover, it shows that **7** can be safely employed as a stopper in the polymerization. The same interactions responsible for dimer formation are observed in the 5_n homopolymer crystal structure (Figure 4.10), where each linear polymeric chain packs against four other antiparallel chains. The charges of the monomers bearing the guest functionality are counterbalanced by triflate anions located in between the lower-rim alkyl chains, close to the methylpyridinium cations. In the

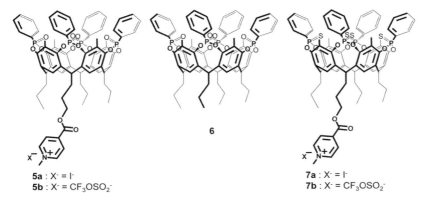

5a : X⁻ = I⁻
5b : X⁻ = CF$_3$OSO$_2$⁻

7a : X⁻ = I⁻
7b : X⁻ = CF$_3$OSO$_2$⁻

Figure 4.8 Structures of monomers **5a,b**, model cavitand **6** and chain stoppers **7a,b**.

Table 4.1 Thermodynamic parameters derived from ITC experiments for the host–guest dimers at 298 K.

Entry	Solvent	ΔH [kJ mol^{-1}]	$T\Delta S$ [kJ mol^{-1}]	ΔG [kJ mol^{-1}]	K_a [M^{-1}]
7b·6	MeOH	-11.9 ± 0.4	16.1 ± 0.5	-28.0 ± 0.4	$(8.2 \pm 0.7)\ 10^4$
7b·6	CH$_2$Cl$_2$	-30.0 ± 0.1	13.4 ± 0.5	-43.6 ± 0.5	$(4.0 \pm 0.1)\ 10^7$
7a·6	CH$_2$Cl$_2$	-26.1 ± 0.2	15.6 ± 0.9	-41.4 ± 0.8	$(1.9 \pm 0.6)\ 10^7$

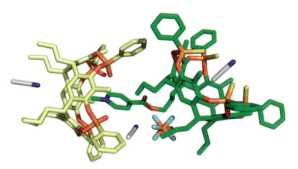

Figure 4.9 Crystal structure of dimer **7b·6**, mimicking the monomer-stopper interaction. Reproduced with permission from Ref. [44].

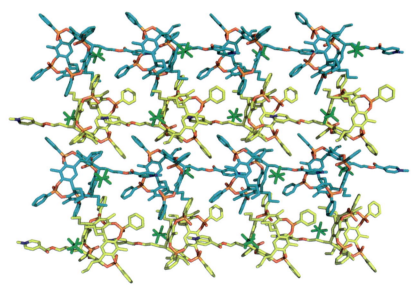

Figure 4.10 Views of the crystal packing of **5b** homopolymer. Reproduced with permission from Ref. [44].

crystal the polymer assumes a straight and extended (all-trans pyridinium linker) conformation.

SLS measurements were performed in a concentration range of 1.04–15.0 g L^{-1}, providing a molecular weight (M_w) upper value of 26 300 g mol^{-1}, corresponding to an average of 18 monomers units linked together. While the stepwise addition of stopper **7a** resulted in a progressive, linear decrease of M_w, the presence of 5,10,15.20-tetrakis(1-methyl-4-pyridinio)porphyrintetra(p-toluene sulfonate) **8** led to a substantial increase in M_w. In the first case, the addition of stopper **7a**, suitable to be complexed but unable to complex, had a detrimental effect on the polymer growth, acting as a monotopic impurity and determining the fall in M_w. In the second case, in contrast, the porphyrin derivative acted as a template molecule, ordering the preformed, linear polymer chains into a star-branched over-structure (Scheme 4.8), as corroborated by the M_w data obtained by SLS (Table 4.2).

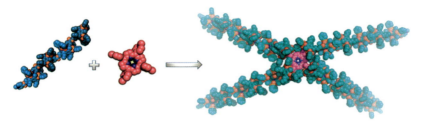

Scheme 4.8 Template-driven switching from linear to star-branched architectures.

The SLS saturation at high monomer concentrations forced us to turn to PGSE as an indirect method to evaluate the average molecular weight as function of the monomer concentration in solution. The results were rather surprising: PGSE showed the presence of two species in solution, both of which increasing their molecular weight with concentration (Figure 4.11). The low M_w species belongs to single polymer chains composed of 4–50 monomer units, while the high M_w species (range 76–380 103 Da) are associated with larger objects in slow exchange with the small ones. These last ones become dominant at monomer concentrations above 70 mM. These results can be interpreted by considering that the polymer chains interact longitudinally to generate bundles. As can be inferred from the crystal structure, the host–guest interaction mode is not the only one present. In solution, multipolar interactions among polymeric chains are also possible: such bundle-

Table 4.2 SLS measurements of the M_w of the linear and corresponding star-branched polymer.

Conc. [g L^{-1}]	Molar Ratio Porphyrin/5a [%]	M_w5a Linear Polymer	M_w5a Star Polymer
1.04	2.5	11 850 ± 150	46 210 ± 580
2.05	2.0	13 490 ± 230	41 600 ± 180
4.10	2.0	15 940 ± 90	49 700 ± 180

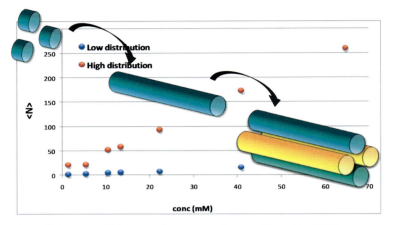

Figure 4.11 Low and high M_w polymer population as evidenced by PGSE measurements. Concentration plot versus number of connected monomeric units.

like structures are in equilibrium with the non-aggregated polymer chains. Due to the lower strength of the interchain interactions, they start to be competitive with the monomer addition when a single polymer chain length approaches six monomer units. Therefore, above a certain chain length, formation of bundles is favored over chain growth. This hypothesis is supported by PGSE measurements in different solvents: since multipolar interactions are very sensitive to the solvent dielectric constant, addition of polar solvents leads to the complete suppression of the aggregation of bundles while retaining linear polymer formation [46].

Reversible disassembly of the homopolymer was achieved by addition of a competitive guest, such as N,N-methylbutylammonium iodide, which is complexed by tetraphosphonate cavitand with an association constant exceeding $10^8\,M^{-1}$ thanks to its ability to establish additional H-bonding between its NH protons and the P=O groups of the host molecule. Subsequent addition of a hindered base such as 1,8-diazabicyclo[5.4.0]undec-7-ene (DBU), too large to be engulfed in the cavity but able to deprotonate the ammonium salt thus canceling its affinity for the cavitand, restored the initial complex with the pyridinium guest and, consequently, the homopolymer 5_n. The whole process was monitored by ^{1}H and ^{31}P NMR spectroscopy in $CDCl_3$, in which the polymer is soluble, and visually in acetonitrile, in which it is insoluble (Figure 4.12).

4.3.6
Homoditopic Cavitands Self-assembled via Host–Guest Interactions

The examples reported until now refer to homopolymeric species formed among self-complementary homo- and heteroditopic single monomers. An alternative approach, leading to the formation of supramolecular copolymers, requires the splitting of the self-assembling functionalities into two different species. Homoditopic complementary partner molecules have recently been reported by our group [47]. Thus, we

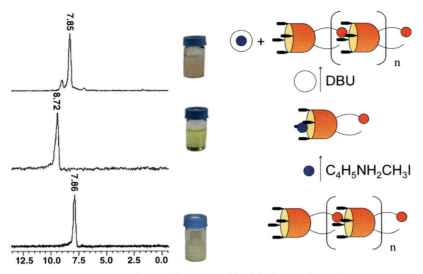

Figure 4.12 Guest triggered assembly–disassembly of the homopolymer **5a**.

synthesized homoditopic hosts (A-A) presenting two tetraphosphonate cavitands covalently attached at their lower rim, and homoditopic guests (B-B), in which flexible ethylene oxide chains of different length are functionalized with two *N*-methylpyridinium end-groups (Scheme 4.9).

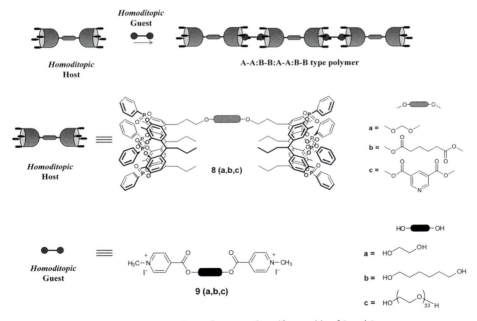

Scheme 4.9 A-A:B-B alternating copolymer formation by self-assembly of **8** and **9**.

The complexation of N-methylpyridinium-based guests inside the host cavity was first determined by a simple ^{1}H NMR titration. Diagnostic up-field shift of the guest signals were observed, as expected for included species that experience the shielding effect of the cavity. In particular, the CH_3-pyridinium moiety moved more than 3 ppm up-field; ortho and meta pyridyl protons featured a lower shift in the same direction, from 8.60 to 8.39 ppm and from 8.51 to about 7.98 ppm respectively. The presence of complexed species was also confirmed by ESI-MS spectroscopy. In order to quantify the thermodynamic parameters $\Delta H°$, K_a, $\Delta G°$, $\Delta S°$ associated with the inclusion process, ITC studies were carried out (Table 4.3). While preliminary experiments performed with both monotopic and ditopic species revealed that no cooperativity, either positive or negative, is present when multiple binding occurs, the influence of the monomer structures on the polymerization thermodynamic parameters clearly appeared from the data analysis. Particularly appealing is the possibility to turn a negative $T\Delta S°$ term into a neutral or positive one by acting on the flexibility of the guest linker. For example, when a really flexible spacer, such as the polyethylene glycol chain in **9c** was exploited, the entropy loss that generally occurs during self-association was reduced, thus leading to a process driven by both enthalpy and entropy. A strong solvent effect was also observed, and a gain of two orders of magnitude for the K_a value was recorded by moving from MeOH to CH_2Cl_2.

The polymer structure and M_W determination were obtained via X-Ray analysis and viscosity measurements. In the case of rigid systems, exploiting, for example, the self-assembly between commercially available methyl viologen $2PF_6^-$ and ditopic host **8c**, X-Ray analysis provided direct physical evidence of the formation of linear supramolecular polymers (Figure 4.13).

In the case of flexible systems, formed by self-assembly of ditopic host **8b** and featuring an adipic spacer connecting the cavitand units and ditopic guest **9c** bearing a polyethyleneglycol linker, viscosimetry investigations accounted for the presence of a ring–chain equilibrium, converging to the formation of linear polymeric species in high concentration regimes.

Thanks to a suitable engineering of the host spacer, structural switches are possible. In our case, for example, the metal-directed conversion of linear polymeric chains into cross-linked supramolecular architectures was achieved via coordination of the pyridine units embedded in the spacer of ditopic host **8c**. In this case, neutral complex $(CH_3CN)_2PdCl_2$ was selected as an orthogonal curing agent, since

Table 4.3 Thermodynamic parameters and K_a values for ITC titration of **8b** with **9a**, **9b**, **9c** in methanol at 25 °C.

MeOH	8b·9a	8b·9b	8b·9c
K_a (M^{-1})	$(1.3 \pm 0.1)\ 10^5$	$(8.4 \pm 0.6)\ 10^4$	$(4.9 \pm 0.2)\ 10^5$
$\Delta H°$ (kJ mol^{-1})	-34.8 ± 0.6	-26.7 ± 0.3	-25.4 ± 0.1
$T\Delta S°$ (kJ mol^{-1})	-5.6 ± 0.5	1.5 ± 0.4	7.1 ± 0.3
$\Delta G°$ (kJ mol^{-1})	$-29.2 \pm 0,2$	-28.2 ± 0.2	-32.5 ± 0.1

[**8b**] = 0.5 mM; [**guest**] = 4 mM. All data fitted well for a 1:1 binding profile.

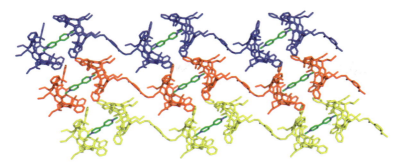

Figure 4.13 Crystal structure of the **8c** methyl viologen A-A:B-B alternating copolymer. Reproduced with permission from Ref. [47].

it easily undergoes ligand exchange of its trans acetonitrile ligands in the presence of pyridine without competing for P=O binding. The whole process was monitored by ^{1}H NMR in CD_3CN.

4.4
Conclusions and Outlook

Now that supramolecular polymers have positioned themselves as a new paradigm in polymer science [36], there is the need for specifically designed stimuli-responsive monomers to bolster the field. Cavitands have shown themselves to be suitable for this purpose. They offer to the practitioners of this research a wide palette of useful properties, summarized here:

i) Synthetic versatility both in terms of number and type of interacting units and their relative positioning in the cavitand scaffold. In particular, different upper and lower rim functionalizations allow for the introduction of two orthogonally interacting motifs like H-bonding, metal coordination, and solvophobic interactions.
ii) In the host–guest mode, they are among the few receptors with large association constants above the 10^5 threshold necessary for the generation of real polymeric systems.
iii) They are amenable to assembly into both neutral and charged supramolecular polymers.
iv) Supramolecular plasticity, namely the ability to switch to a different polymer topology (star-branched, cross-linking) upon addition of a small amount of noncovalent curing agent.
v) Supramolecular control of the degree of polymerization via suitable stoppers.
vi) Guest-triggered disassembly in the host–guest polymerization mode. This last property, specific for host–guest systems, is particularly intriguing for the use of supramolecular polymers in sensing and biomedical applications.

As for the future, several possibilities can be envisioned for the use of cavitand-based supramolecular polymers. Two of them are worth mentioning here. The first is assembling supramolecular polymers on solid supports to generate stimuli-responsive surfaces. As a first step in this direction, we demonstrated hierarchical self-assembly on silicon with cavitands, amenable to polymeric scale-up [48]. The second is embedding cavitand monomers into covalent polymers to induce specific properties such as molecular level miscibility of immiscible polymers [49] or trigger phase segregation upon specific external stimuli [50].

Acknowledgements

Our own work described in this chapter has been supported over the years by the following grants: CNR Nanotechnology Program, EC through the NoE MAGMANet (3-NMP 515767-2), projects FINELUMEN (PITN-GA-2008-215399), and BION (ICT-2007-213219).

References

1 Ciferri, A. (ed.) (2005) *Supramolecular Polymers*, 2nd edn, Francis & Taylor, New York.
2 Lehn, J.-M. (1990) *Angew. Chem. Int. Ed. Engl.*, **29**, 1304–1319.
3 Folmer, B.J.B., Sijbesma, R.P., Versteegen, R.M., van der Rijt, J.A.J., and Meijer, E.W. (2000) *Adv. Mater.*, **12**, 874–878.
4 Zang, L., Che, Y., and Moore, J.S. (2008) *Acc. Chem. Res.*, **41**, 1596–1608.
5 Hartgerink, J.D., Beniash, E., and Stupp, S.I. (2001) *Science*, **294**, 1684–1688.
6 Burattini, S., Colquhoun, H.M., Fox, J.D., Friedmann, D., Greenland, B.W., Harris, P.J.F., Hayes, W., Mackay, M.E., and Rowan, S.J. (2009) *Chem. Commun.*, 6717–6720.
7 Cordier, P., Tournilhac, F., Soulie-Ziakovic, C., and Lieber, L. (2008) *Nature*, **451**, 977–980.
8 Johnson, C.S. Jr. (1999) *Prog. Nucl. Magn. Reson. Spectrosc.*, **34**, 203–256.
9 Schmid, S.A., Abbel, R., Schenning, A.P.H., Meijer, E.W., Sijbesma, R.P., and Herz, L.M. (2009) *J. Am. Chem. Soc.*, **131**, 17696–17704.
10 Samorì, P., Francke, V., Müllen, K., and Rabe, J.P. (1999) *Chem. Eur. J.*, **5**, 2312–2317.
11 Samorì, P., Severin, N., Müllen, K., and Rabe, J.P. (2000) *Adv. Mater.*, **12**, 579–582.
12 Rauwald, U. and Scherman, O.A. (2008) *Angew. Chem., Int. Ed.*, **47**, 3950–3953.
13 Deroo, S., Rauwald, U., Robinson, C.V., and Scherman, O.A. (2009) *Chem. Commun.*, 644–646.
14 Xu, H. and Rudkevich, D.M. (2004) *Chem. Eur. J.*, **10**, 5432–5442.
15 Haino, T., Matsumoto, Y., and Fukazawa, Y. (2005) *J. Am. Chem. Soc.*, **127**, 8936–8937.
16 Pappalardo, S., Villari, V., Slovak, S., Cohen, Y., Gattuso, G., Notti, A., Pappalardo, A., Pisagatti, I., and Parisi, M.F. (2007) *Chem. Eur. J.*, **13**, 8164–8173.
17 Huang, F. and Gibson, H.W. (2005) *Chem. Commun.*, 1696–1698.
18 Wang, F., Zhang, J., Ding, X., Dong, S., Liu, M., Zheng, B., Li, S., Wu, L., Yu, Y., Gibson, H.W., and Huang, F. (2010) *Angew. Chem. Int. Ed.*, **49**, 1090–1094.
19 Castellano, R.K., Rudkevich, D.M., and Rebek, J. Jr. (1997) *Proc. Natl. Acad. USA*, **94**, 7132–7137.
20 Cram, D.J. and Cram, J.M. (1994) *Container Molecules and their Guests*, Royal Society of Chemistry, London.
21 Rudkevich, D.M. and Rebek, J. Jr. (1999) *Eur. J. Org. Chem.*, 1991–2005.

22 Laughrey, Z.R., Gibb, C.L.D., Senechal, T., and Gibb, B.C. (2003) *Chem. Eur. J.*, **9**, 130–139.

23 Pirondini, L. and Dalcanale, E. (2007) *Chem. Soc. Rev.*, **36**, 695–706.

24 Dalcanale, E., Soncini, P., Bacchilega, G., and Ugozzoli, F. (1989) *J. Chem. Soc., Chem. Commun.*, 500–501.

25 Pinalli, R., Suman, M., and Dalcanale, E. (2004) *Eur. J. Org. Chem.*, 451–462.

26 Soncini, P., Bonsignore, S., Dalcanale, E., and Ugozzoli, F. (1992) *J. Org. Chem.*, **57**, 4608–4612.

27 Bianchi, F., Pinalli, R., Ugozzoli, F., Spera, S., Careri, M., and Dalcanale, E. (2003) *New J. Chem.*, **27**, 502–509.

28 Moran, J.R., Ericson, J.L., Dalcanale, E., Bryant, J.A., Knobler, C.B., and Cram, D.J. (1991) *J. Am. Chem. Soc.*, **113**, 5707–5714.

29 Skinner, P.J., Cheetham, A.G., Beeby, A., Gramlich, V., and Diederich, F. (2001) *Helv. Chim. Acta*, **84**, 2146–2153.

30 Cram, D.J., Choi, H.-J., Bryant, J.A., and Knobler, C.B. (1992) *J. Am. Chem. Soc.*, **114**, 7748–7765.

31 Nifantyev, E.E., Maslennikova, V.I., and Merkulov, R.V. (2005) *Acc. Chem. Res.*, **38**, 108–116.

32 Delangle, P., Mulatier, J.-C., Tinant, B., Declercq, J.-P., and Dutasta, J.-P. (2001) *Eur. J. Org. Chem.*, 3695–3704.

33 Kalenius, E., Moiani, D., Dalcanale, E., and Vainiotalo, P. (2007) *Chem. Commun.*, 3865–3867.

34 Melegari, M., Suman, M., Pirondini, L., Moiani, D., Massera, C., Ugozzoli, F., Kalenius, E., Vainiotalo, P., Mulatier, J.-C., Dutasta, J.-P., and Dalcanale, E. (2008) *Chem. Eur. J.*, **14**, 5772–5779.

35 Biavardi, E., Battistini, G., Montalti, M., Yebeutchou, R.M., Prodi, L., and Dalcanale, E. (2008) *Chem. Commun.*, 1638–1640.

36 De Greef, T.F.A., Smulders, M.M.J., Wolffs, M., Schenning, A.P.H.J., Sijbesma, R.P., and Meijer, E.W. (2009) *Chem. Soc. Rev.*, **109**, 5687–5754.

37 Ihm, H., Ahn, J.-S., Lah, M.S., Ko, Y.H., and Paek, K. (2004) *Org. Lett.*, **6**, 3893–3896.

38 Pirondini, L., Stendardo, A.G., Geremia, S., Campagnolo, M., Samorì, P., Rabe, J.P., Fokkens, R., and Dalcanale, E. (2003) *Angew. Chem. Int. Ed.*, **42**, 1384–1387.

39 Pirondini, L., Bonifazi, D., Menozzi, E., Wegelius, E., Rissanen, K., Massera, C., and Dalcanale, E. (2001) *Eur. J. Org. Chem.*, 2311–2320.

40 Kwak, M.-J. and Paek, K. (2007) *Bull. Korean Chem. Soc.*, **28**, 1440–1442.

41 Tancini, F., Rampazzo, E., and Dalcanale, E. (2010) *Aust. J. Chem.*, **63**, 646–652.

42 Beijer, F.H., Sijbesma, R.P., Kooijman, H., Spek, A.L., and Meijer, E.W. (1998) *J. Am. Chem. Soc.*, **120**, 6761–6769.

43 Corbin, P.S. and Zimmermann, S.C. (1998) *J. Am. Chem. Soc.*, **120**, 9710–9711.

44 Yebeutchou, R.M., Tancini, F., Demitri, N., Geremia, S., Mendichi, R., and Dalcanale, E. (2008) *Angew. Chem. Int. Ed.*, **47**, 4504–4508.

45 Menozzi, D., Biavardi, E., Massera, C., Schmidtchen, F.-P., Cornia, A., and Dalcanale, E. (2010) *Supramol. Chem.*, **22**, 768–775.

46 Macchioni, A., Zuccaccia, D., Tancini, F., Yebeutchou, R.M., Geremia, S., and Dalcanale, E.,unpublished results.

47 Tancini, F., Yebeutchou, R.M., Pirondini, L., De Zorzi, R., Geremia, S., Scherman, O.A., and Dalcanale, E. (2010) *Chem. Eur. J.*, **16**, 14313–14321.

48 Tancini, F., Genovese, D., Montalti, M., Cristofolini, L., Nasi, L., Prodi, L., and Dalcanale, E. (2010) *J. Am. Chem. Soc.*, **132**, 4781–4789.

49 For a first example in this direction see: Park, T. and Zimmerman, S.C. (2006) *J. Am. Chem. Soc.*, **128**, 11582–11590.

50 For the electrochemically controlled dissociation of phosphonate cavitand-methypyridinium complexes see: Gadenne, B., Semeraro, M., Yebeutchou, R.M., Tancini, F., Pirondini, L., Dalcanale, E., and Credi, A. (2008) *Chem. Eur. J.*, **14**, 8964–8971.

Part Two
Supramolecular Polymers with Unique Structures

Supramolecular Polymer Chemistry, First Edition. Edited by Akira Harada.
© 2012 Wiley-VCH Verlag GmbH & Co. KGaA. Published 2012 by Wiley-VCH Verlag GmbH & Co. KGaA.

5
Polymers Containing Covalently Bonded and Supramolecularly Attached Cyclodextrins as Side Groups

Helmut Ritter, Monir Tabatabai, and Bernd-Kristof Müller

The design of defined supramolecular structures relies on preferentially selective intermolecular interactions. An ideal and therefore often applied strategy is the formation of host–guest inclusion complexes. Here, cyclodextrins (**CD**) are probably the most promising host molecules. Their exceptional role is based on a combination of some major advantages including low cost, high diversity of potential guest molecules, and biological compatibility [1–5].

Since the late 1960s, CDs have also been applied in the field of polymer chemistry. Starting with highly crosslinked CDs as column materials [6] or as precursors in the preparation of urethane foams [7], the focus quickly shifted to pharmaceutical and biomedical applications [8–11].

CD-containing covalent networks, prepared either by direct cross-linking of CDs or by incorporation of CDs in polymer networks, are still of great interest and widely reviewed [2, 11–14]. Principally linear polymeric structures are being increasingly studied. These structures exhibit better accessibility of the CD-residues and are therefore more suitable for the formation of host–guest complexes. In this context, polymers both with CD as part of the main chain and with CD attached to the backbone are reported [2, 13, 14].

This chapter focuses on these polymers, which contain CDs as side groups. Possible structures are shown in Figure 5.1. In Section 5.1, polymers exhibiting CD covalently bonded to the backbone are reviewed. The next section then deals with CDs which are supramolecularly attached onto side groups and side chains.

5.1
Polymers with Covalently Bonded Cyclodextrins as Side Groups

To obtain linear polymers with CDs as side groups, two major strategies are applicable. On the one hand, monofunctional CD monomers can be employed in a polymerization reaction. Alternatively, the CD moiety can be attached through a polymer analogous reaction.

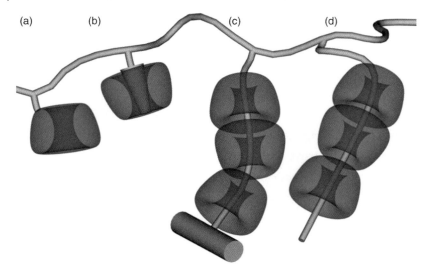

Figure 5.1 Possible structures of polymers with CDs as side groups. (a) CD covalent bonded; (b) CD supramolecular attached onto a side group; (c) CDs supramolecular attached in the manner of a rotaxane; (d) CDs supramolecular attached in the manner of a *pseudo*rotaxane.

5.1.1
Synthesis and Polymerization of Monofunctional Cyclodextrin Monomers

To obtain defined linear polymer structures with CD as side group the best way is to utilize monofunctional CD monomers (*mono*CD-monomer). However, a prerequisite is a convenient synthetic approach to these monomers. Monofunctionality has to be carefully ensured, since multifunctional monomers will lead to branched or crosslinked structures. Up to now, only a small number of *mono*CD-monomers have been reported. The challenge is the large number of hydroxyl groups competing for the modification reagent. Among all methods of monomodification, tosylation of βCD at the 6-position is the most improved reaction. Fairly good yields and purity along with user-friendly reaction conditions make this monofunctional CD derivative a popular precursor for subsequent modification [15]. For this reason, almost all of the *mono*CD-monomers reported so far comprise βCD. The structures of known *mono*CD-monomers are given in Figure 5.2. This figure also provides an overview of the applied monofunctional precursors and the yields obtained in the last step of the synthesis.

Due to the functionalization of the primary hydroxyl group of one of the anhydroglucose units, most of the *mono*CD-monomers are directed to the polymer backbone with their tighter opening. Exceptions are the monomers **1a,b** and **2a,b** of Furue and Harada, which were also the first *mono*CD-monomers of all, published in 1975 [16, 17]. Synthesis starts from native CDs, which are reacted with *m*-nitrophenyl acrylate derivatives under basic conditions at pH 11. The crude product is a mixture of different degrees of substitutions. To obtain the monofunctional derivatives, gel

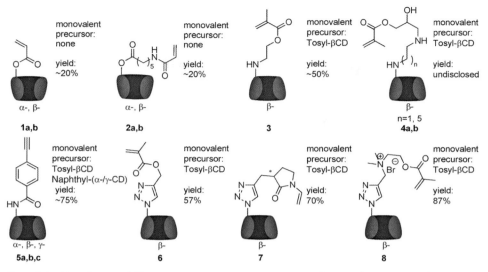

Figure 5.2 Monofunctional CD-monomers (*mono*CD-monomer); Monofunctional precursors and yields of the last synthetic step are annotated. More information can be found in the appendant literature: **1a,b** and **2a,b** [16, 17], **3** [18], **4a** [19, 20], **4b** [19], **5a,b,c** [21], **6** [22], **7** [23], **8** [24].

chromatography over Sephadex G-15 must be carried out, which results in rather low yields of about 20%. However, to obtain αCD-monomers no better procedure is reported.

Only the group of Yashima and coworkers have published another strategy to obtain a monofunctional αCD-monomer. This work also includes a βCD derivative and the only known monofunctional γCD-monomer. The key step for their derivatives is to use mono 6^1-amino-6^1-desoxy-cyclodextrins, which are subsequently reacted with 4-ethinylbenzoic acid chloride. The good yields of the final reaction step, about 75%, are offset by the low yields of the precursor. Monofunctional derivatives of αCD or γCD are obtained after reaction with 2-naphthalenesulfonyl chloride, which is substituted by sodium azide and then reduced to the designated amine. Overall yields for the αCD and γCD derivatives are 9% and 18% respectively. The βCD derivative is obtained in the same way but employs tosyl-βCD as the first intermediate. Overall yield was determined to be 2% [21].

Synthesis of *mono*CD-monomer **3** consists of four steps. Tosyl-βCD is reduced to the mono-5^1-formyl-5^1-deshydroxymethylene-βCD followed by the formation of the appropriate imine and subsequent reduction [18]. **4a,b** are obtained by substitution of tosyl-βCD with the adequate diamine and subsequent reaction with glycidyl methacrylate [19, 20]. For both procedures, overall yields are not reported.

A convenient route to *mono*CD-monomers has been developed by Ritter and coworkers very recently utilizing click-chemistry for coupling mono-6^1-azido-6^1-desoxy-βCD to different propinyl-derivatized monomers. In this way, *mono*CD-monomers **6** [22], **7** [23] and **8** [24] have been synthesized. This method is not in need of laborious purification steps and has an overall yield of at least 25%.

In general, polymerization reactions of the *mono*CD-monomers were carried out in polar solvents as N,N-dimethylformamide or water/methanol-mixtures with AIBN as radical initiator. *mono*CD-monomers were used in copolymerization reactions with NVP [17, 18, 23], NIPAM [19, 25], acrylamide, and acrylic acid [17], yielding copolymers with a number average molecular weight ($\bar{M}_n$) of 15 000–190 000 gmol^{-1}. However, copolymerization parameters or distribution of monomers in the copolymer have not been investigated.

Beside copolymers, homopolymers are also reported, while in most cases $\bar{M}_n$ does not exceed 13 000 gmol^{-1}. Accordingly, a low average degree of polymerization ($\bar{D}P$) of <12 is obtained [16, 17, 20, 22], although in the case of a long flexible spacer even higher $\bar{M}_n$ can be achieved. By use of monomer **8**, a βCD-containing homopolymer with $\bar{M}_n = 59,000$ gmol^{-1} has been prepared by radical polymerization [24].

The sole exception to radical polymerization mechanisms is described in the work of Yashima *et al.* With a rhodium complex as catalyst, monosubstituted acetylene monomers **5a,b** and **c** could be converted to chiral high-molecular-weight poly (acetylenes) with $\bar{M}_n = 111,000-163,000$ gmol^{-1}, which is remarkably high compared to all other CD-containing homopolymers [21].

5.1.2
Polymer-Analogous Reaction with Monofunctional Cyclodextrin

Compared to the polymerization of *mono*CD-monomers, the strategy of bonding CD residues to a polymer in a polymer-analogous reaction is used more often. But also in this case it is of the utmost importance to prevent crosslinking monofunctionality of the CD derivatives. Figure 5.3 gives an overview of monofunctional CD derivatives (***mono*CD**), which have been used for the modification of polymers. Obviously, tosyl-βCD (**9b**) is again an important key intermediate. Six *mono*CDs and two more secondary products are directly accessible from **9b**. αCD derivatives have only been reported for monotosylates [26] (either at C^6 (**9a**) or C^3 (**15**)), monoamines [26] (at C^6 (**22a**) or C^3 (**18**), too) and the monoaldehyde [27] (**12a**). For γCD, just mono-6^I-tosyl-6^I-desoxy-γCD (**9c**) has been used in a polymer-analogous reaction [28]. Besides secondary products from monotosylated CDs, there are only four more *mono*CDs reported, which can be synthesized from native CD in a single step. With the exception of the αCD monoaldehyde (**12a**), all of them are βCD derivatives. Preparation of *mono*CDs **11** [29] and **19** [30] involves extensive purification. In contrast, the procedure for mono deprotonation of a secondary hydroxyl group on C^2 on βCD with lithium hydride is quite simple and high yielding [31]. Likewise, mono oxidation of CD with Dess–Martin periodinane is highly efficient. Both compounds are therefore about as beneficial as **9b**. However, even more *mono*CDs are known, but these have not been used for polymer modification so far [15].

Which *mono*CD is chosen for the polymer-analogous reaction strongly depends on the chemical structure of the target polymer. The first modification of a polymer with CD was reported in 1987 by Seo and coworkers [32]. They attached monotosyl-βCD (**9b**) to poly(allylamine). This substitution reaction on **9b** with a polymeric nucleophile has been applied more often, though with different polymers. As well as poly

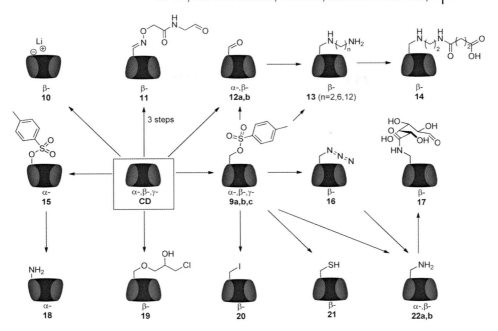

Figure 5.3 Monofunctional CD derivatives which have been used in polymer analogous modification reactions and the pathway of their synthesis. Structures **9a** and **15** have been synthesized only as intermediates. All other monofunctional CDs have been attached to suitable polymers. For experimental details, follow the citations given in the text.

(allylamine) [28, 33–35], poly(vinylamine) [36–40], linear [38] and branched poly(ethylenimine) [38–40] and poly(ethylene glycol-*block*-aspartamide) [41] were used. However, for polymers which do not contain highly nucleophilic amino groups, other strategies had to be developed. These include the second frequently applied modification reaction, the formation of amides from monoamino-CDs (**18**, **22a** and **b**) and poly(carboxylic acid) [26, 42–45] and its copolymers [46, 47]. Guo *et al.* synthesized monoamino-βCD derivatives in which the aminogroups were tethered by alkyl chains of different lengths (**13**). Just like **22**, **13** could be attached to poly(acrylic acid) by formation of amides, but the CD residue is separated from the polymers backbone through a spacer of defined length [48]. Ueno *et al.* confirmed the suitability of amidation for the site-selective derivatization of an aspartic acid residue containing peptide with **22b** [49].

Another frequently used reaction is the reductive amination of amino groups containing polymers and CD monoaldehydes (**12a** and **b**). Yui and coworkers employed this reaction for coupling **12a** and **b** to poly(ε-lysine) [50, 51]. Giammona *et al.* conjugated **12b** to α,β-polyaspartylhydrazide [52]. Likewise, the same authors coupled mono-6-[aminoethyl(4′-carboxybutanamide)]-βCD (**14**) to this polymer. This mild reaction was also applied for attachment of a βCD-monoaldehyde derivative (**11**) to adipic dihydrazide functionalized hyaluronic acid in a site-selective manner [29].

Also by reductive amination, Villalonga et al. grafted **22b** onto some polysaccharides, which were previously activated by periodate oxidation [53]. A rather inconvenient route for polysaccharide modification was chosen by Auzély-Velty and coworkers [54–56]. This route was based on the preparation of a monosubstituted βCD derivative possessing a reducing sugar on the primary face (**17**) followed by its reductive amination with chitosan.

Because of its good accessibility, utilization of monodeprotonated βCD (**10**) for the polymer-analogous reaction has been reported several times. It was shown that a polymer containing acid anhydride residues could be reacted with **10** with high yields – near 30% [31, 57–59].

A rather rarely used modification reaction is the quaternization of tertiary amines. Deratani et al. employed this reaction for coupling 6^{I}-O-(3-chloro-2-hydroxypropyl)-βCD (**19**) to poly(vinylimidazole) for the first time [30]. Tian et al. attached mono-6^{I}-iodo-6^{I}-desoxy-βCD (**20**), which is much easier to prepare, to blockcopolymers, containing a poly(N,N-dimethylaminoethylmethacrylate)-block [60].

Just recently, click chemistry has been introduced to polymer-analogous modification reactions with suitable *mono*CDs. Ritter et al. performed a polymer-analogous coupling of mono-6^{I}-azido-6^{I}-desoxy-βCD (**16**) onto poly(propargyl methacrylate) in a microwave reactor. This reaction turned out to be very efficient. In contrast to homopolymerization of *mono*CD-monomer **6**, high-molecular-weight polymers could be obtained by polymer analogous click reaction [22]. Furthermore, thiol-ene click reaction of mono-6^{I}-thio-6^{I}-desoxy-βCD to maleimide-functionalized dextran has been adopted by Kros et al. [61]. The same authors showed that this rapid reaction is even suitable for *in situ* formation of polymer networks [62].

5.1.3
Structure–Property Relationship of Polymers Containing Cyclodextrins as Side Group

The striking property of CD-containing polymers is their ability to form inclusion complexes with hydrophobic guest molecules of adequate size. Thus, for most of the above introduced polymers, the suitability for complexation has been widely investigated. As guest molecules, both low-molecular compounds and polymers have been employed. Despite the numerous studies, this section is limited to some basic results.

Already the first investigations of polymers with CD-moieties from Harada and coworkers in 1975 and 1976 have disclosed a chelate-type binding behavior with guest molecules possessing two binding sites [17, 63]. Adams et al. studied this complexation phenomenon in detail. They used a model system based on βCD-derivatized poly(allylamine). The synthesized βCD-polymers exhibited a significant change in the complexation behavior depending on the degree of substitution (DS). At high DS (up to 23%) only complexes with 2:1 stoichiometry were observed. This was interpreted as a evidence for intramolecular, chelate-like complexes due to the high local βCD concentration. At low DS (below 5%) 2:1 complexes were formed only intermolecularly. Compared to pure βCD, the overall complexation constant of the

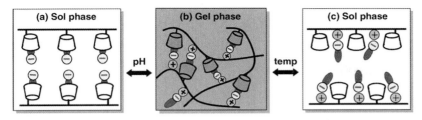

Figure 5.4 A pH- and thermosensitive physical polymer network formed by both ionic and hydrophobic interacting host molecules and a positively chargeable βCD-polymer. Reprinted with permission of reference [64].

βCD-polymers increased by more than two orders of magnitude with increasing DS and was independent of the polymer's molecular weight [34].

Yui and coworkers prepared a positively chargeable polymer with βCD-moieties based on poly(ε-lysine). Supramolecular assemblies were formed in water by addition of negatively charged guest molecules. Ionic and hydrophobic host–guest interactions appeared in a cooperative manner at low temperature, thus provoking formation of a physically crosslinked gel (Figure 5.4). This highly ordered structure was unfavored upon higher temperatures. Hence, the formerly cooperative interactions became competing, leading to a disruption of the weaker host–guest interactions, which was accompanieed by a gel-to-sol transition. Similarly, the gel could be degraded by raising the pH above 8.0, whereby the positive charges were neutralized from the polymer backbone. The reported reversible phase transitions responded very sensitively to minute changes near physiological conditions [64].

Some basic results for the preparation of tailored CD-containing physical networks were reported recently. Guo *et al.* prepared polymers in which the βCD-moiety or an adamantyl group were linked to poly(acrylic acid-*co*-acrylamide) by an alkyl spacer. In this model system, they investigated the influence of the length of the hydrophobic spacers upon hydrogel formation. It turned out that long alkyl spacers could be inter-or intramolecularly complexed by the βCD-moieties, thus competing with the desired host–guest complexes of βCD and adamantyl groups. Additionally, steric effects played a role in the stability of the hydrogel. If the interacting host or guest residues were closely attached to the polymer backbone, complex formation became less favorable. As these effects act contrarily considering the spacer length, a medium spacer length should be used at best [48]. Harada *et al.* showed that the orientation of the wider opening of a CD-residue – either in the direction of the polymer backbone or in the opposite direction – can have an effect on complex stability. They employed a physical network system comprising (a) a poly(acrylic acid) with UV-sensitive azobenzene moieties which were tethered through an alkyl spacer and (b) a polymeric host poly(acrylic acid) with αCD-residues (**18** and **22a**). Due to the chemical structure of the *mono*CDs, αCD was either linked via the wider C^3-site or the narrower C^6-site to the polymer backbone. The trans isomer of the azobenzene unit formed

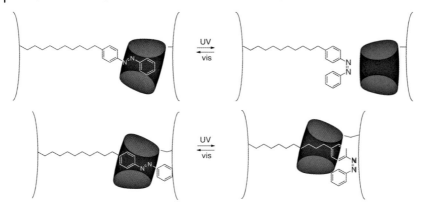

Figure 5.5 Schematic drawing of the interactions of polymer-bonded αCDs and azobenzene moieties upon irradiation with UV and visible light for two polymeric hydrogel systems, differing in the direction αCD is attached to the polymer backbone [26].

complexes with both host polymers. Since the cis isomer is too bulky for the αCD-cavity, UV irradiation results in a decomplexation of the azobenzene structure. Whereas the C^3-linked αCD-polymer network was degraded, the C^6-linked αCD-polymer network hardly lost any of its strength. This phenomenon was caused by the formation of interlocked complexes (Figure 5.5), indicating a tendency of the CD-molecules to slide in the direction of the wider opening [26].

A breakthrough in the theoretical consideration of the dynamics of a hydrogel from associating polymers was reported by Semenov and coworkers. As model system, a mixture of two hyaluronic acid derivatives functionalized either with βCD or adamantly residues was chosen [29, 65]. Above the overlap concentration it was found that these polymer systems generate parallel strands of contrarily derivatized polymers, denoted as 'railway complexes'. So called 'fourfold junction points' cause interlacing of the network. Additional consideration of chemical irregularities enable this model to predict rheological properties fairly precisely [66].

Ma *et al.* reported the successful engineering of core-shell nanospheres from block copolymers comprising two hydrophilic blocks, whereas one comprises βCD. The formation of nanospheres is directed by inclusion of guest substances into the βCD residues. These hydrophobic substances were either small molecules or macromolecules. Depending on their structure, core-shell nanoassemblies of different size and composition were spontaneously formed in an aqueous solution (Figure 5.6) [40].

As was described in Section 5.1, Yashima *et al.* synthesized a helical homopolymer consisting of poly(phenylacetylene) with CD-residues. π-conjugated polymers like poly(acetylenes) are colored substances which can shift their absorption spectrum upon external physical and chemical stimuli. The authors showed that these stimuli could be induced by variation of the solvent or the temperature or by addition of guest molecules. All of these influences caused a specific change of helicity which can reach even a total inversion. Actually, these polymers were sensitive enough to differentiate

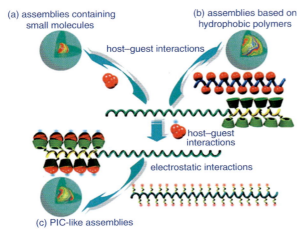

Figure 5.6 Schematic illustration of the formation of various host–guest assemblies: (a) small hydrophobic molecules mediated assemblies, (b) assemblies formed in the presence of a hydrophobic polymer, and (c) PIC-like assemblies. Reprinted with permission of reference [41].

between enantiomers. Finally, cross-linking with epichlorhydrin in the twisted state allowed the handedness of the helix to be fixed. This unique property will make this polymer useful as chiral material for enantiomer separation and as a catalyst [21].

5.2
Side Chain Polyrotaxanes and Poly*pseudo*rotaxanes

Polyrotaxanes are highly interesting and intensively investigated supramolecular materials. One class of polymeric rotaxanes consists of side chain polyrotaxanes (Figure 5.1c) and poly*pseudo*rotaxanes (Figure 5.1a and b). Side chain polyrotaxanes and poly*pseudo*rotaxanes are formed by host–guest interactions of polymer side chains with CDs. Side chain polyrotaxanes are comb-like polymers with CDs in the side chain where they are interlocked by bulky stoppers. Side chain poly*pseudo*rotaxanes are inclusion complexes in which cyclic molecules, in this case CDs, are threaded onto a side chain of the polymer. They are categorized into three types: [14]

1) Polyaxis/rotor: Comb-like polymers interacting with CDs which are not interlocked on the side chain
2) Polyrotor/axis: polymers bearing CDs molecules on the side chain, interacting with guest molecules to form poly*pseudo*rotaxane
3) Polyrotor/polyaxis: polymers bearing covalently bonded CD-moieties interacting with polymers bearing guests in the side chain.

In all these cases poly*pseudo*rotaxanes are present in equilibrium state.

The synthesis of side chain polyrotaxanes is more complicated than the modification of polymers to obtain chain poly*pseudo*rotaxanes. The former are of higher stability compared to poly*pseudo*rotaxanes. Since poly*pseudo*rotaxanes are present in the equilibrium state, they are responsive to external stimuli. Side chain polyrotaxanes and poly*pseudo*rotaxanes have been used in drug release applications, catalysis, and the formation of physical gels.

This section focuses on the synthesis, characterization, and properties of polyrotaxanes and poly*pseudo*rotaxanes (polyaxis/rotor).

5.2.1
Side Chain Polyrotaxanes

The first cyclodextrin side chain polyrotaxanes reported have been synthesized by radical copolymerization of methyl methacrylate and 11-methacroyl aminoundecanoic acid. The latter was activated with ethyl chloroformate in the presence of 2,6-dimethyl-β-cyclodextrins to form the mixed anhydride derived from 4′-triphenylmethyl-4-aminobutaneanilide or 4′-triphenylmethyl-11-aminoundecananilide [67]. The side chain polyrotaxane showed a typical signal widening of the 2,6-anilide protons and a downfield shift of the signals in the ^{1}H NMR spectrum.

Side chain polyrotaxane based on aromatic poly(sulfones) **23** and poly(ether ketones) **24** were also synthesized by Ritter *et al.* [68, 69]. The activation of free acid of poly(sulfone) or poly(ether ketone) (**PEEK**), synthesized under standard condensation conditions with oxalyl chloride and subsequent condensation with 4′-triphenylmethyl-11-aminoundecananilide in the presence of 2,6-dimethyl-βCD, led to side chain polyrotaxanes **23** and **24** (Figure 5.7). Characteristics of the side chain polyrotaxane such as high glass transition temperature (T_g), higher solubility in common organic solvents, concentration-dependent reduced viscosity, a large hydrodynamic volume, and upfield shifted signals in ^{1}H NMR spectra were compared with those of a corresponding guest model compound.

Side chain polyrotaxanes with optically active tandem structures **25** (Figure 5.7) were synthesized by radical copolymerization of 5-[(11-methacryloylamino)-undecanoylamino]isophthaloyl bis(12-aza-dodecanoic acid) and methyl methacrylate in a ratio of 1 : 4.1. The free acid groups of the copolymer were reacted with the free amino groups of 4-amino-N-[4-(triphenylmethyl)phenyl]butaneamide/2,6-dimethyl-βCD complexes in a condensation reaction [70]. After methylation of the residual free OH-group of dimethyl-βCD, T_g decreased [71]. It is postulated that the methylation of residual OH groups reduces hydrogen-bonding interactions, and therefore the mobility of the βCD rings is increased as a side effect.

The synthesis and ^{1}H NMR characterization of a functionalized poly(ether sulfone) bearing about 80% of peracetylated CDs in the side chain was described by Ritter *et al.* [72]. Compared to the characterization data of the model graft polymer without CDs, it was demonstrated that the CD molecules are uniformly located on the side chain, whereas the CDs included the aromatic anilide moiety of the stopper groups from the secondary rim. The glass transition temperature of side chain polyrotaxane is significantly higher (at about 143 °C) than that of the model polymer (111 °C), since rotaxane structures are formed in their side chains.

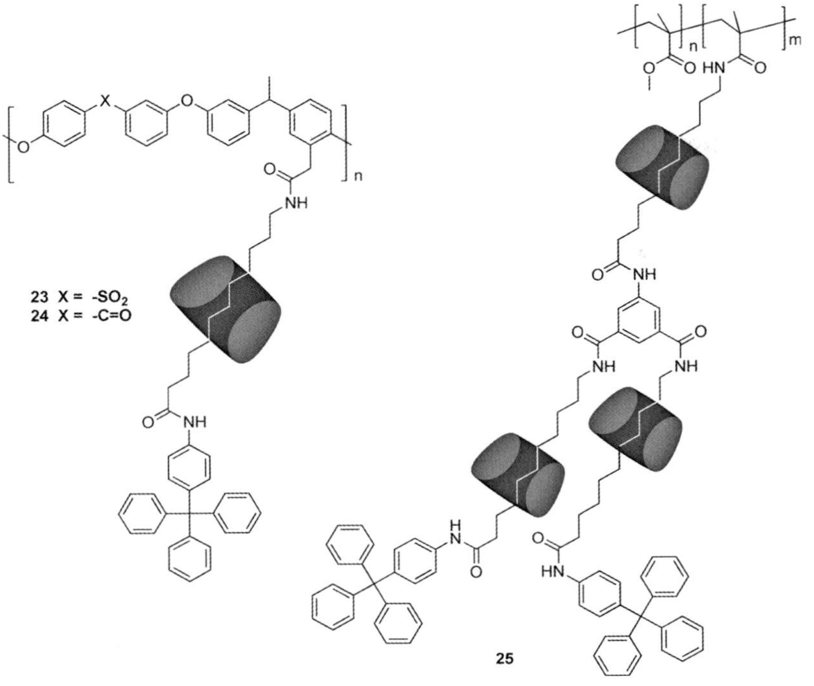

Figure 5.7 Structures of side chain polyrotaxanes **23–25**.

Side chain polyrotaxanes based on poly(benzimidazole) derivatives were obtained by Yamaguchi et al. [73]. NaH promoted deprotonation of the NH groups in poly(p-phenylenebenzimidazole), poly(octamethylenebenzimidazole) and poly[bis(undecamethylene)ether benzimidazole) resulting a deep red solution of polyanions as a first step. The reaction of these polyanions with Br-$(CH_2)_{12}$O (C=O)CH_2C$(C_6H_5)_3$ in the presence of trimethyl-βCD leads to side chain polyrotaxanes **26–28** (Figure 5.8). These side chain polyrotaxanes containing 21–100 mol% CD molecules show higher solubility in organic solvents than that for the basis polymer without CDs. Similar reactions using unprotected βCD resulted in relatively undefined products. Analogously, the reaction of poly(p-phenylenebenzimidazole) anions with 12-bromo-1-dodecanol leads to the N-alkylated polymer with –CH_2ONa end groups in 90% yields. In a subsequent procedure a solution of the above polymer was stirred in the presence of αCD with subsequent addition of 15-crown-5-ether as a stopper molecule. The ^{1}H NMR spectrum of the side chain polyrotaxane indicated the formation of Na^+-crown ether complexes to 100% and inclusion complexes of αCD with polymer side chain to 57%. GPC measurements of the described polyrotaxanes indicated that all the cyclodextrin molecules were incorporated into the side chain. The complexes of Na^+/crown ether were decomposed by addition of water, causing the dethreading of about 37% of αCD from the side chain [74].

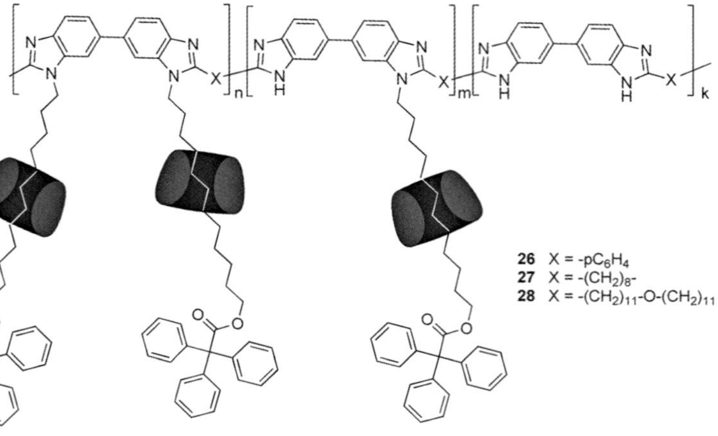

Figure 5.8 Structures of side chain polyrotaxanes **26–28**.

26 X = -pC$_6$H$_4$
27 X = -(CH$_2$)$_8$-
28 X = -(CH$_2$)$_{11}$-O-(CH$_2$)$_{11}$

The chemoenzymatic synthesis of polyrotaxane with noncovalently bonded CD in the side chain was described by Ritter et al. [75]. The copolymerization of methyl methacrylate with macromonomers, which were synthesized by condensation of 12-hydroxylauric acid with 11-methacryloylaminoundecanoic acid using a lipase from *candida antarctica*, led to comb-like polymers bearing free carboxyl groups at the end of oligoesters. These free OH groups were then condensed with the amino groups of *N*-(4-aminobutanoyl)-4-triphenylmethylaniline in the presence of 2,6-dimethyl-βCD. The presence of more than one CD molecule per side chain was proven by NMR spectroscopy.

A water-soluble semi-rotaxane was synthesized by mixing 3-O-(11-acryloylaminoundecanoyl)cholic acid and 2,6-dimethyl-βCD. Free-radical polymerization of the above host–guest complexes in water in the presence of a redox initiator led to a mixture of two different side chain polyrotaxanes. The side chain of the water soluble polymer with a significantly lower degree of polymerization included about 95 mol% of CD. The water-insoluble polyrotaxane with a higher degree of polymerization included more than 40 mol% of CD molecules in the side chain [76].

For the first time, Ritter et al. have described the synthesis of a photosensitive comb-like polymethacrylamide-*co*-methyl methacrylate modified with chalcone (1,3-diphenyl-2-propene-1-one) containing noncovalently attached 2,6-dimethyl-βCD in the side chain (**29**, Figure 5.9) [77]. NMR spectroscopic measurements indicated that the CD rings in the copolymer do not exhibit a fixed position but slide along the polymer side chain. A 10 °C increase in glass transition temperature was observed on comparing the modified polymer to the basic copolymer without CDs. This phenomenon was observed due to the reduced flexibility of the polymer chain caused by the noncovalent attachment of CD molecules. UV irradiation of this side chain polyrotaxane in tetrahydrofuran only indicated *E/Z* isomerization. No other photochemical reactions like [2 + 2] cycloaddition during the irradiations process were observed. The same research group also reported side chain polyrotaxane bearing

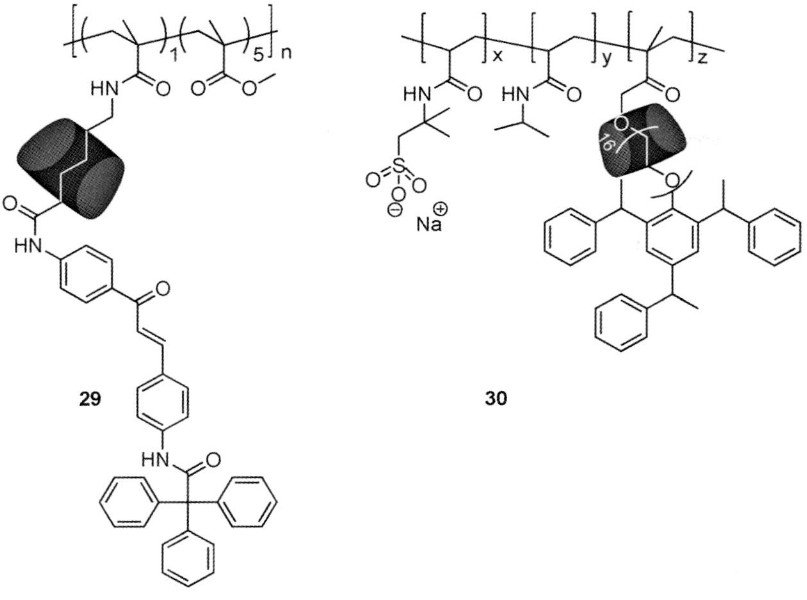

Figure 5.9 Structures of photosensitive side chain polyrotaxanes **29** and **30**.

hydrophobically associative compounds **30** (Figure 5.9) [78]. Copolymerization of the inclusion complex of a macromonomer and βCD with hydrophilic comonomers 2-acrylamido-2-methylpropanesulfonate (AMPS) and N,N-dimethylacrylamide led to hydrophobically associated polymers. These polymers and the formed micelles were characterized by ^{1}H NMR, dynamic light scattering (DLS), and fluorescence spectroscopy using ammonium [8-anilino-1-naphthalene-sulfonate] as fluorescence agent.

A side chain polyrotaxane exhibiting a unidirectional inclusion phenomenon was synthesized by photoisomerization of an azobenzene moiety at the end of the polymer side chain (Figure 5.10) [79]. Harada *et al.* investigated the synthesis of poly(acrylic acid)s modified with azobenzene through hexamethylene (-C_6-) and dodecamethylene (-C_{12}-) and their interactions with αCD. The formation of inclusion complexes was confirmed by UV/Vis and 2D-NOESY analysis. The azobenzene moieties were photoisomerized from trans to cis form. In the case of the polymer with C_6-linker, the UV irradiation caused dissociation of the inclusion complex by trans to cis photoisomerization. On the other hand, 2D NOESY analysis of the inclusion complex **31**-*trans* with C_{12}-linker showed that after UV Irradiation correlation peaks from the CD-moieties with the aliphatic linkers are still apparent. This observation implies that the C_6-linker is too short to form stable side chain polyrotaxanes, in contrast to the polymer with C_{12}-linker. Furthermore, the copolymers of acrylamide and azobenzene-modified acrylamide were prepared by radical polymerization. In the following the characteristics of the inclusion complexes between βCD and the copolymer were examined. The association constant of the copolymer in the

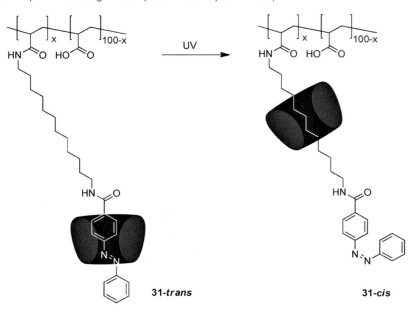

Figure 5.10 Conversion of side chain polypseudorotaxane **31**-trans to stable side chain polyrotaxane **31**-cis by photoisomerization of the azobenzene moiety.

trans form is larger than the one in the cis form [35]. As already discribed in Section 5.1.3 this concept was transfered to prepare physical networks exhibiting a contrast viscosity upon UV irradiation (Figure 5.5).

Based on the previous work by Harada, a side chain hetero polyrotaxane was prepared composed of α-and βCDs with a water-soluble polymer poly(9-decenoic acid-co-maleic anhydride) bearing two recognition sites (Figure 5.11) [80]. The reaction of 4-(phenylazo)benzoic acid and oligo ethylene glycol in the presence of 1-H-benzotriazol-1-yloxytripyrrolidinophosphonium hexafluorophosphate led to N-(3-(2-(3-aminopropoxy)oligoethoxy)propyl)-4-(phenyldiazenyl)benzamide (OEG-Azo). The first step of side chain polyrotaxane synthesis is the formation of the polypseudorotaxane **32**, in which αCD includes the C_7-linker and azobenzene in trans form into the side chain. After UV irradiation of the side chain polypseudorotaxanes prepared in the first step, αCD was interlocked on the side chain because of trans to cis isomerization. The final step was the formation of side chain hetero polyrotaxanes by adding βCD to the product of step two. 2D NOESY NMR of the isolated polyrotaxane confirmed that βCD included the cis-azobenzene moiety and αCD formed a complex with the C_7-linker. This architecture was proposed to extend it to contraction and expansion systems and also to nanotechnological applications such as nanodevices utilizing the combination of the host and guest polymers with CD and hydrophobic moieties in their side chains [80].

CD-based polyrotaxanes supported on gold nanospheres were described by Liu et al. [81]. Gold colloidal dispersions were prepared by an established method and

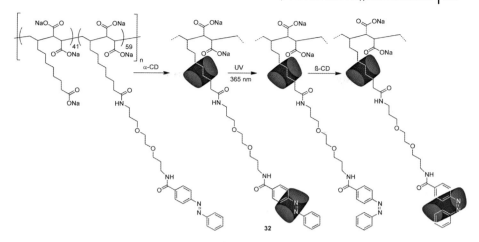

Figure 5.11 Synthesis of a hetero side chain polyrotaxane **32**.

stirred with aqueous thiol-modified ferrocene in the presence of αCD for two days, whereupon a precipitate was formed. The structure of the side chain polyrotaxane was verified by IR-and ^{1}H NMR spectroscopy.

5.2.2
Side Chain Polypseudorotaxane (Polymer (Polyaxis)/Cyclodextrin (Rotor))

In biological systems, recognition of side chains plays an important role in the construction of supramolecular structures which express various functions necessary to maintain living activities. For example, antigens are identified by antibodies through recognition of their side chains. Between 1978 and 1979 Komyama et al. determined the diffusion of α-and βCD in aqueous solution of poly(methacrylic acid), poly(styrene sulfonate) sodium salt, statistical poly(styrene-co-methacrylic acid), and poly(methylbenzyldiallylammonium chloride-SO$_2$). It was found that the diffusion coefficient of CDs in polymeric solutions decreased depending on the polymer concentration, the content of sulfonation and styrene, and the degree of neutralization. These results were interpreted as interaction between CDs and the polymer side chain, but the structures of poly*pseudo*rotaxane were not characterized in detail [82–84].

Harada et al. investigated the interaction of CDs with several guest molecules attached to the polymer side chain as a model system for macromolecular recognition in biological systems in detail. Polymers having guests on the side chain have been prepared by copolymerization of acrylamide with methacrylates or acrylates of different alcohols. It was found that CDs interact more selectively with polymers bearing alkyl guests than with low-molecular-weight model compounds. When αCD was added to a suspension of insoluble copolymers bearing n-butyl groups as side chains, the insoluble polymers turned water soluble. In contrast to these findings, no change was observed in the presence of βCD. When βCD was added to a suspension

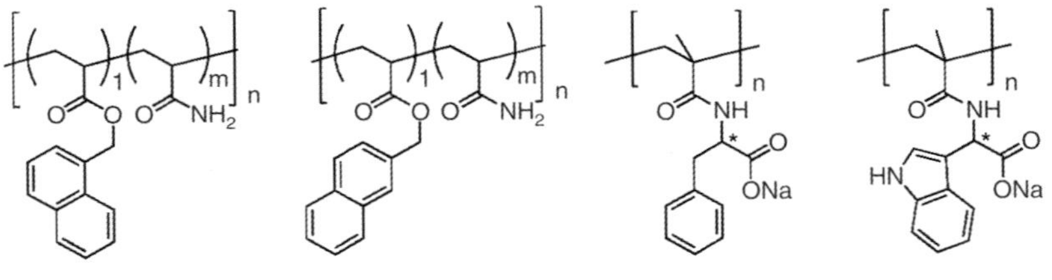

Figure 5.12 Structures of polymers **33a–36**.

of poly(acrylamide-co-tert-butyl methacrylate), the formerly insoluble polymer became soluble in water and formed a clear homogeneous solution. The analogous suspension containing αCD instead of βCD did not show changes in the solubility behavior of the polymer. These results indicated that αCD formed inclusion complexes with n-butyl side chain while the tert-butyl methacrylate only formed side chain poly*pseudo*rotaxanes with βCD [85]. As an extension of the research on the recognition of polymer side chains, the work group mentioned above has chosen (1-naphthyl)methyl (**1Np**) and (2-naphthyl)methyl (**2Np**) groups as guest molecules [86]. They investigated the interactions of different CDs with 1Np and 2Np groups attached to the poly(acrylamide) backbone (polymer **33** and **34**, Figure 5.12) by several techniques like circular dichroism and steady-state fluorescence. No interaction was observed by the use of α-and γCD s in combination with polymer side chains. When βCD was added to the 1Np- or 2Np-modified poly(acrylamide) solutions the absorbance did not change, while the fluorescence intensity increased significantly. Furthermore, the increase in fluorescence intensity is larger for 2Np-modified polymer than for 1Np-poly(acrylamide). The circular dichroism spectrum of CD/1NP-poly(acrylamide) showed no significant signals, while a negative induced circular dichroism signal was observed for CD/2Np-poly(acrylamide). This could be ascribed to a deep inclusion of the 2Np groups in the side chain of the polymer into the βCD cavity. In contrast, 1Np groups exhibited a shallow inclusion into βCD because of their steric hindrance. In another example, tryptophan and phenylalanine were attached to poly(acrylamide), and the interaction of CDs with poly(acrylamide)-bearing hydrophobic amino acid residues was investigated as a simple model system of macromolecular recognition of proteins (Figure 5.12) [87]. The interaction of cyclodextrins with polymer side chains has been studied by ^{1}H NMR and NOESY, and the association constant (**K**) was estimated. K-values for the model compound, the sodium salt of tryptophan, were found to be 43, 59 and 12 M^{-1} (for α-, β-, and γCD). Nearly identical K-values (16 M^{-1} for αCD, 69 M^{-1} for βCD and 3 M^{-1} for γCD) were found for the respective sodium salts of phenylalanine. However, there is a significant difference in the apparent K-values roughly estimated by ^{1}H NMR for poly(N-methacryloylphenylalanine) **35**. The K-values are considerably smaller (by about 10 M^{-1}) than the values for poly(N-methacryloyltryptophan) **36** (30 M^{-1} for αCD, 83 M^{-1} for βCD and 11 M^{-1} for γCD). Previous ^{1}H NMR and 2D NOESY studies [46] indicated that a significant moiety of aromatic rings in **35** and **36** were located in close proximity to the polymer main chain, presumably because of hydrophobic and CH interactions. Since the benzyl group in poly(N-methacryloyl-phenylalanine) **35** is smaller and more hydrophobic than the (3-indolyl)methyl group in poly(N-methacryloyltryptophan) **36**, interactions with the polymer main chain are stronger than these of (3-indolyl)methyl groups in poly(N-methacryloyltryptophan). This demonstrated that (3-indolyl)methyl groups in poly(N-methacryloyltryptophan) were readily recognized by CDs.

Copolymers of acrylic acid and N-dodecylacrylamide were synthesized from poly (acrylic acid) and dodecylamine in the presence of N,N-dicyclohexylcarbodiimide (DCC). The degree of modification was determined by elemental analysis and ^{1}H NMR spectroscopy. When the molecular weights of the copolymers were high and the content of alkyl side chains were moderate, interpolymer aggregates could be formed. Such aggregates were readily formed in water when the solution of the copolymer was highly concentrated, due to intramolecular interactions of the C_{12}-side chain. A drastic decrease in viscosity was observed by addition of αCD to this hydrogel. Creep recovery experiments were also carried out for the mixture of above-described copolymer in combination with oligo(αCD). Steady shear viscosity increased with increasing concentrations of the oligo(αCD). The copolymer behaved like a hydogel at [oligo(αCD)] ≥ 10 gL^{-1}, whereas η_0 was too large to be determined [88]. This sol-to-gel transition system was extended to photoresponsive hydrogel systems by adding photoresponsive [89] or redoxresponsive (Figure 5.13) [90] competitive guests to a binary mixture of αCD and PAA-co-polydodecylacrylamide. Addition of αCD to the gel-like solution of the copolymer induced a drastic decrease of η_0 of more than

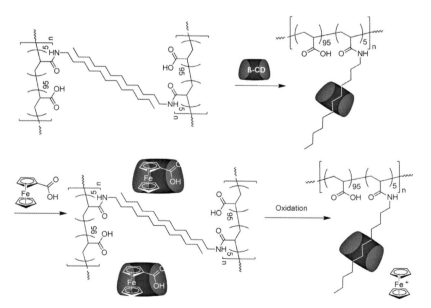

Figure 5.13 Redox responsible hydrogel system by combination of dodecyl modified poly(acrylic acid), βCD and the redox-responsive guest ferrocenecarboxylic acid.

6 magnitudes. This was caused by the formation of inclusion complexes of αCD and the hydrophobic alkyl side chains, thus preventing the side chains from aggregating. By addition of *trans*-4,4'-azodibenzoic acid (ADA) to the binary sol mixture, η_0 increased, since αCD interacts more favorably with *trans*-ADA than with the alkyl side chains. Thus, sol-to-gel transition was observed. After UV irradiation, ADA was isomerized from the trans to the cis configuration, and the mixture underwent a gel-to-sol transition. In this case αCD s interact more favorably with dodecyl side chains than with *cis*-ADA. When the sol ternary mixture was irradiated with visible light, the sol was gradually converted to the gel-like mixture [89]. Analogously, a redox-responsive hydrogel system was constructed by the use of water soluble ferrocene-carboxylic acid (FCA) as a redox-responsive competitive guest (Figure 5.13). In this case, the addition of FCA to the sol system induced the dissociation of the inclusion complexes of βCD with the alkyl side chains. Oxidation of FCA with sodium hypochlorite in turn led to the dissociation of the FCA-αCD complexes, and the gel-to-sol transition was observed [90].

Poly(*isobutene-alt*-maleic anhydride) (**PiBMA**, $\bar{M}_w = 60 \text{ kg mol}^{-1}$) was substituted with *p-tert*-butylphenyl or adamantyl groups respectively. The interaction of guest polymers with βCD in water was investigated by means of capillary electrophoresis. The inclusion of the hydrophobic substituents into the βCD cavity reduced the electrophoretic mobility of the polymers. For both sets of polymers the decrease in electrophoretic mobility upon complex formation was increased with a higher degree of substitution. On the other hand, in case of the underivatized polymer, no significant change of mobility was observed after addition of excess βCD. The higher electrophoretic mobility of the free polymer was explained by the formation of compact unimer globules. The hydrophobic derivatized polymers in turn formed extended random coils [91]. The interaction of βCD with poly(maleic anhydride)-*alt*-isobutene substituted with 4-*tert*-butylanilide groups was further investigated by isothermal titration calorimetry (ITC) [57].

Harada *et al.* published numerous papers about the interaction of CDs with hydrophobically modified water-soluble polymers (HMASPs), most of them dealing with the change of association properties of HMASPs in the presence of CDs. Steady-state fluorescence and sedimentation equilibrium experiments confirmed that water-soluble polymers with high contents of hydrophobic groups, for example, dodecyl groups, formed micelle-like aggregates caused by the association of the hydrophobic groups. The formation of inclusion complexes of CD with hydrophobic guest groups led to dissociation of intermolecular aggregates in cases where that the *pseudo*-rotaxane structures were favored. Based on these findings the interaction of CD with a water-soluble poly(sodium maleate)-*alt*-dodecyl acrylate) was investigated. The formation of micelle-like aggregates in the aqueous phase was analyzed by ^{1}H NMR and 2D-NOESY NMR under basic conditions. A significant interaction of αCD with the dodecyl side chains was observed, while βCD and γCD did not interact. These results indicated a major tendency of the dodecyl groups towards αCD, since the interaction of αCD with the dodecyl groups competes with intramicellar hydrophobic associations. The hydrodynamic size of the micelles examined by pulse field gradient spin echo (PGSE) NMR was confirmed to be 3.0 or 2.0 nm in absence or presence of

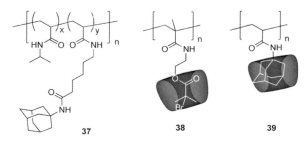

Figure 5.14 Structures of polymer **37** and side chain polypseudorotaxanes **38** and **39**.

CD, respectively [92]. The same research group investigated the effect of the molecular weight of these polymers on the stability of their complexes with αCD. The binding isotherms obtained from ^{1}H NMR data exhibited sigmoidal-shaped curves, which is indicative of cooperative complexation of αCD and the polymers. With increasing molecular weight, the complexation was found to be even more cooperative. Analysis of the binding isotherms utilizing a proposed model based on one-dimensional lattice theory indicated that the molecular-weight dependence of the cooperative complexation was due to molecular-weight-dependent attractive interactions between free dodecyl chains and adjacent complexing αCD molecules [93].

Several studies have shown that the lower critical solution temperature (LCST) of thermally responsive polymer systems can be influenced by supramolecular interaction with CDs. The copolymerization of N-isopropylacrylamide (NIPAM) and adamantyl-bonded acrylamide via a C_6-linker was chosen as a model system (Figure 5.14). The LCST behavior of copolymer **37** was studied in the absence and presence of 2,6-dimethyl-βCD [94]. The LCST drastically decreased down to 40 °C by addition of 2,6-dimethyl-βCD and strongly depended on its concentration. These observations indicated that side chain poly*pseudo*rotaxanes were formed. Dissociation of the poly*pseudo*rotaxanes and loss of turbidity intensity appeared when potassium 1-adamantane carboxylate was added to the solution of the cyclodextrin/polymer mixture. The workgroup of Ritter also reported the synthesis of a polymer inclusion complex **38** consisting of 2,6-dimethyl-βCD and a poly(methacrylamide), which showed a reversible phase transition in aqueous solution as a result of a dissociation/association process of the side chain polyrotaxane [95]. Since this system indicated low chemical stability, 1-adamantylacrylamide and 6-acryloylaminohexanoic acid-1-adamantylamide were chosen as suitable guest monomers. Additionally, the synthesis of side chain poly*pseudo*rotaxanes **39** and **40** were also reported [96]. The incorporation of a flexible hydrophobic spacer between the polymer main chain and the adamantyl moiety strongly affected the thermosensitivity of the supramolecular complex. Figure 5.15 shows the transmittance of an aqueous solution of polymer **40** as a function of temperature, in which the transmittance of the solution drops from 100 to 0% within a temperature range of about 1–2 °C around the cloud point at 38.6 °C. A phase separation was apparent at higher temperatures, in which the side chain poly*pseudo*rotaxanes were dissociated. The same research group described the synthesis of a silyl-protected monomer

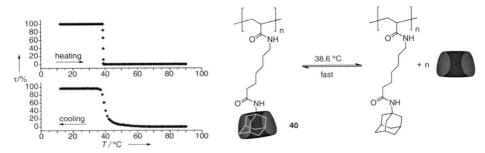

Figure 5.15 Transmittance as a function of temperature for an aqueous solution of polymer/βCD-complex **40** at heating/cooling rate of 1 K min^{-1}.

4-(E)-2-{4-[(triisopropylsilyl)oxy]phenyl}ethenyl)-1-(vinylbenzyl)pyridinium chloride, a polymerizable Reichards dye, its copolymerization with NIPAM and the solvatochromism of the copolymer **41** upon temperature change and addition of 2,6-dimethyl-βCD (Figure 5.16) [97]. A characteristic bathochromic shift from orange to dark red was observed when the aqueous polymer solution at a pH > 7 was heated above the LCST at about 31 °C. It was found that a bathochromic shift due to a negative polarity change was also visible after addition of an excess of the βCD derivative to the polymer solution below its LCST due to the formation of a side chain poly*pseudo*rotaxane.

Ogoshi et al. prepared side chain poly*pseudo*rotaxanes constituted of heteromacrocyclic receptors of αCD, γCD, and cucurbit[7]uril (**CB7**). A vinyl monomer comprising a hydrophobic decyl chain attached to a viologen moiety was copolymerized with N-isopropylacrylamide via radical polymerization with 2,2′-azobis(isobutyronitrile) (**AIBN**) to yield the thermosensitive copolymer **42** (Figure 5.16). The host–guest interaction of copolymer **42** with CDs and CB7 was investigated by ^{1}H NMR measurements, which indicated that CDs formed host–guest complexes with the decyl moieties in **42** whereas CB7 interacted with the viologen moieties.

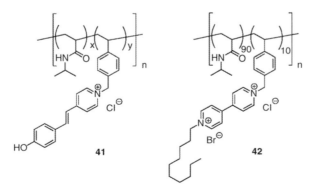

Figure 5.16 Structure of compound **41** and thermosensitive copolymer **42**.

The LCST of the copolymer with αCD or CB7 was increased, while that with γCD was decreased [98].

The inclusion complexation of comb-like poly(ethylene oxide) grafted polymers **43**, P(PEOMA)s, with αCD was studied by He *et al*. **43** was prepared by atom transfer radical homopolymerization (ATRP) of the macromonomers containing poly(ethylene oxide) (**PEOMA**) of different molecular weights ($\bar{M}_n = 300, 475,$ and 1100). The grafted polymers containing poly(ethylene oxide) side chains of $\bar{M}_n = 1100$ gmol^{-1} and $\bar{M}_n = 475$ gmol^{-1} formed crystalline inclusion complexes with αCD, while the polymer with the PEOMA of molecular weight $\bar{M}_n = 300$ gmol^{-1} showed no interaction with αCD. The formation of side chain poly*pseudo*rotaxanes was confirmed by powder X-ray diffraction (XRD), differential scanning calorimetry (DSC) and solide-state ^{13}C CP/MAS NMR analysis. The results of XRD indicated that the obtained inclusion complexes exhibited channel-type crystalline structures. ^{1}H NMR proved that the ratio of EO units to αCD of the side chain poly*pseudo*rotaxanes was higher than 2 : 1 [99]. But also the complexation of double-grafted polymer brushes (PBIEM-*g*-P(PEOMA)) **44** with αCD was described. Poly(ethylene glycol) side chains again had a molecular weight of $\bar{M}_n = 300$ gmol^{-1} and $\bar{M}_n = 475$ gmol^{-1} [100]. The formation of inclusion complexes was confirmed by XRD, ^{1}H NMR, electron diffraction (ED), scanning electron microscopy (SEM), DSC, and TGA. The aqueous solution of the polymer brushes (PBIEM-*g*-P(PEOMA)475) formed hydrogels after addition of αCD. The hydrogel was a nearly transparent liquid with a light blue tint. When the hydrogel system was heated above 40 °C it dissipated but degenerated after cooling to room temperature gradually. This reversible process demonstrated that the side chain poly*pseudo*rotaxanes exhibited a thermosensitive sol-to-gel transition due to physical crosslinking. XRD analysis of dried gels demonstrated a channel type structure. Polymer brushes (PBIEM-*g*-P(PEOMA)300) formed fine hexagonal crystalline inclusion complexes with αCD (EO/αCD, 1 : 1).

A supramolecular self-assembling injectable hydrogel was reported by He *et al*. They employed poly(ethylene glycol)-grafted poly(α,β-malic acid) **45** as a polyaxis and added αCD, which threaded onto the poly(ethylene glycol) side chains. The hydrogel structure was characterized by ^{1}H NMR, XRD, DSC, TGA, and SEM. The results showed that the poly*pseudo*rotaxanes of αCD s/mPEG-*g*-PMA acted as physical crosslink sites in the hydrogel. The anti-tumor drug doxorubicin hydrochloride (**DOX**) was loaded in the hydrogel. The release and anti-tumor effect were studied *in vitro* [101].

Photo-regulated sol-to-gel transition of azobenzene functionalized hydropropyl methylcellulose (**Azo-HPMC**) in aqueous solution was reported. The effect of αCD on its aggregation behavior was surveyed [102–104]. Azo-HPMC polymers were synthesized by an esterification of hydroxyl groups of HPMC with 4-phenylazobenzyl chloride. The authors have demonstrated that by incorporation of a small number of azobenzene moieties onto HPMC polymers, the sol-to-gel transition behavior of HPMC in aqueous solutions can be regulated by photoirradiation, which can be observed by rheological measurements at different temperatures. In the absence of αCD, the gelation temperature increased after UV irradiation (26.5 to 36.0 °C) because of trans to cis photoisomerization of the azobenzene side chain. In the

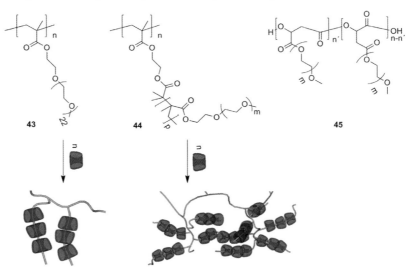

Figure 5.17 Structures of polymer brushes **43** and double grafted polymer brushes **44** and schematic drawing of their interaction with αCD; structure of copolymer **45**.

presence of αCD, the azobenzene moieties in trans conformation formed inclusion complexes with αCD, and the gelation temperature increased from 26.5 to 57 °C. After UV irradiation of this side-chain poly*pseudo*rotaxane, the inclusion complexes dispersed resulting in an aggregation of Azo-HPMC. Hence, the gelation temperature decreased to 49 °C. Gelation temperatures were dependent on the azobenzene content and αCD concentration. ITC, UV-visible spectroscopy, ^{1}H NMR, FT-Raman spectroscopy and FT-IR were used to prove the supramolecular interaction of αCD and Azo-HPMC in aqueous solutions.

Cho *et al.* reported the preparation of the hydrophobic block copolymer **46** containing adamantyl polyphosphazene (**PN**) and polystyrene (**PS**) via controlled living cationic polymerization (Figure 5.18). The complexation of adamantyl groups on a phosphazene block with βCD resulted in a significant change of the block copolymer character from hydrophobic to amphiphilic. The resulting amphilphilic PN-PS block copolymers self-aggregated to form micelles in aqueous media.

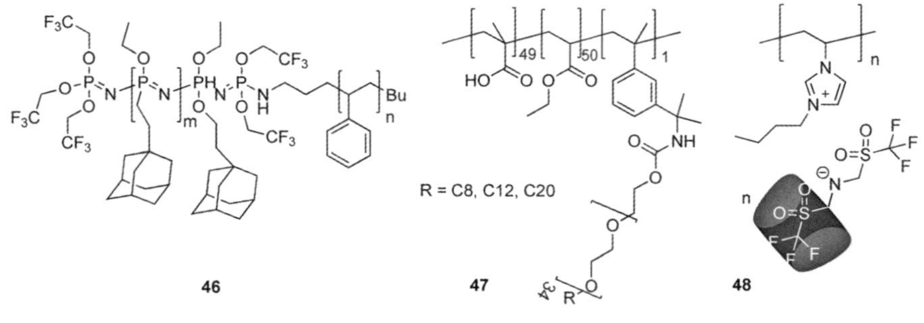

Figure 5.18 Structures of compounds **46–48**.

Thereby, the hydrophobic PS segments were incorporated into the micelle core. The micelle behavior of these complexes was monitored using fluorescence techniques, transmission electron microscopy (TEM), and DLS. Compared to other synthetic amphiphilic block copolymers, a lower critical micelle concentration (**cmc**) of 0.925 mgL^{-1} was found for the block copolymer complex with βCD. This demonstrated that the highly hydrophilic shell of the micelles is due to the hydrophilic outer surface of CD. It was also possible to control the micelle formation by changing the amount of βCD while maintaining the concentration of the block copolymer [105].

Yang et al. used methyl-αCD to prevent the hydrophobic association of hydrophobically modified alkali-soluble emulsion (**HASE**) polymers. Thus, the authors were able to determine the weight average molecular weight ($\bar{M}_w$), hydrodynamic radius (R_h), and radius of gyration (R_g) of monodisperse solute HASE polymers. Generally, the characterization of HASE polymers is difficult due to their hydrophobic aggregation. HASE polymers **47** were synthesized by emulsion copolymerization of methacrylic acid, a nonassociative acrylate monomer like ethyl acrylate, and an associative macromonomer (Figure 5.18). The side chain poly*pseudo*rotaxanes were prepared by addition of a basic solution of potassium hydroxide and potassium chloride to methyl-αCD and the polymer [106].

Ritter et al. described the synthesis and the *pseudo*-LCST effect of side chain poly*pseudo*rotaxanes **48** based on a hydrophobic polyelectrolyte (Figure 5.18) [107, 108]. Radical polymerization of the ionic liquid 1-vinyl-3-butylimidazolium bis(trifluoromethylsulfonyl)imide (**[vbim][Tf2N]**) was carried out in water in presence of 2,6-dimethyl-βCD and LiCl. TGA analysis was performed to prove the thermal stability of polymeric [vbim][Tf2N]. It was found that the complex between the polyelectrolyte and 2,6-dimethyl-βCD had a significantly lower thermal stability than that of pure polymeric [vbim][Tf2N] and 2,6-dimethyl-βCD. The obtained solid poly(ionic liquid) with an $\bar{M}_w$ of 8.000 gmol^{-1} showed a '*pseudo*'-LCST effect in the presence of 2,6-dimethyl-βCD. On heating the clear aqueous solution of the complexed polyelectrolyte poly[vbim][Tf2N] above 45 °C, it clouded suddenly. The turbid dispersion became completely transparent again when the solution was cooled, indicating the temperature-dependent dissociation and association of the side chain poly*pseudo*rotaxane. The LCST increased significantly up to 62 °C when the concentration of CD derivative was increased. R_h was measured before and after reaching the LCST by DLS. A value of 1.3 nm was measured for R_h at 56 °C, when the solution was clear. R_h at the LCST at 57 °C increased to 8.8 nm due to intermolecular aggregation, and the solution became turbid.

The applications of single-walled carbon nanotubes (**SWNTs**) are extremely limited because of their low solubility in solvents. Therefore, solubilization of SWNTs has become a hot topic in recent years. Ogoshi et al. [109] reported supramolecular SWNT hydrogels by attaching pyrene-modified βCD (**Py β CD**) on the surface of SWNT. From atomic force microscopic (AFM) images, it could be seen that nanotubes which were sandwiched between Py βCD were formed that had an average size of about 2.5 nm. The nanotubes were completely soluble in aqueous media. Supramolecular SWNT hydrogels were synthesized by host–guest interactions between the cavity of βCDs in Py βCD/SWNT hybrids and dodecyl groups of poly(acryl acid-*co*-dodecylacrylate) (containing 2 mol% dodecyl groups). The supramolecular SWNT hydrogel exhibited gel-to-sol transition by adding competitive guests and hosts. When sodium adamantane carboxylate was added to the hydrogel, gel-to-sol transition was observed. This transition was also observed upon addition of αCD as a competitive host. These results indicated the formation of side chain poly*pseudo*rotaxanes of modified poly(acrylic acid)s with CD, since dodecyl moieties form complexes with αCD more favorably than with βCD.

References

1. Saenger, W. (1980) Cyclodextrin inclusion-compounds in research and industry. *Angew. Chem. Int. Ed. Engl.*, **19** (5), 344–362.
2. Szejtli, J. and Osa, T. (eds) (1996) *Cyclodextrins, Comprehensive Supramolecular Chemistry*, vol. 3, Elsevier Science Ltd., Oxford.
3. Szejtli, J. (1998) Introduction and general overview of cyclodextrin chemistry. *Chem. Rev.*, **98** (5), 1743–1753.
4. Wenz, G. (2000) An overview of host-guest chemistry and its application to nonsteroidal anti-inflammatory drugs. *Clin. Drug Invest.*, **19** (2), 21–25.
5. Del Valle, E.M.M. (2004) Cyclodextrins and their uses: a review. *Process Biochem.*, **39** (9), 1033–1046.
6. Wiedenhof, N. (1969) Properties of cyclodextrins.4.features and use of insoluble cyclodextrin-epichlorohydrin-resins. *Starke*, **21** (6), 163–166.
7. Gramera, R.E. and Caimi, R.J. (1968) Cyclodextrin polyethers.
8. Bekers, O., Uijtendaal, E.V., Beijnen, J.H., Bult, A., and Underberg, W.J.M. (1991) Cyclodextrins in the pharmaceutical field. *Drug Dev. Ind. Pharm.*, **17** (11), 1503–1549.
9. Loftsson, T. and Brewster, M.E. (1996) Pharmaceutical applications of cyclodextrins.1. drug solubilization and stabilization. *J. Pharm. Sci.*, **85** (10), 1017–1025.
10. Uekama, K., Hirayama, F., and Irie, T. (1998) Cyclodextrin drug carrier systems. *Chem. Rev.*, **98** (5), 2045–2076.
11. Davis, M.E. and Brewster, M.E. (2004) Cyclodextrin-based pharmaceutics: Past, present and future. *Nat. Rev. Drug Discov.*, **3** (12), 1023–1035.
12. Li, J. (2009) Cyclodextrin inclusion polymers forming hydrogels, in *Inclusion Polymers, Advances in Polymer Science*, vol. 222, Springer-Verlag Berlin, Berlin, pp. 79–112.
13. van de Manakker, F., Vermonden, T., van Nostrum, C.F., and Hennink, W.E. (2009) Cyclodextrin-based polymeric materials: Synthesis, properties, and pharmaceutical/biomedical applications. *Biomacromolecules*, **10** (12), 3157–3175.
14. Harada, A., Hashidzume, A., Yamaguchi, H., and Takashima, Y. (2009) Polymeric rotaxanes. *Chem. Rev.*, **109** (11), 5974–6023.
15. Khan, A.R., Forgo, P., Stine, K.J., and D'Souza, V.T. (1998) Methods for selective modifications of cyclodextrins. *Chem. Rev.*, **98** (5), 1977–1996.
16. Furue, M., Harada, A., and Nozakura, S.I. (1975) Preparation of cyclodextrin-containing polymers and their catalysis in ester-hydrolysis. *J. Polym. Sci. Pol. Lett.*, **13** (6), 357–360.
17. Harada, A., Furue, M., and Nozakura, S. (1976) Cyclodextrin-containing polymers.1. preparation of polymers. *Macromolecules*, **9** (5), 701–704.
18. Ren, S.D., Chen, D.Y., and Jiang, M. (2009) Noncovalently connected micelles based on a beta-cyclodextrin-containing polymer and adamantane end-capped poly(epsilon-caprolactone) via host-guest interactions. *J. Polym. Sci. Pol. Chem.*, **47** (17), 4267–4278.

19 Liu, Y.Y., Fan, X.D., and Gao, L. (2003) Synthesis and characterization of beta-cyclodextrin based functional monomers and its copolymers with n-isopropylacrylamide. *Macromol. Biosci.*, **3** (12), 715–719.

20 Wang, J. and Jiang, M. (2006) Polymeric self-assembly into micelles and hollow spheres with multiscale cavities driven by inclusion complexation. *J. Am. Chem. Soc.*, **128** (11), 3703–3708.

21 Maeda, K., Mochizuki, H., Watanabe, M., and Yashima, E. (2006) Switching of macromolecular helicity of optically active poly(phenylacetylene)s bearing cyclodextrin pendants induced by various external stimuli. *J. Am. Chem. Soc.*, **128** (23), 7639–7650.

22 Munteanu, M., Choi, S., and Ritter, H. (2008) Cyclodextrin methacrylate via microwave-assisted click reaction. *Macromolecules*, **41** (24), 9619–9623.

23 Trellenkamp, T. and Ritter, H. (2010) Poly(n-vinylpyrrolidone) bearing covalently attached cyclodextrin via click-chemistry: Synthesis, characterization, and complexation behavior with phenolphthalein. *Macromolecules*. **43**, 5538–5543.

24 Gonsior, N. and Ritter, H. (2011) Ucst-behavior of cyclodextrin containing poly(pseudo-betaines) based on supramolecular structures. *macromelcular chemistry and physics, accepted.*

25 Maciollek, A., Munteanu, M., and Ritter, H. (2010) New generation of polymeric drugs: Copolymer from nipaam and cyclodextrin methacrylate containing supramolecular-attached antitumor derivative. *Macromol. Chem. Phys.*, **211** (2), 245–249.

26 Tomatsu, I., Hashidzume, A., and Harada, A. (2006) Contrast viscosity changes upon photoirradiation for mixtures of poly(acrylic acid)-based alpha-cyclodextrin and azobenzene polymers. *J. Am. Chem. Soc.*, **128** (7), 2226–2227.

27 Huh, K.M., Tomita, H., Lee, W.K., Ooya, T., and Yui, N. (2002) Synthesis of alpha-cyclodextrin-conjugated poly(epsilon-lysine)s and their inclusion complexation behavior. *Macromol. Rapid Commun.*, **23** (3), 179–182.

28 Ruebner, A., Statton, G.L., and James, M.R. (2000) Synthesis of a linear polymer with pendent gamma-cyclodextrins. *Macromol. Chem. Phys.*, **201** (11), 1185–1188.

29 Charlot, A., Heyraud, A., Guenot, P., Rinaudo, M., and Auzely-Velty, R. (2006) Controlled synthesis and inclusion ability of a hyaluronic acid derivative bearing beta-cyclodextrin molecules. *Biomacromolecules*, **7** (3), 907–913.

30 Deratani, A., Popping, B., and Muller, G. (1995) Linear cyclodextrin-containing polyelectrolytes.1. synthesis of poly(1-vinylimidazole)-supported beta-cyclodextrin-effect of pH and ionic strength on the solution behavior. *Macromol. Chem. Physic.*, **196** (1), 343–352.

31 Weickenmeier, M. and Wenz, G. (1996) Cyclodextrin sidechain polyesters-synthesis and inclusion of adamantane derivatives. *Macromol. Rapid Commun.*, **17** (10), 731–736.

32 Seo, T., Kajihara, T., and Iijima, T. (1987) The synthesis of poly(allylamine) containing covalently bound cyclodextrin and its catalytic effect in the hydrolysis of phenyl-esters. *Macromol. Chem. Phys.*, **188** (9), 2071–2082.

33 Seo, T., Kajihara, T., Miwa, K., and Iijima, T. (1991) Catalytic hydrolysis of phenyl-esters in cyclodextrin poly(allylamine) systems. *Makromol. Chem.*, **192** (10), 2357–2369.

34 Hollas, M., Chung, M.A., and Adams, J. (1998) Complexation of pyrene by poly(allylamine) with pendant beta-cyclodextrin side groups. *J. Phys. Chem. B*, **102** (16), 2947–2953.

35 Takashima, Y., Nakajama, T., Miyauchi, M., Kawaguchi, Y., Yamaguchi, H., and Harada, A. (2004) Complex formation and gelation between copolymers containing pendant groups and cyclodextrin polymers. *Chem. Lett.*, **33** (7), 890–891.

36 Martel, B., Leckchiri, Y., Pollet, A., and Morcellet, M. (1995) Cyclodextrin-poly(vinylamine) systems.1. synthesis, characterization and conformational

properties. *Eur. Polym. J.*, **31** (11), 1083–1088.

37 Crini, G., Torri, G., Guerrini, M., Martel, B., Lekchiri, Y., and Morcellet, M. (1997) Linear cyclodextrin-poly (vinylamine): Synthesis and nmr characterization. *Eur. Polym. J.*, **33** (7), 1143–1151.

38 Pun, S.H., Bellocq, N.C., Liu, A.J., Jensen, G., Machemer, T., Quijano, E., Schluep, T., Wen, S.F., Engler, H., Heidel, J., and Davis, M.E. (2004) Cyclodextrin-modified polyethylenimine polymers for gene delivery. *Bioconjug. Chem.*, **15** (4), 831–840.

39 Suh, J.H., Lee, S.H., and Zoh, K.D. (1992) A novel host containing both binding-site and nucleophile prepared by attachment of beta-cyclodextrin to poly (ethylenimine). *J. Am. Chem. Soc.*, **114** (20), 7916–7917.

40 Zhang, J.X., Sun, H.L., and Ma, P.X. (2010) Host-guest interaction mediated polymeric assemblies: Multifunctional nanoparticles for drug and gene delivery. *Acs Nano*, **4** (2), 1049–1059.

41 Zhang, J.X. and Ma, P.X. (2009) Polymeric core-shell assemblies mediated by host-guest interactions: Versatile nanocarriers for drug delivery. *Angew. Chem. Int. Ed.*, **48** (5), 964–968.

42 Guo, X.H., Abdala, A.A., May, B.L., Lincoln, S.F., Khan, S.A., and Prud'homme, R.K. (2005) Novel associative polymer networks based on cyclodextrin inclusion compounds. *Macromolecules*, **38** (7), 3037–3040.

43 Guo, X.H., Abdala, A.A., May, B.L., Lincoln, S.F., Khan, S.A., and Prud'homme, R.K. (2006) Rheology control by modulating hydrophobic and inclusion associations in modified poly (acrylic acid) solutions. *Polymer*, **47** (9), 2976–2983.

44 Li, L., Guo, X.H., Wang, J., Liu, P., Prud'homme, R.K., May, B.L., and Lincoln, S.F. (2008) Polymer networks assembled by host-guest inclusion between adamantyl and beta-cyclodextrin substituents on poly(acrylic acid) in aqueous solution. *Macromolecules*, **41** (22), 8677–8681.

45 Li, L., Guo, X.H., Fu, L., Prud'homme, R.K., and Lincoln, S.F. (2008) Complexation behavior of alpha-, beta-, and gamma-cyclodextrin in modulating and constructing polymer networks. *Langmuir*, **24** (15), 8290–8296.

46 Hashidzume, A. and Harada, A. (2005) Steady-state fluorescence and nmr study on self-association behavior of poly (methacrylamides) bearing hydrophobic amino acid residues. *Polymer*, **46** (5), 1609–1616.

47 Hashidzume, A., Tomatsu, I., and Harada, A. (2006) Interaction of cyclodextrins with side chains of water soluble polymers: A simple model for biological molecular recognition and its utilization for stimuli-responsive systems. *Polymer*, **47** (17), 6011–6027.

48 Guo, X.H., Wang, J., Li, L., Pham, D.T., Clements, P., Lincoln, S.F., May, B.L., Chen, Q.C., Zheng, L., and Prud'homme, R.K. (2010) Tailoring polymeric hydrogels through cyclodextrin host-guest complexation. *Macromol. Rapid Comm.*, **31** (3), 300–304.

49 Hossain, M.A., Hamasaki, K., Takahashi, K., Mihara, H., and Ueno, A. (2001) Guest-induced diminishment in fluorescence quenching and molecule sensing ability of a novel cyclodextrin-peptide conjugate. *J. Am. Chem. Soc*, **123** (30), 7435–7436.

50 Huh, K.M., Ooya, T., Lee, W.K., Sasaki, S., Kwon, I.C., Jeong, S.Y., and Yui, N. (2001) Supramolecular-structured hydrogels showing a reversible phase transition by inclusion complexation between poly (ethylene glycol) grafted dextran and alpha-cyclodextrin. *Macromolecules*, **34** (25), 8657–8662.

51 Choi, H.S., Huh, K.M., Ooya, T., and Yui, N. (2003) pH-and thermosensitive supramolecular assembling system: Rapidly responsive properties of beta-cyclodextrin-conjugated poly(epsilon-lysine). *J. Am. Chem. Soc.*, **125** (21), 6350–6351.

52 Giammona, G., Cavallaro, G., Maniscalco, L., Craparo, E.F., and Pitarresi, G. (2006) Synthesis and characterization of novel chemical

conjugates based on alpha,beta-polyaspartylhydrazide and beta-cyclodextrins. *Eur. Polym. J.*, **42** (10), 2715–2729.
53. Ramirez, H.L., Valdivia, A., Cao, R., Torres-Labandeira, J.J., Fragoso, A., and Villalonga, R. (2006) Cyclodextrin-grafted polysaccharides as supramolecular carrier systems for naproxen. *Bioorg. Med. Chem. Lett.*, **16** (6), 1499–1501.
54. Auzely-Velty, R. and Rinaudo, M. (2001) Chitosan derivatives bearing pendant cyclodextrin cavities: Synthesis and inclusion performance. *Macromolecules*, **34** (11), 3574–3580.
55. Auzely-Velty, R. and Rinaudo, M. (2002) New supramolecular assemblies of a cyclodextrin-grafted chitosan through specific complexation. *Macromolecules*, **35** (21), 7955–7962.
56. Charlot, A., Auzely-Velty, R., and Rinaudo, M. (2003) Specific interactions in model charged polysaccharide systems. *J. Phys. Chem. B*, **107** (32), 8248–8254.
57. Weickenmeier, M., Wenz, G., and Huff, J. (1997) Association thickener by host guest interaction of a beta-cyclodextrin polymer and a polymer with hydrophobic side-groups. *Macromol. Rapid Comm.*, **18** (12), 1117–1123.
58. Renard, E., Volet, G., and Amiel, C. (2005) Synthesis of a novel linear water-soluble beta-cyclodextrin polymer. *Polym. Int.*, **54** (3), 594–599.
59. Volet, G. and Amiel, C. (2009) pH sensitive supramolecular assembling system between a new linear water soluble beta-cyclodextrin terpolymer and an amphiphilic poly(ethylene oxide). *Eur. Polym. J.*, **45** (3), 852–862.
60. Tian, W., Fan, X.D., Kong, J., Liu, Y.Y., Zhang, W.H., Cheng, G.W., and Jiang, M. (2009) Amphiphilic hyperbranched polymers containing two types of beta-cyclodextrin segments: Synthesis and properties. *Macromol. Chem. Phys.*, **210** (24), 2107–2117.
61. Peng, K., Tomatsu, I., and Kros, A. (2010) Light controlled protein release from a supramolecular hydrogel. *Chem. Commun.*, **46** (23), 4094–4096.
62. Peng, K., Tomatsu, I., Korobko, A.V., and Kros, A. (2010) Cyclodextrin-dextran based in situ hydrogel formation: a carrier for hydrophobic drugs. *Soft Matter*, **6** (1), 85–87.
63. Harada, A., Furue, M., and Nozakura, S. (1976) Cyclodextrin-containing polymers.2. cooperative effects in catalysis and binding. *Macromolecules*, **9** (5), 705–709.
64. Choi, H.S. and Yui, N. (2006) Design of rapidly assembling supramolecular systems responsive to synchronized stimuli. *Prog. Polym. Sci.*, **31** (2), 121–144.
65. Charlot, A. and Auzely-Velty, R. (2007) Synthesis of novel supramolecular assemblies based on hyaluronic acid derivatives bearing bivalent beta-cyclodextrin and adamantane moieties. *Macromolecules*, **40** (4), 1147–1158.
66. Semenov, A., Charlot, A., Auzely-Velty, R., and Rinaudo, M. (2007) Rheological properties of binary associating polymers. *Rheol. Acta*, **46** (5), 541–568.
67. Born, M. and Ritter, H. (1991) Comb-like rotaxane polymers containing noncovalently bound cyclodextrins in the side-chains. *Makromol. Chemie. Rapid Commun.*, **12** (8), 471–476.
68. Born, M., Koch, T., and Ritter, H. (1994) Side-chain polyrotaxanes.2. functionalized polysulfone with noncovalently anchored cyclodextrins in the side-chains. *Acta Polym.*, **45** (2), 68–72.
69. Born, M., Koch, T., and Ritter, H. (1995) Side-chain polyrotaxanes.3. synthesis, characterization and enzymatically catalyzed degradation of noncovalently anchored cyclodextrines in the side-chains of poly(ether-ether-ketone)s. *Macromol. Chem. Physic.*, **196** (5), 1761–1767.
70. Born, M. and Ritter, H. (1995) Side-chain polyrotaxanes with a tandem structure based on cyclodextrins and a polymethacrylate main-chain. *Angew. Chem. Int. Edit.*, **34** (3), 309–311.
71. Born, M. and Ritter, H. (1996) Pseudo-polymer analogous reactions: Methylation of alcohol groups of non-covalently anchored 2,6-dimethyl-beta-

cyclodextrin components located in branched side chains of a poly(tandem-rotaxane). *Adv. Mater.*, **8** (2), 149.

72 Born, M. and Ritter, H. (1996) Topologically unique side-chain polyrotaxanes based on triacetyl-beta-cyclodextrin and a poly(ether sulfone) main chain. *Macromol. Rapid Comm.*, **17** (4), 197–202.

73 Yamaguchi, I., Osakada, K., and Yamamoto, T. (1997) Introduction of a long alkyl side chain to poly (benzimidazole)s. n-alkylation of the imidazole ring and synthesis of novel side chain polyrotaxanes. *Macromolecules*, **30** (15), 4288–4294.

74 Yamaguchi, I., Osakada, K., and Yamamoto, T. (2000) A novel crown ether stopping group for side chain polyrotaxane. preparation of side chain polybenzimidazole rotaxane containing alkyl side chain ended by crown ether-ona group. *Macromolecules*, **33** (7), 2315–2319.

75 Noll, O. and Ritter, H. (1997) New side-chain poly(methacryl-rotaxanes) bearing cyclodextrins as non-covalently anchored ring components. chemoenzymatic synthesis and degradation. *Macromol. Rapid Comm.*, **18** (1), 53–58.

76 Noll, O. and Ritter, H. (1998) Synthesis of new side-chain polyrotaxanes via free radical polymerization of a water-soluble semi-rotaxane monomer consisting of 2,6-dimethyl-beta-cyclodextrin and 3-0-(11-acryloylaminoundecanoyl)cholic acid. *Macromol. Chem. Phys.*, **199** (5), 791–794.

77 Goretzki, C. and Ritter, H. (2002) Photosensitive polyrotaxanes: New comb-like polymethacrylamides containing 1,3-diphenyl-2-propen-1-one (chalcone) functions and non-covalently bound cyclodextrin in the side-chain. *E-Polymers*, **019**, 1–17.

78 Pang, A. and Ritter, H. (2006) Novel side chain polyrotaxane with cyclodextrin: Synthesis and study of water-soluble copolymers bearing hydrophobically associative components. *Macromol. Chem. Physic.*, **207**, 201–208.

79 Tomatsu, I., Hashidzume, A., and Harada, A. (2006) Cyclodextrin-based side-chain polyrotaxane with unidirectional inclusion in aqueous media. *Angew. Chem. Int. Ed.*, **45** (28), 4605–4608.

80 Taura, D., Li, S.J., Hashidzume, A., and Harada, A. (2010) Formation of side-chain hetero-polypseudorotaxane composed of alpha-and beta-cyclodextrins with a water-soluble polymer bearing two recognition sites. *Macromolecules*, **43** (4), 1706–1713.

81 Liu, J., Xu, R.L., and Kaifer, A.E. (1998) In situ modification of the surface of gold colloidal particles. preparation of cyclodextrin-based rotaxanes supported on gold nanospheres. *Langmuir*, **14** (26), 7337–7339.

82 Iijima, T., Uemura, T., Tsuzuku, S., and Komiyama, J. (1978) Diffusion of cyclodextrins in aqueous polymer solution. *J. Polym. Sci.: Pol. Phys. Edit.*, **16**, 793–802.

83 Moro, T., Kobayashi, S., Kainuma, K., Uemura, T., Harada, T., Komiyama, J., and Iijima, T. (1979) Interaction mit cycloamyloses with polymers. *Carbohydr. Res.*, **75**, 345–348.

84 Uemura, T., Moro, T., Komiyama, J., and Iijima, T. (1979) Cooperative binding and diffusion of cyclodextrins in aqueous polymer solution. *Macromolecules*, **12**, 737–739.

85 Harada, A., Adachi, H., Kawaguchi, Y., and Kamachi, M. (1997) Recognition of alkyl groups on a polymer chain by cyclodextrins. *Macromolecules*, **30** (17), 5181–5182.

86 Harada, A., Ito, F., Tomatsu, I., Shimoda, K., Hashidzume, A., Takashima, Y., Yamaguchi, H., and Karmtori, S. (2006) Spectroscopic study on the interaction of cyclodextrins with naphthyl groups attached to poly(acrylamide) backbone. *J. Photochem. Photobio. A*, **179** (1–2), 13–19.

87 Hashidzume, A. and Harada, A. (2006) Macromolecular recognition by cyclodextrins. interaction of cyclodextrins with polymethacrylamides bearing hydrophobic amino acid residues. *Polymer*, **47** (10), 3448–3454.

88 Tomatsu, I., Hashidzume, A., and Harada, A. (2005) Gel-to-sol and sol-to-gel transitions utilizing the interaction of alpha-cyclodextrin with dodecyl side chains attached to a poly(acrylic acid) backbone. *Macromol. Rapid Commun.*, **26** (10), 825–829.

89 Tomatsu, I., Hashidzume, A., and Harada, A. (2005) Photoresponsive hydrogel system using molecular recognition of alpha-cyclodextrin. *Macromolecules*, **38** (12), 5223–5227.

90 Tomatsu, I., Hashidzume, A., and Harada, A. (2006) Redox-responsive hydrogel system using the molecular recognition of beta-cyclodextrin. *Macromol. Rapid Commun.*, **27** (4), 238–241.

91 Ravoo, B.J. and Jacquier, J.C. (2002) Host-guest interaction between beta-cyclodextrin and hydrophobically modified poly(isobutene-alt-maleic acid) studied by affinity capillary electrophoresis. *Macromolecules*, **35** (16), 6412–6416.

92 Taura, D., Hashidzume, A., and Harada, A. (2007) Macromolecular recognition: Interaction of cyclodextrins with an alternating copolymer of sodium maleate and dodecyl vinyl ether. *Macromol. Rapid Commun.*, **28** (24), 2306–2310.

93 Taura, D., Hashidzume, A., Okumura, Y., and Harada, A. (2008) Cooperative complexation of alpha-cyclodextrin with alternating copolymers of sodium maleate and dodecyl vinyl ether with varying molecular weights. *Macromolecules*, **41** (10), 3640–3645.

94 Ritter, H., Sadowski, O., and Tepper, E. (2003) Influence of cyclodextrin molecules on the synthesis and the thermoresponsive solution behavior of n-isopropylacrylamide copolymers with adamantyl groups in the side-chains. *Angew. Chem. Int. Ed.*, **42** (27), 3171–3173.

95 Schmitz, S. and Ritter, H. (2005) Unusual solubility properties of polymethacrylamides as a result of supramolecular interactions with cyclodextrin. *Angew. Chem. Int. Ed.*, **44** (35), 5658–5661.

96 Kretschmann, O., Steffens, C., and Ritter, H. (2007) Cyclodextrin complexes of polymers bearing adamantyl groups: Host-guest interactions and the effect of spacers on water solubility. *Angew. Chem. Int. Ed.*, **46** (15), 2708–2711.

97 Koopmans, C. and Ritter, H. (2007) Color change of n-isopropylacrylamide copolymer bearing Reichardts dye as optical sensor for lower critical solution temperature and for host-guest interaction with beta-cyclodextrin. *J. Am. Chem. Soc.*, **129** (12), 3502.

98 Ogoshi, T., Masuda, K., Yamagishi, T., and Nakamoto, Y. (2009) Side-chain polypseudorotaxanes with heteromacrocyclic receptors of cyclodextrins (cds) and cucurbit[7]uril (cb7): Their contrasting lower critical solution temperature behavior with alpha-CD, gamma-CD, and cb7. *Macromolecules*, **42** (21), 8003–8005.

99 He, L.H., Huang, J., Chen, Y.M., and Liu, L.P. (2005) Inclusion complexation between comblike peo grafted polymers and alpha-cyclodextrin. *Macromolecules*, **38** (8), 3351–3355.

100 He, L.H., Huang, J., Chen, Y.M., Xu, X.J., and Liu, L.P. (2005) Inclusion interaction of highly densely peo grafted polymer brush and alpha-cyclodextrin. *Macromolecules*, **38** (9), 3845–3851.

101 He, B., Zeng, J., Nie, Y., Ji, L., Wang, R., Li, Y., Wu, Y., Li, L., Wang, G., Luo, X.L., Zhang, Z.R., and Gu, Z.W. (2009) In situ gelation of supramolecular hydrogel for anti-tumor drug delivery. *Macromol. Biosci.*, **9** (12), 1169–1175.

102 Hu, X., Zheng, P.J., Zhao, X.Y., Li, L., Tam, K.C., and Gan, L.H. (2004) Preparation, characterization and novel photoregulated rheological properties of azobenzene functionalized cellulose derivatives and their at-cd complexes. *Polymer*, **45** (18), 6219–6225.

103 Zheng, P.J., Hu, X., Zhao, X.Y., Li, L., Tam, K.C., and Gan, L.H. (2004) Photoregulated sol-gel transition of novel azobenzene-functionalized hydroxypropyl methylcellulose and its

alpha-cyclodextrin complexes. *Macromol. Rapid Comm.*, **25** (5), 678–682.

104 Zheng, P.J., Wang, C., Hu, X., Tam, K.C., and Li, L. (2005) Supramolecular complexes of azocellulose and alpha-cyclodextrin: Isothermal titration calorimetric and spectroscopic studies. *Macromolecules*, **38** (7), 2859–2864.

105 Cho, S.Y. and Allcock, H.R. (2009) Synthesis of adamantyl polyphosphazene-polystyrene block copolymers, and beta-cyclodextrin-adamantyl side group complexation. *Macromolecules*, **42** (13), 4484–4490.

106 Islam, M.F., Jenkins, R.D., Bassett, D.R., Lau, W., and Ou-Yang, H.D. (2000) Single chain characterization of hydrophobically modified polyelectrolytes using cyclodextrin/hydrophobe complexes. *Macromolecules*, **33** (7), 2480–2485.

107 Amajjahe, S. and Ritter, H. (2008) Supramolecular controlled pseudo-lcst effects of cyclodextrin-complexed poly (ionic liquids). *Macromolecules*, **41** (9), 3250–3253.

108 Amajjahe, S. and Ritter, H. (2008) Anion complexation of vinylimidazolium salts and its influence on polymerization. *Macromolecules*, **41** (3), 716–718.

109 Ogoshi, T., Takashima, Y., Yamaguchi, H., and Harada, A. (2007) Chemically-responsive sol-gel transition of supramolecular single-walled carbon nanotubes (swnts) hydrogel made by hybrids of swnts and cyclodextrins. *J. Am. Chem. Soc.*, **129** (16), 4878.

6
Antibody Dendrimers and DNA Catenanes

Hiroyasu Yamaguchi and Akira Harada

6.1
Molecular Recognition in Biological Systems

Molecular recognition by biomacromolecules such as DNAs and proteins plays an important role, for example, in translation and transcription of the genetic code, and substrate specificities of enzyme and antigen-antibody reactions in life processes. Selective recognition among these biomacromolecules is achieved by noncovalent bonds with a large number of weak bonding interactions including hydrogen bonds, van der Waals, dipole, and/or electrostatic interactions and so on. In recent years, much attention has been focused on supramolecular science (science of noncovalent assembly) because of the recognition of the importance of specific noncovalent interactions in biological systems and in chemical processes [1]. Many supramolecular architectures, for example, supramolecular dendrimers [2–12], rotaxanes [13–24], and catenanes [25–34], have been synthesized, and a huge number of papers have been published. Especially, the artificial supramolecular complexes constructed by bio-functional molecules as building blocks can be expected to have a novel property that is never seen in naturally occurring systems or to improve the property of each component. In this chapter, we focus on supramolecular architectures composed of antibodies or DNAs.

6.1.1
Supramolecular Complex Formation of Antibodies

The first topic in this chapter is supramolecular complexes of antibodies [35–39]. The immune system has an ability to generate antibodies against virtually any molecule of interest. Antibodies can recognize a large and complicated compound with high specificity and high affinity. Recently, much attention has been directed toward antibodies not only in the field of biology but also in the field of chemistry because of their unique structures and functions. Antibodies (immunoglobulins) have been studied as sensors [37, 40–44], diagnostics [45, 46], drug delivery systems (DDS) [47], catalysts (catalytic antibodies) [48–52], and components for nano-technology [53–61].

Supramolecular Polymer Chemistry, First Edition. Edited by Akira Harada.
© 2012 Wiley-VCH Verlag GmbH & Co. KGaA. Published 2012 by Wiley-VCH Verlag GmbH & Co. KGaA.

With the advent of cell technology [62], it has become possible to prepare individual antibodies ('monoclonal antibodies') in large amounts and in homogeneous form. Especially monoclonal antibodies have been widely used as an efficient reagent to detect a target molecule. Based on the principle that a monoclonal antibody specifically reacts with an antigen, several procedures have been developed in immunosorbent assay [63, 64]. Labeled antibodies or antigens are used for the detection, localization and quantification of biological constituents. More recently, an optical technique based on surface plasmon resonance (SPR) [65–68] or a microgravimetric quartz-crystal-microbalance (QCM) [69–71] technique has been found to be useful for measuring and characterizing macromolecular interactions in the increasingly expanding area of biosensor technology. The use of biosensors based on SPR has made it possible to determine kinetic parameters in real time and without any labeling of biomacromolecules for detection. However, as the SPR response reflects a change in mass concentration at the detector surface as molecules bind or dissociate, the specific sensing of substrates with low molecular weight is difficult. In such a case, functional molecules with a high molecular weight such as antibodies have a great potential for amplification of the response signals expressing molecular recognition events. We describe here the design and preparation of supramolecular complexes of antibodies and their application for a highly sensitive detection method. The complex formation between antibodies, for example immunoglobulin G (IgG) and multivalent antigens is investigated. When an antibody solution is mixed with divalent antigen, linear or cyclic supramolecular complexes are formed. These supramolecular formations are utilized for the amplification of detection signals on the biosensor.

The structure of the typical antibody molecule, IgG, is shown in Figure 6.1a. IgG is generated in a final stage of immunization with a highly selectivity for antigens. Two identical heavy chains of about 50 000 daltons and two identical light chains of about 25 000 daltons are cross-linked to each other by disulfide bonds. The molecule adopts a conformation that resembles the letter Y [72–74]. There are two identical binding sites at the top of the Fab fragments of IgG (see Figure 6.1), which are bound by

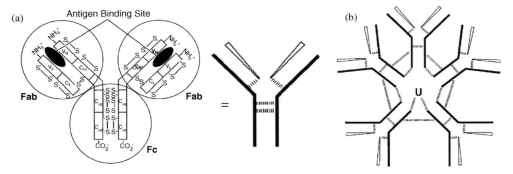

Figure 6.1 The schematic representation of the structure of (a) immunoglobulin G (IgG) and (b) IgM.

flexible hinges with a single constant stem (Fc). The complementarity-determining regions in the antigen binding site differ in length and sequence among different antibodies and are mainly responsible for the specificity (recognition) and affinity (binding) of the antibodies to their target molecules. Fc does not bind antigen, but it has other important biological activities, including the mediation of responses termed effector functions. Antibodies are divided into five classes, IgG, IgM, IgA, IgE, and IgD, on the basis of the number of Y-like units and the type of heavy-chain polypeptide they contain. Each type has common characteristic sequences and variable sequences characteristic of antibodies of the five types. IgM ($M_w = 960\,000$, Figure 6.1b) has a pentameric structure of IgG and ten antigen binding sites in a single molecule [75]. It binds large multivalent antigens strongly (avidity) because of their multivalent structures. However, IgM is the first class of antibody to appear in the serum after exposure to an antigen, and it is unmatured and less specific for the antigen than IgG.

Sensitivity and specificity are required and important in the construction of excellent sensing materials. In order to design an antibody system with a high specificity and a high affinity, a combination of the functions of both IgG and IgM seems to be important. We have designed and prepared dendritic antibody supramolecules in which IgM is placed in a core and many IgGs are bound around the IgM as branches.

6.1.2
Supramolecular Complexes Prepared by DNAs

The second topic is the direct observation of topological structures of DNA supramolecules [76–78]. Catenated DNA molecules [79–83], that is, multiply linked DNA rings, are the products of DNA replication and recombination involving circular genomes and the substrate for topoisomerases [84]. Despite the importance of catenated DNA, little is known about its structure in solution. It is crucial to determine the complex structure of catenanes in order to provide important information about the processes that generate them and the structure of DNA. [n]Catenanes can be expected to have novel physicochemical properties.

6.1.3
Observation of Topological Structures of Supramolecular Complexes by Atomic Force Microscopy (AFM)

AFM [85] is a useful technique for imaging biorelated materials on a nanometer scale under conditions close to their natural environments without complicated sample preparation techniques, such as staining, used in transmission electron microscopy [86–89]. Since the specimens of AFM do not need to be electrically conductive, it is the preferred choice for biological applications. Thus AFM can be potentially applied for investigating microscopic structures of the complicated supramolecular complexes of biomacromolecules by taking advantage of the height information [90–94].

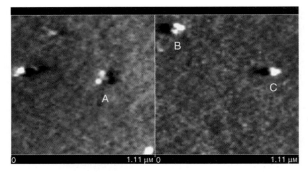

Figure 6.2 AFM images of the individual antibody (IgG) molecules.

6.2
Antibody Supramolecules

6.2.1
Structural Properties of Individual Antibody Molecules

A typical antibody molecule (IgG) was observed by AFM at room temperature depositing from a solution onto a freshly cleaved HOPG surface. Figure 6.2 shows AFM images of the IgG molecule, whose molecular weight is 150 000. Characteristic T or Y shape molecules [95] can be seen. The overall lateral dimension is 40–50 nm as shown A, B, and C in Figure 6.2, which is in good agreement with the expected values [96]. Although these molecules are monoclonal antibodies, their antibody molecular images are not identical. There are two fragments whose heights are the same and one fragment whose height is different from the other two fragments in each molecular image. Fab fragments and an Fc fragment can be differentiated by comparison of their heights [97]. The angles between the Fab fractions of each immunoglobulin are different. Each antibody molecules takes on a somewhat different shape. This is probably due to the flexibility of its hinge region. Figure 6.3 shows IgM images. Two kinds of images are evident, although the object is a single species. The bigger one is a flat pentamer and the smaller one has a higher center. The results are consistent with those obtained by the cryo-AFM measurement [98, 99].

6.2.2
Supramolecular Formation of Antibodies with Multivalent Antigens

In this study, methyl viologen is selected as one of the target molecules to be detected. Viologens are well-known functional molecules as a herbicide and an electron acceptor, although they are harmful. Methyl viologen has been suggested as a potential etiological factor in Parkinson's disease because of its structural similarity to the known dopaminergic neurotoxicant, 1-methyl-4-phenyl-1,2,3,6-tetrahydropyridine [100, 101]. It is important to use the anti-viologen antibodies for

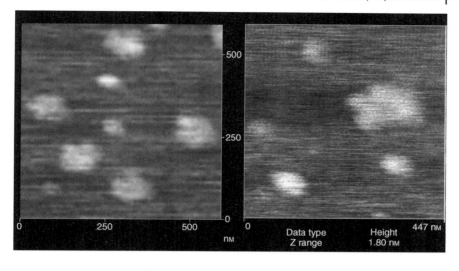

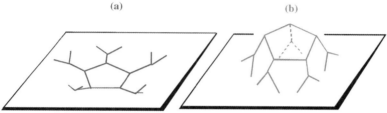

Figure 6.3 The molecular images of the monoclonal IgM on a cleaved highly oriented pyrolytic graphite (HOPG) surface and schematic representation of the images for IgM. A flat pentagram (a) and a smaller object with higher center (b).

the sensitive and specific detection of viologens by the SPR biosensor because methyl viologen is a charged substrate with low molecular weight. However, the detection of viologens at low concentrations is difficult owing to the low sensitivity (small response) in a common SPR biosensor technique using corresponding antibodies. To improve the sensitivity, it is important to detect methyl viologen as a large response signal caused by the antibody bindings. A solution of this problem is thought to be the inhibition of the supramolecular formation between the antiviologen antibody and viologen dimer by methyl viologen. We investigated the complex formation of one of the antibodies, 10D5, with methyl viologen or viologen dimer **2** (Figure 6.4).

6.2.2.1 Supramolecular Formation of Antibodies with Divalent Antigens

The specific binding of the antibody (divalent) and antigen dimer (divalent) produces the supramolecules such as linear or cyclic antigen–antibody oligomers [102, 103]. Step-wise additions of antigen dimer and antibody to the sensor

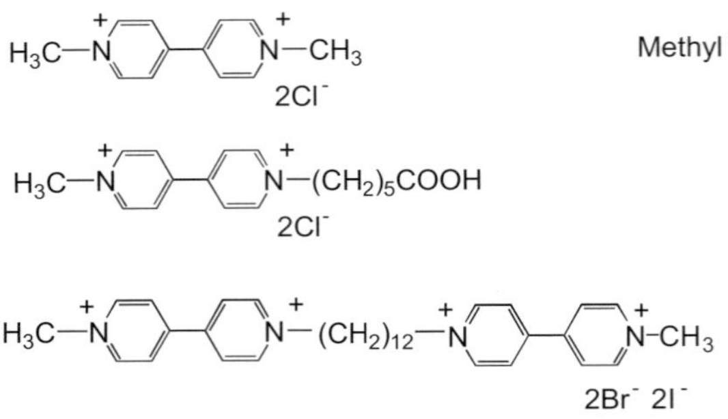

$H_3C-N^+\text{⟨py⟩}-N^+\text{⟨py⟩}-CH_3$ Methyl Viologen

2Cl$^-$

$H_3C-N^+\text{⟨py⟩}-N^+\text{⟨py⟩}-(CH_2)_5COOH$ **1**

2Cl$^-$

$H_3C-N^+\text{⟨py⟩}-N^+\text{⟨py⟩}-(CH_2)_{12}-N^+\text{⟨py⟩}-N^+\text{⟨py⟩}-CH_3$ **2**

2Br$^-$ 2I$^-$

Figure 6.4 The structures of methyl viologen, the hapten molecule **1** (4,4′-bipyridinium, 1-(carboxypentyl)-1′-methyl-dichloride), and divalent antigen **2** prepared in this study.

chip of the biosensor, whose surface is modified with the antibody, leads to the enhancement of the response signal intensity in SPR due to the complex formation of antibodies with antigens. When methyl viologen is added to the flow cell instead of viologen dimer, the binding sites of the antibody at the end of the supramolecule are occupied by methyl viologen and can not bind viologen dimer that acts as a connector of the antibodies in the supramolecular formation. The addition of methyl viologen may bring a reduction in the response signal enhancement. In this method, the binding of methyl viologen to the antibody affects the next growing step in supramolecular formation between the antibody and viologen dimer. The amount of methyl viologen ($M_w = 257$) is expressed as the amount of the antibody ($M_w = 150\,000$) that cannot form the supramolecules between the viologen dimer and the antibody.

The antibody 10D5 (IgG$_1$) binds hapten **1** with the dissociation constant $K_d = 2.0 \times 10^{-7}$ M. The dissociation constant between antibody 10D5 and methyl viologen was found to be 2.0×10^{-7} M. The antibody 10D5 recognizes the bipyridinium moiety with high specificity. Figure 6.5 shows the result of competition, enzyme-linked immunosorbent assay (ELISA), between antibody 10D5 and viologen dimer **2**. At low concentrations of viologen dimer, the absorbance of the product

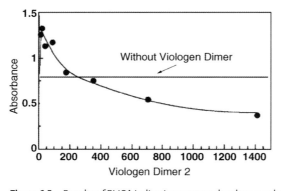

Figure 6.5 Results of ELISA indicating supramolecular complex formation between antibody 10D5 and viologen dimer 2. A number of antibody molecules are immobilized to the ELISA plate at lower concentrations of viologen dimer 2.

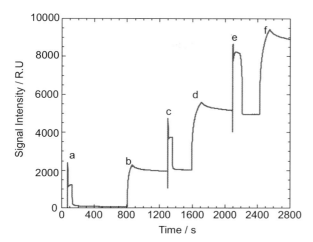

Figure 6.6 The sensorgram of the repeated injection of the aqueous viologen dimer **2** (a, c, e) and the antibody (b, d, f) solutions. [Viologen dimer **2**] = 2.0 μM and [antibody] = 2.0 μM in phosphate borate buffer. The injection period: 60 s for (a)–(c) and 120 s for (d)–(f). A solution of viologen dimer **2** or the antibody passes over the surface of the sensor chip for 60 or 120 s at a constant flow rate of 20 μL min^{-1}. The surface of the sensor chip was subsequently washed with buffer.

of enzymatic reactions is higher than that in the absence of viologen dimer. It is suggested that a number of antibody molecules are immobilized to the ELISA plate in the presence of viologen dimer. The signal intensity on the SPR biosensor also increased on the addition of an aqueous solution of antibody 10D5 to the sensor chip on which the viologen dimer–antibody complex was pre-coated. Figure 6.6 shows the sensorgram of the repeated injection of the aqueous viologen dimer and antibody solutions. The signal intensity is enhanced by the binding of antibody to the viologen dimer–antibody complex. The viologen dimer molecule is considered to act as a connector between antibodies. It is suggested that the phenomena observed in the ELISA and SPR measurements are ascribable to the higher-order complex formation between antibodies and viologen dimer.

6.2.2.2 Direct Observation of Supramolecular Complexes of Antibodies with Porphyrin Dimers

The structure of complexes between one of the monoclonal antibodies for water-soluble porphyrins and porphyrin dimers was observed by AFM. Figure 6.7a shows one of the images of the complex obtained by mixing the IgG with the divalent antigen (TCPP dimer shown in Figure 6.7d). This image suggests the presence of the supramolecular structures of the antibodies connected by the divalent antigens, for example, dimer, trimer, pentamers, and so on. The molecular images of trimer and pentamer are clearly observed by picking up each image and expanding them as shown in Figure 6.7b and c, respectively. From the 3D image, it is suggested that this pentamer is a precursor of a cyclic pentamer.

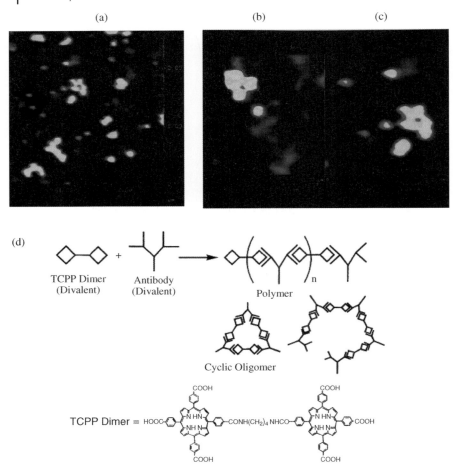

Figure 6.7 The molecular images of antibody–divalent antigen complexes (a), trimer (b), and pentamer (c). Schematic representation of supramolecular complex formation between monoclonal antibodies and divalent antigens (d).

6.2.2.3 Applications for the Highly Sensitive Detection Method of Small Molecules by the Supramolecular Complexes between Antibodies and Multivalent Antigens

We found that the supramolecular formation of antibodies and viologen dimer 2 brings an SPR signal enhancement of the biosensor. The additional binding of the antibody to the viologen dimer–antibody complex gives a remarkable increase of signal intensities. On the other hand, the addition of methyl viologen (viologen monomer) instead of viologen dimer is expected to block the antigen binding sites and to inhibit the additional antibody binding. A small amount of methyl viologen can be detected from a decrease of signal enhancement due to the inhibition of the complex formation between viologen dimer and antibody by methyl viologen, compared with the signal intensity of complete supramolecular formation between

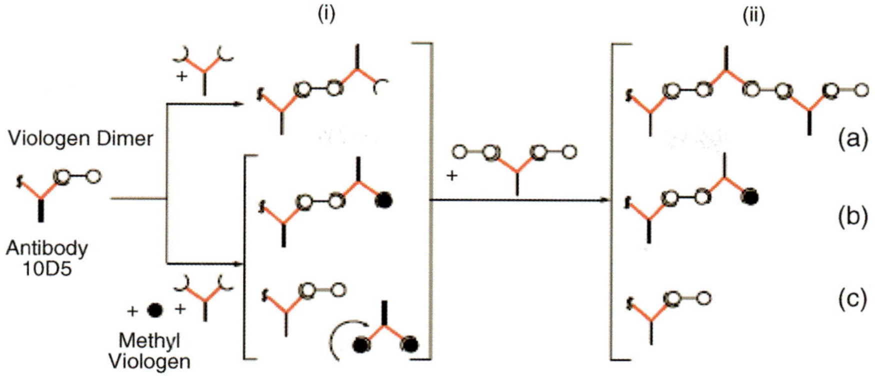

Scheme 6.1 Strategy for the highly sensitive detection of viologen based on the SPR biosensor technique. Inhibition of the complex formation of the antibody with viologen dimer by methyl viologen and the signal enhancement due to the supramolecular formation between the antibody and viologen dimer. Antibody 10D5–viologen dimer complex is immobilized onto the surface of the sensor chip. An aqueous solution of antibody 10D5 and a sample including methyl viologen or one without methyl viologen is injected into the flow cell (i) before the addition of the complex between antibody 10D5 and viologen dimer 2 (ii). The supramolecular formation between antibody 10D5 and viologen dimer 2 without methyl viologen (a), and that in the presence methyl viologen (b) and (c). [Methyl viologen] < [antibody combining site]: (b) and [methyl viologen] ≫ [antibody combining site]: (c).

the antibody and viologen dimer. Scheme 6.1 shows the strategy for the amplification of detection signals for methyl viologen based on the signal enhancement by the supramolecular formation between the antibody and viologen dimer using the biosensor technique. This system includes a two-step procedure as follows: (i) the injection of the aqueous solution of anti-viologen antibody with methyl viologen to the sensor chip whose surface is modified with the antibody-viologen dimer complex, and then (ii) the addition of antibody-viologen dimer complex to the previous state. The changes of the signal intensities in the presence of methyl viologen are compared with those in the absence of methyl viologen.

The amount of antibody immobilized to the sensor chip decreased with increase in the concentration of methyl viologen. To enlarge the difference in the signal intensities in the presence of a small amount of methyl viologen, a solution of the antibody-viologen dimer complex was added to the previous state (i). The differences in signal intensities due to the binding of the additional antibody (0.9 μM) in the presence of methyl viologen ranging in concentration from 0.2 to 1.1 μM is slight in step (i). However, further addition of viologen dimer 2 and antibody solutions in step (ii) causes a clear difference in the response signal intensities in the same concentration range of methyl viologen. The total changes in the signal intensities were found to have a linear relationship to the logarithm of the concentration of methyl viologen. The sensitivity in this system is 140-fold larger than that in the simple addition of methyl viologen to the antibody immobilized to the surface of the sensor chip. It is clear that this system can be utilized for the detection of methyl viologen. Amplification of methyl viologen sensing processes is realized by the inhibition of complex formation between the antibody and viologen dimer–antibody complex and signal enhancement due to the supramolecular formation of the antibody and viologen dimer.

6.2.3
Supramolecular Dendrimers Constructed by IgM and Chemically Modified IgG

6.2.3.1 Preparation of Antibody Dendrimers and their Topological Structures

A novel antibody supramolecule is designed and prepared by using immunoglobulin M (IgM) as a core and chemically modified IgGs as branches as shown in Scheme 6.2. The characteristic binding ability and specificity of IgG were found to remain during the chemical modification of IgG with 3MPy1C (Figure 6.8). When IgM for 3MPy1C is treated with IgG covalently bound cationic porphyrin, IgM binds the cationic porphyrin attached on the IgG to give a dendritic antibody supramolecule 'antibody dendrimer' (**G1** in Scheme 6.2). The dendritic supramolecules are composed of proteins with molecular weight of about 2 million and constructed by noncovalent bonds. The structural observation of the antibody dendrimer was carried out by using AFM. The sample surface was observed under conditions such that any damage caused by scanning the cantilever is minimized and that any nonspecific assembly among antibodies does not occur. Figure 6.9 shows an AFM image of the dendrimer. This image was twice as large as that of starting IgM. Some branches (IgGs) can be seen outside of the IgM core. Such an assembled structure was not observed in a chemically-modified IgG solution or an IgM solution alone.

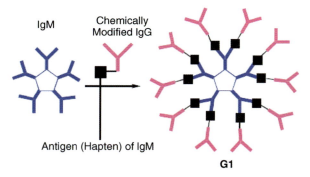

Scheme 6.2 The synthetic route of the complete antibody dendrimers. An ideal structure of the dendritic supramolecule is shown as **G1**.

6.2.3.2 Binding Properties of Antibody Dendrimers for Antigens

The binding property of the antibody dendrimer (**G1**) with a cationic or anionic porphyrin was measured by ELISA. Figure 6.10a shows the binding properties of the IgG, IgM, and **G1** with the cationic porphyrin, *meso*-tetrakis(4-methylpyridyl)por-

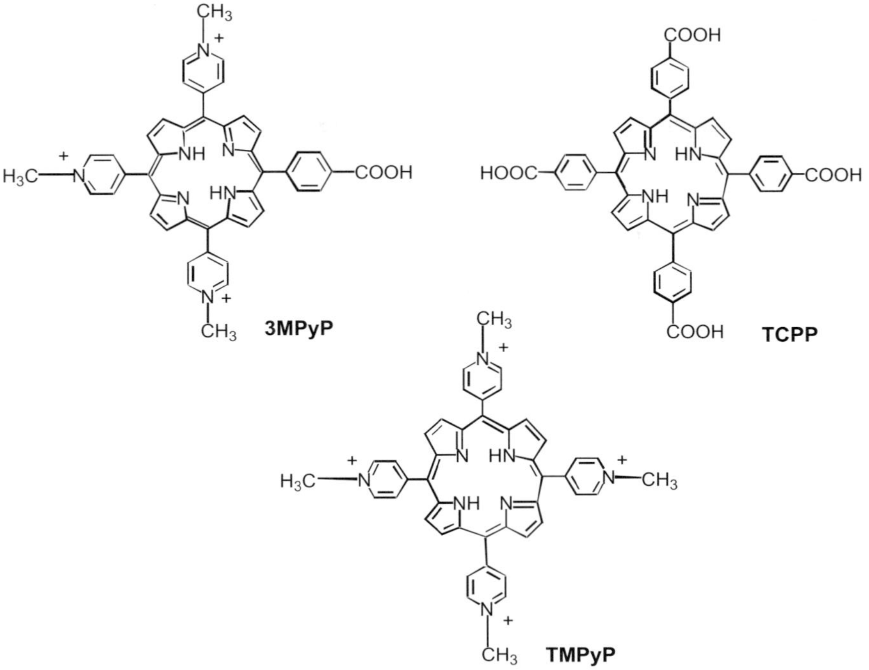

Figure 6.8 Anti-porphyrin antibodies were elicited for [5-(4-carboxyphenyl)-10,15, 20-tris-(4-methyl pyridyl)]porphine (**3MPy1C**) or meso-tetrakis(4-carboxyphenyl)porphine (**TCPP**). meso-Tetrakis(4-methyl pyridyl) porphine (**TMPyP**) was used to investigate the specificity of the antibody dendrimer for porphyrins.

phyrin (TMPyP). Although IgG did not bind the cationic porphyrin and IgM did, the dendrimer did not bind TMPyP. These results show that the cationic porphyrin attached to IgG occupies the binding sites of IgM in the dendrimer, and thus there are no free binding sites against TMPyP on IgG. Figure 6.10b shows the binding of IgG, IgM, and **G1** to TCPP. The IgM used in this study can bind both anionic and cationic

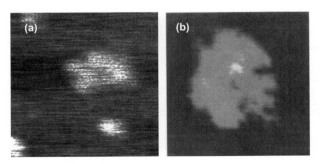

Figure 6.9 The AFM images of IgM (a) and the antibody dendrimer (b). A total 2 μL of solutions of the antibodies (3.0×10^{-9} M) in 0.1 M phosphate borate buffer (pH 9.0) was deposited onto the surface of HOPG and air dried.

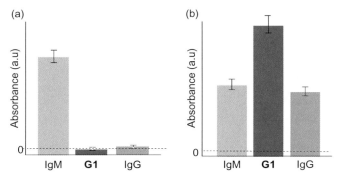

Figure 6.10 Binding affinities of IgG, IgM, and the antibody dendrimer (G1) with the cationic porphyrin (TMPyP, a) and those with the anionic porphyrin (TCPP, b) estimated by ELISA.

porphyrins, due to the low specificity of IgM against porphyrins. Both IgM and IgG bound TCPP, while **G1** bound TCPP more efficiently than IgM or IgG. The increase in affinity of **G1** for the anionic porphyrin indicates that many IgG molecules attach to the surface of the IgM molecule.

The biosensor technique based on SPR shows that the antibody dendrimer has an advantage of the amplification of detection signals for antigens. A solution of **G1** was added to the sensor chip on which TCPP was pre-coated by the coupling with hexamethylenediamine as a spacer. The total concentration of the antibody was fixed at 0.2 mg mL^{-1} (0.1 mg of IgM + 0.1 mg of chemically modified IgG in 1 mL buffer for **G1**). The sensorgram for the binding of the antibody dendrimer to TCPP was compared with that of IgG to TCPP, as shown in Figure 6.11. The signal intensity

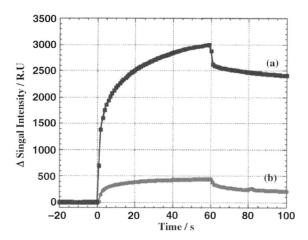

Figure 6.11 The sensorgrams for the binding of the antibody dendrimer (a) or IgG (b) to the anionic porphyrin immobilized onto the surface of the sensor chip. Phosphate borate buffer (0.1 M, pH 9.0) was used. TCPP was immobilized via hexamethylenediamine spacer onto the sensor chip and then a solution of IgG or the dendrimer was injected into the flow cell. Sixty seconds after the injection of the antibody solutions, the flow cell was filled with buffer.

increased by the injection of the antibody dendrimer was sufficiently larger than that of simple addition of IgG. Taking into account the change of the binding property of the antibody dendrimer for porphyrins with the increase in the amount of bound antibody to the anionic porphyrin on the SPR biosensor, the antibody dendrimer has many IgG molecules successively bound to IgM molecule.

6.3
DNA Supramolecules

6.3.1
Imaging of Individual Plasmid DNA Molecules

The two main methods used for determining the topology of DNA catenanes are agarose gel electrophoresis and AFM. Although only AFM gives the complete stereostructure of individual molecules, electrophoresis provides a more rapid and quantitative overview of the population. Therefore, it is best to use both methods in combination. Figure 6.12a shows electrophoresis of plasmid DNA pBR 322 using 1.0% agarose gel. Two bands are clearly separated. The upper and lower bands correspond to form I (closed-circular DNA) and form II (open-circular DNA), respectively. The ratio of illumination intensities of form I band and form II band is ca. 4. To prepare samples for visualization, a total 4 µL of DNA solution (0.5 ng µL^{-1} in 10 mM Tris-HCl, 1 mM EDTA, 10 mM MgCl$_2$, 10 mM HEPES) was placed onto freshly cleaved mica for 15 min, rinsed with deionized water, and then dried for 6 h in a desiccator with CaCl$_2$. The sample surface was observed by tapping mode AFM. All

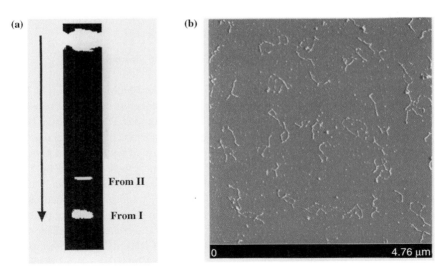

Figure 6.12 Agarose gel electrophoresis (a) and the tapping mode AFM image (b) of plasmid DNA pBR322. A downward-pointing arrow shows the direction of electrophoresis.

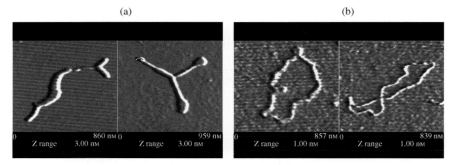

Figure 6.13 The single molecular images of plasmid DNA extracted from 'form I' (closed-circular DNA, a) and 'form II' (open-circular DNA, b).

scales were calibrated against a standard sample (5 μm × 5 μm grid). Figure 6.12b shows the images of plasmid DNA pBR322 on freshly cleavaged mica. The buffer contains 10 mM Mg^{2+} to bind individual DNA molecule on substrate under conditions where the buffer salts can be washed off. Each DNA molecule can be clearly observed by a tapping mode AFM under these conditions. The single molecular images of DNA molecules extracted from form I band and those of form II band are shown in Figure 6.13a and b, respectively. There are two types of topological structure of plasmid DNA. The ratio of the number of molecules in Figure 6.13a and b is equal to that of intensities of form I and form II bands in agarose gel electrophoresis. The structural information obtained by AFM measurement corresponds to that of electrophoresis. The AFM images of DNA molecules are indicated to show the DNA structures in solution. The topological structures of plasmid DNA are considered to be remaining during the extraction procedure and the sample preparation for AFM.

6.3.2
Preparation of Nicked DNA by the Addition of DNase I to Plasmid DNA

Nicked DNA was prepared by the addition of DNase I (in 10 mM Tris-HCl (pH 7.5), 10 mM $MgCl_2$, 10 mM $CaCl_2$ and 50% glycerol) to the aqueous solution of pBR322. DNase I can cut one of the two chains on double-stranded DNA. Plasmid DNA becomes an expanded circular molecule by the enzymatic reaction. The increase in the concentration of DNase I brings about the production of linear DNA or partially separated smaller-DNA molecules. The reaction products were analyzed by 1.0% agarose gel electrophoresis using Tris/acetate/EDTA buffer at 5 V/cm for 90 min. Gels were stained with ethidium bromide (20 mg mL^{-1}) and photographed using ultraviolet illumination (312 nm). Figure 6.14 shows the electrophoresis of pBR322 in the presence of various concentrations of DNase I. The concentration of form I DNA decreased by the increase in the concentration of DNase I. The more rapidly migrated bands are obtained as shown in line 9 and lane 10 at [DNase I] $> 2.0 \times 10^{-6}$ units μL^{-1}. The DNase I concentration dependence on the topological structure of nicked DNA is also monitored by AFM using a tapping mode. Figure 6.15a shows the

6.3 DNA Supramolecules | 141

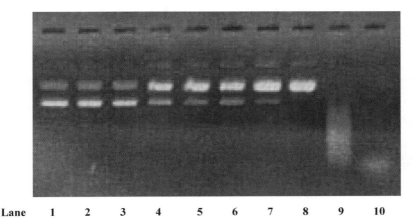

Lane   1   2   3   4   5   6   7   8   9   10

Figure 6.14 Electrophoresis of the reaction products of plasmid DNA pBR 322 with DNase I, showing the effect of the concentration of DNase I on the structures of nicked DNAs. The concentration of DNA was fixed at 20 ng µL^{-1}. [DNase I] (units µL^{-1}): 0 (lane 1), 10^{-8} (lane 2), 10^{-7} (lane 3), 1.0 × 10^{-6} (lane 4), 2.0 × 10^{-6} (lane 5), 4.0 × 10^{-6} (lane 6), 8.0 × 10^{-6} (lane 7), 10^{-5} (lane 8), 10^{-4} (lane 9), 10^{-3} (lane 10).

AFM image of the DNA after the enzymatic reaction of 1.0×10^{-6} units/µL DNase I. The topological structure of pBR322 is mainly form II (open-circular structure) compared with the original structure of plasmid DNA (mixture of open and closed circular DNA). Linear and shorter DNAs are observed by AFM in the presence of higher concentration of DNase I (Figure 6.15b and c). The structural information obtained by electrophoresis is confirmed by AFM measurement.

6.3.3
Catenation Reaction with Topoisomerase I

Catenation reaction was performed as shown in Scheme 6.3. A open-circular DNA was obtained by mixing pBR322 DNA (12.5 ng µL^{-1}) and DNase I (7.4 × 10^{-6} ng) in

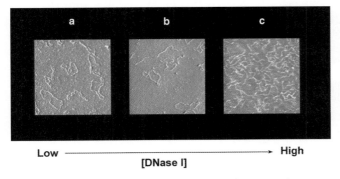

Low ────────────────→ High
              [DNase I]

Figure 6.15 AFM images of the reaction products of pBR322 with DNase I used in Figure 6.9, lane 4 (a), lane 7 (b), and lane 9 (c).

6 Antibody Dendrimers and DNA Catenanes

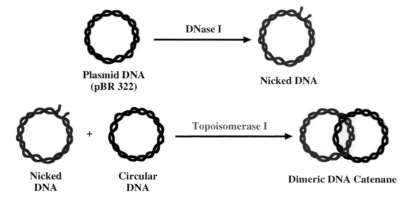

Scheme 6.3 Preparation of dimeric DNA catenane by the reaction of topoisomerase I to the nicked DNA in the presence of the circular DNA.

16 μL of solution containing 40 mM Tris-HCl (pH 7.5), 6 mM $MgCl_2$, and 6 mM $CaCl_2$ at 37 °C for 30 min. Reactions were terminated by heating at 65 °C for 5 min. Catenation reactions between nicked DNA and circular DNA were carried out at 37 °C for 4 h in 10 μL of solution containing 20 mM Tris-HCl (pH 7.5), 20 mM KCl, 5 mM spermidine, 2 mM dithiothreitol, 50 μg mL bovine serum albumin, 10% glycerol, 12.5 ng μL pBR322 DNA, and 10 ng topoisomerase I. The enzymatic reaction was terminated by heating at 65 °C for 5 min. The reaction of topoisomerase I was monitored by electrophoresis. Figure 6.16 shows the results of electrophoresis of the reaction products. There are only two bands (lanes 1 and 2) in the presence of 0.05 units/mL of topoisomerase I. However, a new band has been observed in the presence of double the amounts of topoisomerase I compared to the previous

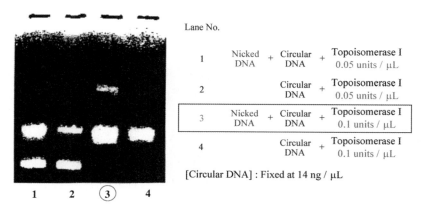

Figure 6.16 Agarose gel electrophoresis of the reaction mixture between nicked DNA and circular DNA in the presence of topoisomerase I. [Topoisomerase I]: 0.05 units $μL^{-1}$ (lanes 1 and 2), 0.1 units $μL^{-1}$ (lanes 3 and 4). Lane 1 and lane 3 include nicked DNA. The concentration of circular DNA was fixed at 14 ng $μL^{-1}$.

conditions. The more slowly migrated band appearing in lane 3 is considered to be a [2]catenane.

6.3.4
AFM Images of DNA Catenanes

DNA catenane was extracted from the band of 1.0% agarose gel by using a centrifugal filter device, Ultrafree-DA (MILLIPORE), spinning at 6000 × g for 10 min. Figure 6.17a, c and d show the DNA molecular images extracted from the more slowly migrated band appearing in lane 3 in Figure 6.16. Parts of two circles fused over a distance of ∼ 90 nm can be seen. Each circle is assigned to be pBR 322 plasmid,

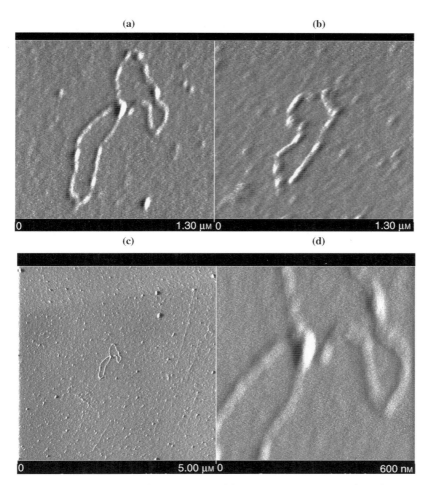

Figure 6.17 The tapping mode AFM images of dimeric DNA catenane (a), plasmid DNA (DNA monomer) (b), an image of wide area (5 μm × 5 μm, c), and an enlarged image of the fused part of catenated DNA molecule (d).

whose length is about 1.7 μm – comparable in length to the 4361 bps. This molecular image can be ascribed to dimeric DNA catenane for the following reasons; (1) this fused DNA is constituted by two plasmid DNA molecules, and (2) the molecules seen in Figure 6.17a were extracted in the higher molecular weight band by electrophoresis and inclued no DNA monomer (Figure 6.17b). The interlocked structure is clearly observed by expanding the fused part as shown in Figure 6.17d. The overpassing and underpassing strands at crossings can be seen.

6.3.5
DNA [n]Catenanes Prepared by Irreversible Reaction with DNA Ligase

DNA catenanes were prepared by the addition of T4 DNA ligase to the mixture of linear DNA and nicked DNA as shown in Scheme 6.4. The linear and nicked DNAs were prepared by the addition of EcoRI or DNase I to a solution of pBR322 plasmid DNA, respectively. The advantage of using T4 DNA ligase is an irreversible recombination between the complementary ends of DNA, in contrast with a reversible one for topoisomerase I. Therefore we can expect the increase of the yield of DNA catenanes. Many more slowly migrated bands were observed in the electrophoresis by the addition of ligase to the mixture of linear and nicked DNA. These bands were found to include DNA catenanes, large circular DNAs, and linear DNA oligomers by extracting DNAs from each band in agarose gel electrophoresis. Figure 6.18b shows a tapping mode AFM image of a large circular DNA. In comparison with the chain length of plasmid DNA (monomer), this cyclic molecule turned out to be a tetramer. However, side reactions such as intermolecular recombination of linear DNAs and cyclization of the linear-DNA oligomer occur, and we found that stable [n]catenanes (n > 2) were also produced by the addition of T4 DNA ligase to an aqueous solution of concentrated linear DNA in the presence of nicked DNA. Figure 6.18a, c, and d show the topological structure of [2], [3] and [4]catenanes observed by AFM. Cozzarelli *et al.*

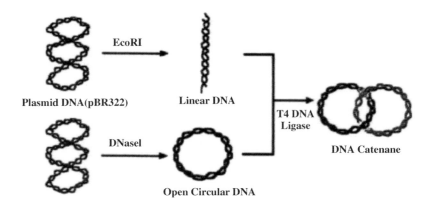

Scheme 6.4 DNA catenanes prepared by the addition of T4 DNA ligase to the mixture of linear and nicked (open circular) DNAs.

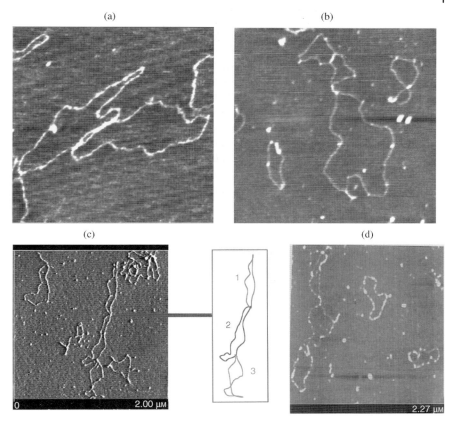

Figure 6.18 The tapping mode AFM images of [2]catenane (a), the large circular DNA (b), [3] and [4]catenanes ((c) and (d), respectively) prepared by the reaction of ligase. Specimen: a total 4 μL of DNA solution including reaction products (0.5 ng mL^{-1} in 10 mM Tris-HCl, 1 mM EDTA, 10 mM MgCl$_2$, 10 mM HEPES) was placed onto freshly cleaved mica for 15 min, rinsed with deionized water, and then dried for 6 h in a desiccator with CaCl$_2$.

reported on the topological structures of DNA catenanes by using an electron microscope [79–83]. However, the resolution of the segments at the junction is poor in electron microscopic examination of conventionally strained and shadowed DNA. To overcome this problem, they used a protein (RecA) coating method to enhance visualization of the crossings. Our experiment is the first example of the direct observation of DNA catenanes [104–106].

6.4 Conclusions

The enhancement of SPR signal intensity was observed by the addition of the antibody to the divalent antigen-antibody complex immobilized onto the surface of the sensor chip, indicating the formation of linear supramolecules. An amplification method of the detection signals for a target molecule has been devised by using the signal enhancement in the supramolecular assembly of anti-viologen antibodies and divalent antigens. Target substrate added to the flow cell of SPR can be detected quantitatively by monitoring the total amount of the antibody bound to the surface of the sensor chip. The sensitivity in this system was found to be two orders larger than that in the simple addition of target substrate to the antibody immobilized to the surface of the sensor chip. This method can be potentially applied for many compounds to be detected with high sensitivity and specificity by using corresponding antibodies and dimers of the target molecule (divalent antigen).

New antibody dendrimers were designed and prepared by a combination of IgG and IgM, that is using IgM as a core and IgG as branches. Many binding sites of IgG were arranged radially on the surface of one object, and the resulting artificial antibodies bound antigens more selectively than IgM and more strongly than IgG. The characteristic features of the antibody dendrimers are (i) they are composed of proteins, (ii) they are of large size, with a molecular weight of about 2 million, (iii) they are constructed by noncovalent bonds, and (iv) they bind antigen strongly with high specificity. The antibody dendrimers will be used as functionalized materials for sensitive detection of many kinds of chemicals, for diagnosis, and for DDS.

DNA catenanes were prepared by the addition of topoisomerase I to the mixture of nicked DNA and plasmid DNA or by the addition of T4 DNA ligase to the mixture of linear DNA and nicked DNA. The single molecular images of *naked* DNA catenanes are clearly observed by AFM using a tapping mode at room temperature and in an ambient atmosphere.

References

1 Lehn, J.-M. (1990) *Angew. Chem. Int. Ed. Engl.*, **29**, 1304–1319.
2 Tomalia, D.A. and Majoros, I. (2000) Dendritimeric Supramolecular and Supramacromolecular Assemblies, in *Supramolecular Polymers* (ed. A. Ciferri), Marcel Dekker, Inc., New York, pp. 359–434.
3 Baars, M.W.P.L. and Meijer, E.W. (2000) *Top. Curr. Chem.*, **210**, 131–182.
4 Smith, D.K. and Diederich, F. (2000) *Top. Curr. Chem.*, **210**, 183–227.
5 Zimmerman, S.C. and Lawless, L.J. (2001) *Top. Curr. Chem.*, **217**, 95–120.
6 Frechet, J.M.J. (2002) *Proc. Natl. Acad. Sci. USA*, **99**, 4782–4787.
7 Pyun, J., Zhou, X.-Z., Drockenmuller, E., and Hawker, C.J. (2003) *J. Mater. Chem.*, **13**, 2653–2660.
8 Smith, D.K. (2006) *Chem. Commun.*, 34–44.
9 Ribaudo, F., van Leeuwen, P.W.N.M., and Reek, J.N.H. (2006) *Top. Organomet. Chem.*, **20**, 39–59.
10 Hammond, M.R. and Mezzenga, R. (2008) *Soft Matter*, **4**, 952–961.
11 Palmer, L.C. and Stupp, S.I. (2008) *Acc. Chem. Res.*, **41**, 1674–1684.
12 Astruc, D., Boisselier, E., and Ornelas, C. (2010) *Chem. Rev.*, **110**, 1857–1959.
13 Harada, A., Hashidzume, A., Yamaguchi, H., and Takashima, Y. (2009) *Chem. Rev.*, **109**, 5974–6023.
14 Faiz, J.A., Heitz, V., and Sauvage, J.-P. (2009) *Chem. Soc. Rev.*, **38**, 422–442.
15 Harada, A., Takashima, Y., and Yamaguchi, H. (2009) *Chem. Soc. Rev.*, **38**, 875–882.

16 Crowley, J.D., Goldup, S.M., Lee, A.-L., Leigh, D.A., and McBurney, R.T. (2009) *Chem. Soc. Rev.*, **38**, 1530–1541.
17 Balzani, V., Credi, A., and Venturi, M. (2009) *Chem. Soc. Rev.*, **38**, 1542–1550.
18 Stoddart, J.F. (2009) *Chem. Soc. Rev.*, **38**, 1802–1820.
19 Bodis, P., Panman, M.R., Bakker, B.H., Mateo-Alonso, A., Prato, M., Buma, W.J., Brouwer, A.M., Kay, E.R., Leigh, D.A., and Woutersen, S. (2009) *Acc. Chem. Res.*, **42**, 1462–1469.
20 Silvi, S., Venturi, M., and Credi, A. (2009) *J. Mater. Chem.*, **19**, 2279–2294.
21 Wenz, G. (2009) *Adv. Polym. Sci.*, **222**, 1–54.
22 Yui, N., Katoono, R., and Yamashita, A. (2009) *Adv. Polym. Sci.*, **222**, 55–77.
23 Tonelli, A.E. (2009) *Adv. Polym. Sci.*, **222**, 115–173.
24 Haenni, K.D. and Leigh, D.A. (2010) *Chem. Soc. Rev.*, **39**, 1240–1251.
25 Niu, Z. and Gibson, H.W. (2009) *Chem. Rev.*, **109**, 6024–6046.
26 Faiz, J.A., Heitz, V., and Sauvage, J.-P. (2009) *Chem. Soc. Rev.*, **38**, 422–442.
27 Coronado, E., Gavina, P., and Tatay, S. (2009) *Chem. Soc. Rev.*, **38**, 1674–1689.
28 Stoddart, J.F. (2009) *Chem. Soc. Rev.*, **38**, 1802–1820.
29 Puddephatt, R.J. (2008) *Chem. Soc. Rev.*, **37**, 2012–2027.
30 Sauvage, J.-P., Collin, J.-P., Faiz, J.A., Frey, J., Heitz, V., and Tock, C. (2008) *J. Porphyrins & Phthalocyanines*, **12**, 881–905.
31 Kay, E.R. and Leigh, D.A. (2008) *Pure Appl. Chem.*, **80**, 17–29.
32 Griffiths, K.E. and Stoddart, J.F. (2008) *Pure Appl. Chem.*, **80**, 485–506.
33 Fang, L., Olson, M.A., Benitez, D., Tkatchouk, E., Goddard, W.A. III, and Stoddart, J.F. (2010) *Chem. Soc. Rev.*, **39**, 17–29.
34 El-Sagheer, A.H. and Brown, T. (2010) *Chem. Soc. Rev.*, **39**, 1388–1405.
35 Harada, A., Yamaguchi, H., Oka, F., and Kamachi, M. (1999) *Proc. SPIE Int. Soc. Opt. Eng.*, **3607**, 126–135.
36 Yamaguchi, H. and Harada, A. (2002) *Chem. Lett.*, 382–383.
37 Yamaguchi, H. and Harada, A. (2002) *Biomacromolecules*, **3**, 1163–1169.
38 Harada, A., Yamaguchi, H., Tsubouchi, K., and Horita, E. (2003) *Chem. Lett.*, **32**, 18–19.
39 Yamaguchi, H. and Harada, A. (2003) *Top. Curr. Chem.*, **228**, 237–258.
40 Skottrup, P.D., Nicolaisen, M., and Justesen, A.F. (2008) *Biosens. Bioelectron.*, **24**, 339–348.
41 Byrne, B., Stack, E., Gilmartin, N., and O'Kennedy, R. (2009) *Sensors*, **9**, 4407–4445.
42 Wagner, M.K., Li, F., Li, J., Li, X.–F., and Le, X.C. (2010) *Anal. Bioanal. Chem.*, **397**, 3213–3224.
43 Choi, Y.–E., Kwak, J.–W., and Park, J.W. (2010) *Sensors*, **10**, 428–455.
44 Yamaguchi, H., Ogoshi, T., and Harada, A. (2011) Sensor Development Using Existing Scaffolds, in *Chemosensors: Principles, Strategies, and Applications* (eds. B. Wang and E. V. Anslyn), John Wiley & Sons, Inc., Hoboken, NJ, USA, pp. 211–226.
45 Borrebaeck, C.A.K. (2000) *Immunol. Today*, **21**, 379–382.
46 Poetz, O., Hoeppe, S., Templin, M.F., Stoll, D., and Joos, T.O. (2009) *Proteomics*, **9**, 1518–1523.
47 Kaneda, Y. (2000) *Adv. Drug Deliv. Rev.*, **43**, 197–205.
48 Harada, A., Fukushima, H., Shiotsuki, K., Yamaguchi, H., Oka, F., and Kamachi, M. (1997) *Inorg. Chem.*, **36**, 6099–6102.
49 Yamaguchi, H., Tsubouchi, K., Kawaguchi, K., Horita, E., and Harada, A. (2004) *Chem. Eur. J.*, **10**, 6179–6186.
50 Yamaguchi, H., Hirano, T., Kiminami, H., Taura, D., and Harada, A. (2006) *Org. Biomol. Chem.*, **4**, 3571–3573.
51 Yamaguchi, H. and Harada, A. (2008) *Chem. Lett.*, **37**, 1184–1189.
52 Heinisch, T. and Ward, T.R. (2010) *Curr. Opin. Chem. Biol.*, **14**, 184–199.
53 Harada, A., Okamoto, K., Kamachi, M., Honda, T., and Miwatani, T. (1990) *Chem. Lett.*, 917–918.
54 Harada, A., Okamoto, K., and Kamachi, M. (1991) *Chem. Lett.*, 953–956.
55 Harada, A., Fukushima, H., Shiotsuki, K., and Kamachi, M. (1993) *Supramol. Chem.*, **2**, 153–155.
56 Harada, A., Shiotsuki, K., Fukushima, H., Yamaguchi, H., and

Kamachi, M. (1995) *Inorg. Chem.*, **34**, 1070–1076.

57 Shenton, W., Davies, S.A., and Mann, S. (1999) *Adv. Mater.*, **11**, 449–452.

58 Harada, A., Yamaguchi, H., Okamoto, K., Fukushima, H., Shiotsuki, K., and Kamachi, M. (1999) *Photochem. Photobiol.*, **70**, 298–302.

59 Yamaguchi, H., Kamachi, M., and Harada, A. (2000) *Angew. Chem. Int. Ed.*, **39**, 3829–3831.

60 Onji, T., Ohara, H., Yamaguchi, H., Ikeda, N., and Harada, A. (2006) *Chem. Lett.*, **35**, 1126–1127.

61 Yamaguchi, H., Onji, T., Ohara, H., Ikeda, N., and Harada, A. (2009) *Bull. Chem. Soc. Jpn.*, **82**, 1341–1346.

62 Köhler, G. and Milstein, C. (1975) *Nature*, **256**, 495–497.

63 Yamaguchi, H., Harada, A., and Kamachi, M. (1998) *React. Polym.*, **37**, 245–250.

64 Ueda, H. (2002) *J. Biosci. Bioeng.*, **94**, 614–619.

65 Löfas, S. and Johnsson, B. (1990) *J. Chem. Soc. Chem. Commun.*, 1526–1528.

66 Jönsson, U., Fägerstam, L., Ivarsson, B., Johnsson, B., Karlsson, R., Lundh, K., Löfas, S., Persson, B., Roos, H., Ronnberg, I., Sjölander, S., Stenberg, E., Stahlberg, R., Urbaniczky, C., Östlin, H., and Malmqvist, M. (1990) *Biotechniques*, **11** 620–627.

67 Malmqvist, M. (1993) *Nature*, **361**, 186–187.

68 Rich, R.L. and Myszka, D.G. (2008) *J. Mol. Recognit.*, **21**, 355–400.

69 Niikura, K., Nagata, K., and Okahata, Y. (1996) *Chem. Lett.*, 863–864.

70 Patolsky, F., Lichtenstein, A., and Willner, I. (2000) *J. Am. Chem. Soc.*, **122**, 418–419.

71 Mattiasson, B., Teeparuksapun, K., and Hedström, M. (2009) *Trends Biotechnol.*, **28**, 20–27.

72 Terry, W.D., Matthews, B.W., and Davies, D.R. (1968) *Nature*, **220**, 239–241.

73 Davies, D.R., Padlan, E.A., and Segal, D.M. (1975) *Annu. Rev. Biochem.*, **44**, 639–667.

74 Silverton, E.W., Navia, M.A., and Davies, D.R. (1977) *Proc. Nat. Acad. Sci. USA*, **74**, 5140–5144.

75 Tomalia, D.A., Naylor, A.M., and Goggard, W.A. III (1990) *Angew. Chem. Int. Ed. Engl.*, **29**, 138–175.

76 Yamaguchi, H., Kubota, K., and Harada, A. (2000) *Proc. SPIE Int. Soc. Opt. Eng.*, **3922**, 228–235.

77 Yamaguchi, H., Kubota, K., and Harada, A. (2000) *Chem. Lett.*, 384–385.

78 Yamaguchi, H., Kubota, K., and Harada, A. (2000) *Nucleic Acids Symp. Ser.*, **44**, 229–230.

79 Brown, P.O. and Cozzarelli, N.R. (1981) *Proc. Natl. Acad. Sci. USA*, **78**, 843–847.

80 Krasnow, M.A. and Cozzarelli, N.R. (1982) *J. Biol. Chem.*, **257**, 2687–2693.

81 Krasnow, M.A., Stasiak, A., Spengler, S.J., Dean, F., Koller, T., and Cozzarelli, N.R. (1983) *Nature*, **304**, 559–560.

82 Droge, P.D. and Cozzarelli, N.R. (1992) *Methods Enzymol.*, **212**, 120–130.

83 Levene, S.D., Donahue, C., Boles, T.C., and Cozzarelli, N.R. (1995) *Biophys. J.*, **69**, 1036–1045.

84 Wang, J.C. (1985) *Annu. Rev. Biochem.*, **54**, 665–697.

85 Binnig, G., Quate, C.F., and Gerber, Ch. (1986) *Phys. Rev. Lett.*, **56**, 930–933.

86 Blackford, B.L., Jericho, M.H., and Mulhern, P.J. (1991) *Scanning Microsc.*, **5**, 907–918.

87 Butt, H.-J., Guckenberger, R., and Rabe, J.P. (1992) *Ultramicroscopy*, **46**, 375–393.

88 Hoh, J.H. and Hansma, P.K. (1992) *Trends Cell Biol.*, **7**, 208–213.

89 Yang, J., Tamm, L.K., Somlyo, A.P., and Shao, Z. (1993) *J. Microsc.*, **171**, 183–198.

90 Scheuring, S. (2006) *Curr. Opin. Chem. Biol.*, **10**, 387–393.

91 Jena, B.P. (2006) *Ultramicroscopy*, **106**, 663–669.

92 Kitagishi, H., Oohora, K., Yamaguchi, H., Sato, H., Matsuo, T., Harada, A., and Hayashi, T. (2007) *J. Am. Chem. Soc.*, **129**, 10326–10327.

93 Muller, D.J. (2008) *Biochemistry*, **47**, 7986–7998.

94 Kitagishi, H., Kakikura, Y., Yamaguchi, H., Oohora, K., Harada, A., and Hayashi, T. (2009) *Angew. Chem. Int. Ed.*, **48**, 1271–1274.

95 Leatherbarrow, R.J., Stedman, M., and Wells, T.N.C. (1991) *J. Mol. Biol.*, **221**, 361–365.

96 Mazeran, P.-E., Loubet, J.-L., Martelet, C., and Theretz, A. (1995) *Ultramicroscopy*, **60**, 33–40.

97 Harada, A., Yamaguchi, H., and Kamachi, M. (1997) *Chem. Lett.*, 1141–1142.

98 Han, W., Mou, J., Sheng, J., Yang, J., and Shao, Z. (1995) *Biochemistry*, **34**, 8215–8220.

99 Zhang, Y., Sheng, S., and Shao, Z. (1996) *Biophys. J.*, **71**, 2168–2176.

100 Liou, H.H., Tsai, M.C., Chen, C.J., Jeng, J.S., Chang, Y.C., Chen, S.Y., and Chen, R.C. (1997) *Neurology*, **48**, 1583–1588.

101 Brooks, A.I., Chadwick, C.A., Gelbard, H.A., Cory-Slechta, D.A., and Federoff, H.J. (1999) *Brain Res.*, **823**, 1–10.

102 Yamaguchi, H., Oka, F., Kamachi, M., and Harada, A. (1999) *Kobunshi Ronbunsyu*, **56**, 660–666.

103 Bilgiçer, B., Thomas, S.W., Shaw, B.F. III, Kaufman, G.K., Krishnamurthy, V.M., Estroff, L.A., Yang, J., and Whitesides, G.M. (2009) *J. Am. Chem. Soc.*, **131**, 9361–9367.

104 Ercolini, E., Valle, F., Adamcik, J., Witz, G., Metzler, R., Rios, P.D.L., Roca, J., and Dietler, G. (2007) *Phys. Rev. Lett.*, **98**, 058102(1)–058102(4).

105 Deffieux, A. and Schappacher, M. (2009) *Cell. Mol. Life Sci.*, **66**, 2599–2602.

106 Schappacher, M. and Deffieux, M. (2009) *Angew. Chem. Int. Ed.*, **48**, 5930–5933.

7
Crown Ether-Based Polymeric Rotaxanes
Terry L. Price Jr. and Harry W. Gibson

7.1
Introduction

Supramolecular chemistry encompasses bi- or multi-molecular systems bound together by noncovalent interactions. Looking more closely at this concept, in its simplest form, noncovalent binding can be defined as a bi-molecular system in which one molecule, H, is considered the host and another is considered the guest, G. Hydrogen bonding (possible the most important), ion pairing, dipole–dipole interactions, charge transfer, hydrophobic–hydrophilic interactions, and van der Waals interactions constitute the individual intermolecular interactions possible. The host is generally the larger of the two molecules and provides a seat or cavity in which the smaller guest molecule may reside.

Crown ethers, e.g., dibenzo-24-crown-8 (**1**), have been known for some time to form complexes with cations, both metallic and organic. In regard to crown ether

1

complexes, the crown ether functions as a host. In this system interactions between the host and guest rely upon attractive forces between the electron-rich crown ether and an electron-poor ammonium salt, that is, H-bonding, charge transfer, and dipole–dipole interactions. Once complexation between the host and guest has occurred, a pseudorotaxane may be formed in a reversible manner (see Figure 7.1). If the complex is locked in place so that dissociation cannot take place, a rotaxane is formed. Rotaxanes are normally held in place with some type of blocking group which sterically hinders dissociation. Rotaxanes are, however, different from catenanes, which consist of two or more interlocked rings. The strength of association between a

Supramolecular Polymer Chemistry, First Edition. Edited by Akira Harada.
© 2012 Wiley-VCH Verlag GmbH & Co. KGaA. Published 2012 by Wiley-VCH Verlag GmbH & Co. KGaA.

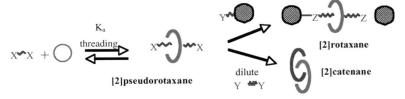

Figure 7.1 Threading equilibrium to form a pseudorotaxane, an intermediate for rotaxanes and catenanes.

host and guest molecule can be translated into an association constant which is obtained empirically by NMR, luminescence or UV-Vis spectroscopies, or isothermal titration calorimetry (ITC). Association constants, being dependent on the interactions between a host and guest pair, may be greatly affected by a range of stimuli such as: temperature, solvent, pH, common ions, and so on.

Complexation between a host and a guest can occur in many different fashions for polymeric systems and includes the following categories: supramolecular polymers, main chain polyrotaxanes and pseudorotaxanes, side chain polyrotaxanes and polypseudorotaxanes, and polymers that end in either a host or guest functionality (see Figure 7.2). Supramolecular polymers are polymeric structures produced from small molecules or oligomers, brought about and held together by noncovalent forces. Subgrouping of supramolecular polymers gives the following categories: poly[2]rotaxanes (as well as daisy chains which are a subset of poly[2]rotaxanes), poly[3]rotaxanes, and di-functional pseudorotaxanes (being either an AA-BB or AB host–guest system). Main chain polyrotaxanes and pseudorotaxanes are produced by incorporating either a host or guest moiety into the backbone of the polymer to give a complexed repeat unit in the main chain. Side chain polyrotaxanes and polypseudorotaxanes contain either the host or guest moiety as a pendent group which is used for complexation. Polymers that end with either a host or guest functionality may be used for chain extension, formation of block copolymers, or to produce star polymers. As for dendritic rotaxanes and dendronized polymers, they are generally formed from a host or guest core that is allowed to associate with the corresponding dendritic host or guest. To avoid confusion, throughout this chapter hosts will appear as red and guests will appear as blue.

The different types of polymer complexations are shown as cartoons in Figure 7.2; blue boxes represent guests and red ellipses represent hosts, while the black circles correspond to blocking groups. In each cartoon the equivalent pseudorotaxane may be depicted by the absence of the blocking group. Main chain polyrotaxanes and pseudorotaxanes normally contain either a host or guest incorporated into the repeat unit of the polymer; this gives complexation directly on the backbone of the polymer. However, side chain polyrotaxanes and pseudorotaxanes contain their complexes in the pendent groups. Poly[2]rotaxanes and poly[3]rotaxanes (as well as the analogous pseudorotaxanes) are built from smaller pieces that self assemble to form the backbone of a polymeric structure. In addition to these complexations a polymer

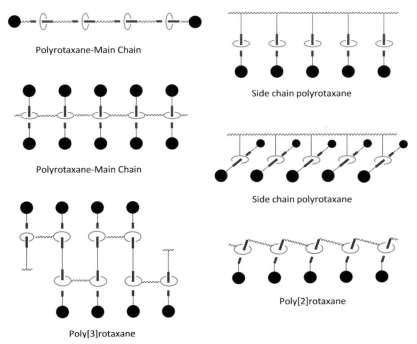

Figure 7.2 Classes of crown ether-based polyrotaxanes.

can also end in either a host or guest moiety to bring about a single association to yield some form of chain extension.

This review is meant to give an account of crown ether-containing polymers (both traditional and supramolecular) which undergo a form of complexation to yield either a rotaxane, pseudorotaxane, or catenane, and encompasses only those papers which have been published since our previous review on the subject [1] and those of other workers [2–5] (i.e., 2004–2010).

7.2
Daisy Chains

In addition to the well-known uses for host–guest chemistry regarding polymers such as chain extensions, crosslinking, and production of star polymers, the area of supramolecular polymers shows great promise. If a small molecule is made to be heterobifunctional, or heteroditopic in the supramolecular jargon, with a host–guest motif, the molecules can organize themselves into linear head-to-tail arrays. The name 'daisy chains' has been given to those systems formed via rotaxane or pseudorotaxane association. Two possibilities exist for producing a daisy chain. The first possibility requires two small molecules, one of which is a homoditopic host

(H-H) and the other a homoditopic guest (G-G); upon complexation, in principle, a daisy chain forms. The second possibility requires a molecule which is heteroditopic, having one end as a host (H) and the other as a guest (G), enabling formation of poly(H-G). In both cases the length of the array is dependent upon the association constant of the host–guest system and the concentration; to form long oligomers or polymers, it is crucial that the host–guest pairs have extremely high association constants [2]. In addition to their dependence upon association constants, daisy chain formation can be hindered by the tendency to form cyclic dimers as side products [3, 6–9].

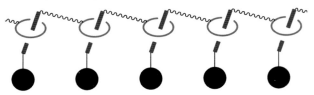

Poly[2]rotaxane

The heteroditopic molecule **2** has been heavily studied for its use in producing daisy chains. Stoddart and coworkers reported that the trifluoroacetate salt of **2** led to the formation of the cyclic dimer, via self-complexation [7, 8]. An X-ray crystal structure of the cyclic dimer was provided, and MS results revealed only the dimer [7, 8]. However, Gibson et al. reported **2** (as the PF_6 salt) and showed that oligomers could be formed from this compound [9]. MS results indicated complexation up to the hexamer in the positive ion MALDI-TOF MS spectra and up to the pentamer in FAB MS. With ample proof that **2** would auto-assemble into a daisy chain to some extent, they proceeded to produce **3** in an attempt to increase the association constant and solubility. This was done because **3** contains an alcohol moiety that will allow for hydrogen bonding, as reports by Busch et al. have demonstrated [10]. Compound **3**, however, led to the cyclic dimer as the major product, but MS results again provided proof for self-assembly up to the pentamer [9].

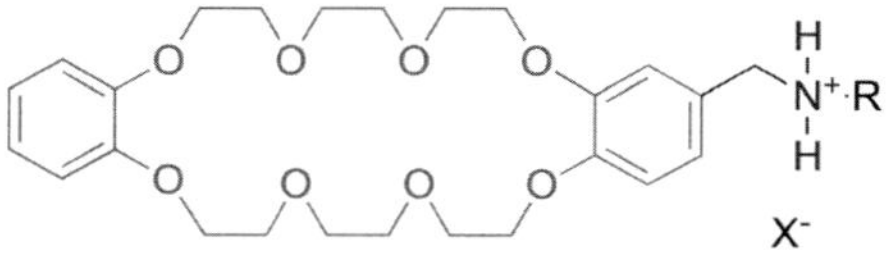

2 $R = CH_2C_6H_5$

3 $R = (CH_2)_6OH$

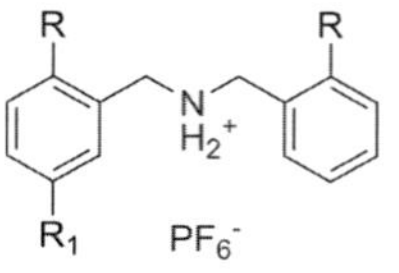

4 $R = CH_3, R_1 = H$

5 $R = R_1 = OCH_3$

In addition to the formation of daisy chains, dibenzo-24-crown-8 (DB24C8) has potential for forming rotaxanes. Upon observing the cavity size of DB24C8 and the bulkiness of the phenyl groups, Gibson *et al.* proceeded to conduct a model study geared toward designing a system in which thermal energy could produce a rotaxane via a process known as slippage [9]. Keeping the crown motif of DB24C8 they tested complexation and slippage of the crown with guests **4** and then **5**. NMR results showed that a 1:1 solution of DB24C8 and **4** gave the pseudorotaxane structure and a 1:1 solution of DB24C8 and **5** gave no complexation at room temperature. Lack of complexation was attributed to the small cavity size of DB24C8 relative to **5**. They proceeded to show that if ample thermal energy was applied to the system a rotaxane could be formed by slippage. Results showed that a solution of DB24C8 with **5** after 11 days at 53 °C reached an equilibrium that was 8% complexed as a rotaxane; a melt of DB24C8 with **5** at 150 °C for 30 min produced the rotaxane at a lower degree of complexation. The 'rotaxane' from DB24C8 and **5** was found to have a half-life of 66 ± 10 h at room temperature in a 1:1 mixture of chloroform and DMSO, which leads to the conclusion that methoxy substituents on the guest have the potential for synthesizing rotaxanes via the slippage method. However, much more efficient, that is, higher association constant, systems are required to produce true polymers (defined as containing 100 or more repeat units) by this approach.

Currently daisy chains seem to be making a niche for themselves in the area of molecular muscles; one such system showing potential in this area has been developed by Stoddart and coworkers [5, 7, 8, 11]. The daisy chain they developed (see Scheme 7.1) makes use of the DB24C8 moiety with two different guests (paraquat PF_6 and an ammonium PF_6 salt) placed in a linear sequence and held in place with a functional blocking group. A click reaction was then used to link the rotaxanes together to form a doubly threaded daisy chain containing two different guests for each host. When both salts are present, the DB24C8 has a preference to complex with the ammonium PF_6 salt; however, if the ammonium PF_6 salt is deprotonated, the DB24C8 moiety will complex with the adjacent paraquat unit. The act of the DB24C8 moiety changing its locus from the ammonium salt to the paraquat salt results in an extension of the daisy chain. The process is reversible through the addition of an acid and then a base (trifluoroacetic acid and 1,4-diazabicyclo[2.2.2]octane), allowing the daisy chain to undergo reversible contractions and extensions, forming a type of molecular muscle. Structure **6** depicts the system developed by Stoddart and coworkers [11]. The outcome is a daisy chain that responds to external stimuli, in this case a change in pH. UV-Vis spectrophotometric titrations were used for determining the contraction and extension mechanisms; interestingly it was observed that the contraction process (due to protonation) was slower than extension (due to deprotonation). The difference in speeds for contraction and extension of the daisy chain was attributed to the latter process proceeding through intermediates that are less stable than those of the contraction process.

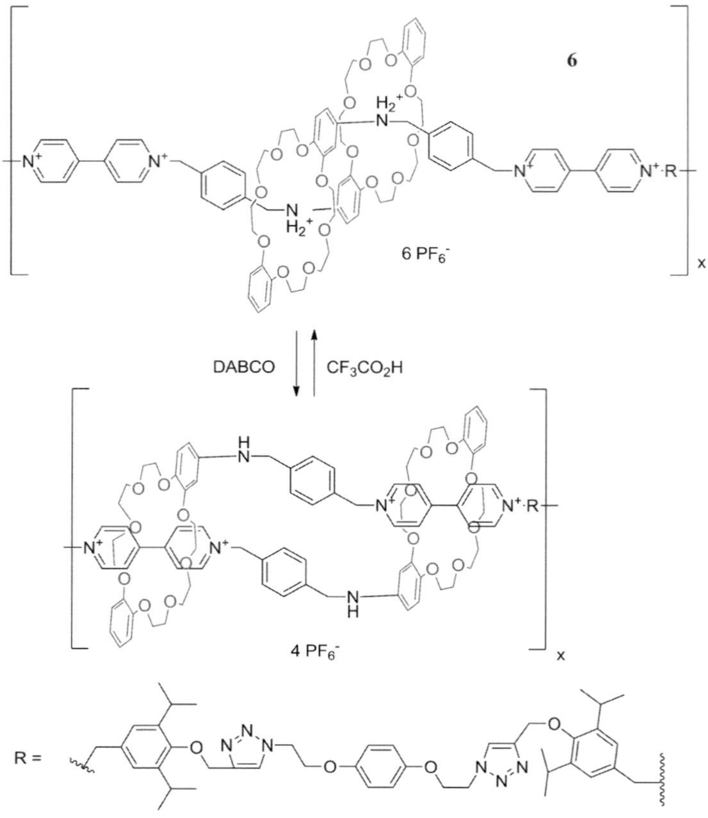

Scheme 7.1 Daisy chain capable of reversible contractions and extensions, developed by Stoddart and coworkers [11].

7.3
Supramolecular Polymers

Traditionally, polymers are formed by either step-growth or chain-growth reactions, in which covalent bonds are made to form the desired polymer. Through the use of host–guest chemistry another type of polymerization is possible in which small molecules associate in a step-growth manner to form a supramolecular polymer; this is possible whenever a small molecule is given a host–guest functionality of 2 or more. Huang and coworkers extended the concept of supramolecular chemistry to provide a supramolecular polymer in which the backbone was constructed from pseudorotaxane units and also contained an independent pseudorotaxane moiety [12]. This was made possible by the bis(*p*-phenylene)-34-crown-10 (BPP34C10) and DB24C8 moieties, which preferentially associate with paraquat and dibenzylammonium cations, respectively. To bring about the formation of this polymeric structure,

7.4 Dendritic Rotaxanes

two BPP34C10 molecules were attached to a DB24C8 center to form molecule **7**. A difunctional paraquat PF$_6$ (**8**) along with dibenzylammonium PF$_6$ (**9**) were then introduced to the host to form the desired supramolecular structure. The dibenzylammonium PF$_6$ selectively associated with the DB24C8 to form the central pseudorotaxane, and at high concentrations the difunctional paraquat PF$_6$ associated with the BPP34C10 to form a linear polymer, while at low concentrations it formed cyclic species. It is also notable that it was possible for Huang and coworkers to draw a fiber from a concentrated equimolar solution of the host with the guests **8** and **9** (conducted in a 1 : 1 solution of chloroform and acetonitrile).

7.4
Dendritic Rotaxanes

A one-pot synthesis reaction for the production of mechanically interlocked dendrimers has been reported by Stoddart and coworkers for dendrimers up to the third generation [13, 14]. This one-pot synthesis capitalizes on a two-part association, an ether-amine and dialdehyde with a dialkylammonium PF$_6$ salt, that react to form a rotaxane. Initially four dendrimers were reported by Stoddart and coworkers, each using the same 3-armed dialkylammonium PF$_6$ core with a rotaxane varying in the dendritic generation from zero to three based on the host component; with three rotaxanes complexed to the core, dendrimers **10a–d** resulted. It is noteworthy that these single-pot reactions are fast as well as efficient; once the reagents are mixed, equilibrium with **10c** is reached after 15 min in the nitromethane and 20 min in acetonitrile. High yields were possible for this system because the reversible, thermodynamically driven reactions bring about the formation of the imine-containing rotaxane. The imine synthesis being reversible allows for the formation of products which represent the highest stability; for this system this was the formation of products **10a–d**. Molecular weights for the dendrimers were 3193 for **10a**, 3997 for **10b**, 5606 for **10c**, and 8824 for **10d**, with each reaction giving a quantitative yield. Additionally, it was observed by ESI-MS that upon mixing different generations of **10**, the dendrimer arms rearrange themselves onto the cores to give a new combination of dendrimers. However, due to the instability of the imine linkages in **10a–d**, the dendrimers required further modification to be of practical use; for this reason the imines were reduced to form the more robust set of amine dendrimers **11a–c**. Yields for the reduction of **10a–c** to **11a–c** were reported as 90–91%. However, reduction of all imine groups contained within **10d** was not achievable. The lack of reactivity for **10d** was attributed to the high degree of steric hindrance produced by the third-generation dendritic groups.

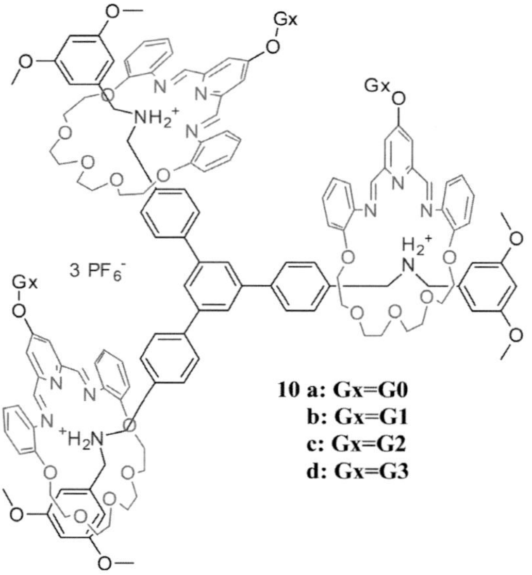

10 a: Gx=G0
b: Gx=G1
c: Gx=G2
d: Gx=G3

7.5
Dendronized Polymers

Dendronized polymers are polymers that contain one or more dendritic groups located at either the end of the polymer or along the main chain as pendent groups. In comparison to a straight-chain counterpart, the properties of the dendronized polymer are very different. This allows for the polymers to be used for different applications and appeal to new research areas. Host–guest chemistry has the potential to further the field by noncovalent formation of dendronized polymers from two separate structures, a normal polymer and a dendritic molecule, each fitted with either a host or guest functionality.

Several reversible dendronized polymers have been produced by Stoddart and coworkers, all of which take advantage of the DB24C8 and secondary ammonium

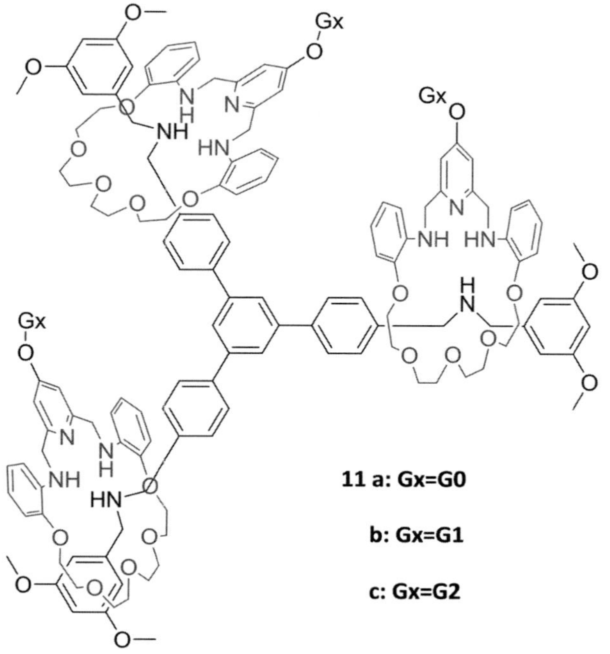

11 a: Gx=G0
b: Gx=G1
c: Gx=G2

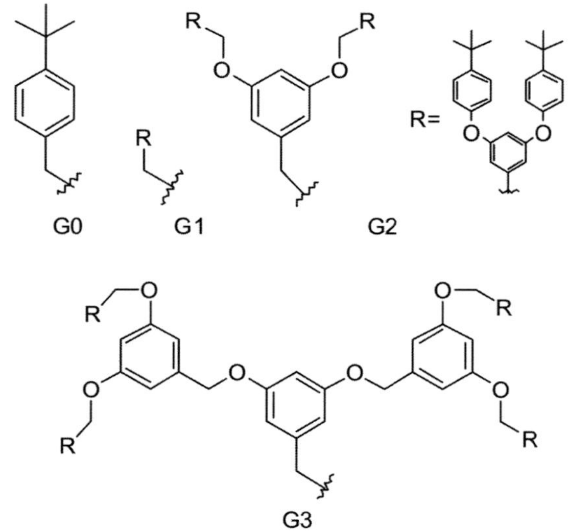

host–guest system [15]. Linear polystyrene and polyacetylene were functionalized with DB24C8 as host pendent groups, molecules **13** and **14**. Guests included dialkylammonium PF$_6$ salts with a blocking group attached, these being made in three different generations, molecules **12a–c**. Upon introduction of a guest to one of the host polymers, association between the host and guest occurred to produce a dendronized polymer. The resulting polymers were found to be rigid and rod-like as well as stable over time in the solvent dichloromethane. It was further found that the formation of the dendronized polymer from the linear polymer could be made to be reversible by either deprotonation or protonation of the dialkylammonium through the addition of either an base or acid. Although the system is reversible, there is a finite number of times that the system may undergo a switch from the dendronized polymer to the linear polymer. This is due to the buildup of salts that hinder recomplexation. After five acid-base cycles, hindrance by the salt for interaction of **14** with **12a** was reported to give a loss in complexation to 89%. Likewise **13** with **12a** gave a drop in the formation of the dendronized polymer to 77%, which suggests that **14** with **12a** is more stable towards the acid-base cycle used.

7.6
Main chain Rotaxanes Based on Polymeric Crowns (Including Crosslinked Systems)

Main chain polyrotaxanes are found as two types: either the polymer backbone can be composed of host moieties or it will contain the guest functionalities (see Figure 7.2). For this class, however, either the host or the guest must be contained within the backbone. When the host is contained within the backbone two methods may be used for preparing the polyrotaxane. First, the host-containing monomer may be allowed to complex with a guest, forming a pseudorotaxane, which is blocked to form a rotaxane monomer that is then used for polymerization. Or alternatively, the host-containing monomer can be polymerized and then guests are introduced to form polypseudorotaxanes that are subsequently blocked to form polyrotaxanes. As for polyrotaxanes containing the guest functionality in the repeat unit, they too are produced in one of two ways. They can be made by allowing the guest-containing monomer to complex with a host species, forming a pseudorotaxane monomer, followed by polymerization to form a polypseudorotaxane that is then blocked to form the polyrotaxane. Or they can be synthesized by initially forming a polymer that has the guest moiety in the repeat unit and allowing the host molecules to thread themselves onto the polymer, a blocking group being added afterwards to form the desired polyrotaxane.

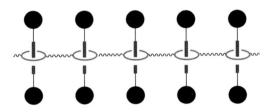

Main Chain Polyrotaxane

Recently, a new class of polyaramides has been synthesized by Gibson et al. that contain the crown ether motif in the repeat unit [16]. The class is synthesized using a step growth reaction with a given bis(5-carboxy-1,3-phenylene)crown ether (BCP) and a diamine. This family of polyamides was studied because of the ability of the crown ethers to associate with the amine and amide units by hydrogen bonding. Results were reported using crown ether macrocycles of various sizes: BCP32C10, BCP26C8,

BCP20C6, and BCP14C4 with the diamines 4,4′-oxydianiline (ODA) and [4-(m-aminophenoxy)phenyl]phenylphosphine oxide (m-BAPPO). The resulting polymers are shown as **15 (a–d)** and **16 (a–b)**. This class of polyaramides proved to be very interesting. The polycondensation reaction of BCP32C10 with ODA resulted in a polymer that was insoluble in all attempted solvents. The reaction, being the product of two difunctional monomers (in a covalent sense), should have yielded a linear polymer; however, the resulting polymer was highly crosslinked. By alteration of the crown cavity size and diamine used for the polycondensation (synthesis of polymers **15b–d** and **16a–b**) it was determined that crosslinking occurred in **15a** due to hydrogen bonding between the crown and amine or amide moieties, which led to threading as the reaction proceeded. Gibson et al. went on to show that the extent of threading of the crown moieties by amines or amides was highly dependent upon the cavity size of the crown. BCP14C4 with ODA resulted in the formation of a polymer with a unimodal molecular weight distribution (linear polymer); BCP26C8 and BCP20C6 with ODA and m-BAPPO resulted in bimodal molecular weight distributions. As a result of the tendency of polyamides themselves to form cyclics, the polymers produced in this study contained polypseudorotaxane, polyrotaxane, and polycatenene units, depending upon the size of the crown used. This, coupled with previous work by the authors [17], provides the understanding needed to control the amount of branching based on crown size, feed ratio, and solvent. This work also suggests that with proper design this system could be directed towards the synthesis of polycatenanes.

15 a: n=3 c: n=1
b: n=2 d: n=0

16 a: n=3
b: n=2

Seeking to assess the possibilities that could come about as a result of its novel structure, Takata and coworkers developed a main chain polyrotaxane with a backbone consisting of a DB24C8 threaded by a dialkylammonium PF_6 salt [18]. Synthesis of the polyrotaxane was carried out by allowing the ammonium salt to complex the polymer containing DB24C8 moieties, followed by the addition of a blocking group to convert the polypseudorotaxane to a polyrotaxane. The polymer

7.6 Main chain Rotaxanes Based on Polymeric Crowns (Including Crosslinked Systems)

used for this study was the low molecular weight oligomer **17**, having $M_n = 4.0$ kDa and PDI $= 1.68$ and consisting of two isomers. Although the crown oligomer was of a low molecular weight, results seemed to indicate that the molecular weight of the oligomer/polymer was insignificant in regard to complexation; this conclusion is based on the observtion that increasing the molecular weight of **17** to $M_n = 10.0$ kDa resulted in only a very small alteration to the percent pseudo-rotaxane formed. An attempt to characterize the molecular weight of **18** was unsuccessful due to its lack of solubility in common GPC solvents as a result of its ionic character. Acetylation of **18** ($M_n = 2.4$ kDa) resulted in **19**, which had $M_n = 4.4$ kDa.

Once they were satisfied that a rotaxane had been formed, the authors set out to use the system to generate a network in which crosslinking was due to rotaxanation. Mechanically crosslinked network **20** was successfully prepared by reacting **17** at elevated concentrations (i.e., 0.65M produced gelation, while 0.20 M did not) with MDI at 25 °C using chloroform as the solvent and di(n-butyl)tin dilaurate as the catalyst. The resulting network **20** was a clear gel having a low degree of crosslinking; rotaxane formation was calculated to be 30% by ^{1}H NMR. Results from **18** and **19** allowed for the conclusion that the system is efficient for producing main chain polyrotaxanes, and **20** showed that the system could be used for producing networks with a low degree of crosslinking.

Rotaxane formation offers a unique method for cross-linking polymers, providing the ability to cross-link one polymer chain with another without forming a covalent bond between the two chains. Cross-linking a polymer through the formation of a rotaxane can be accomplished when macrocycles are present. Cross-linked rotaxanes can by formed through a statistical process in which a polymerizing chain inserts itself through a large macrocycle or through a directed process accomplished by host–guest chemistry.

A unique protocol has been developed by Takata and coworkers for cross-linking main chain crown ether polymers via rotaxane formation; this has been accomplished through the use of both host–guest chemistry and a reversible disulfide reaction [19, 20]. The aim is the production of cross-linked polymers which can eventually undergo a reverse cross-linking reaction for the purpose of recycling the original polymers. The reported cross-linking utilizes a DB24C8-containing main chain polymer **21** and a bis(dialkylammonium) PF$_6$ salt (**22**, linked by a disulfide bond). Scheme 7.2 depicts the cross-linking reaction that occurs when **22**

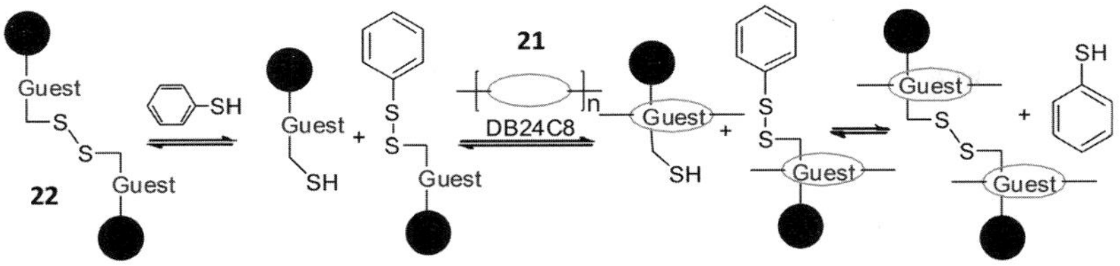

7.6 Main chain Rotaxanes Based on Polymeric Crowns (Including Crosslinked Systems)

Scheme 7.2 Protocol for cross-linking main chain crown ether polymers, developed by Takata and coworkers [19, 20].

is introduced to the crown ether polymer along with benzenethiol. Benzenethiol reversibly reacts with the dithiol bond in **22** causing the molecule to split into two halves, both of which are available to complex with a DB24C8 ring of polymer **21** to form a pseudorotaxane. When the recombination of the two halves occurs, if both have complexed with DB24C8, the two pseudorotaxanes become rotaxanes, forming a mechanical cross-link. The cross-linking protocol was assessed and shown to be successful for DB24C8 polymers **23** (M_n 2.4 kDa, PDI 3.7), **24** (M_n 5.5 kDa, PDI 1.5), **25** (M_n 2.9 kDa, PDI 1.7) and an array of block copolymers with general structure **26** varying in the feed ratio of x to z (x/z: 1/0.5, M_n 14.6 kDa, PDI 4.9; 1/1, M_n 25.7 kDa, PDI 4.1, 1/1.8, M_n 18.0 kDa, PDI 3.6, 1/2, M_n 45.7 kDa, PDI 3.2 and 1/3, M_n 49.5 kDa, PDI 3.0). All the reported polymers underwent cross-linking by this new protocol and produced gels. It was further shown that an alteration of the amount of **22** for cross-linking resulted in a change of properties of the gel, such as swelling, viscoelasticity, and so on. Additionally, success resulting from the block copolymers indicates that this protocol may be applied to a range of polymers, provided that a DB24C8 host functionality is present.

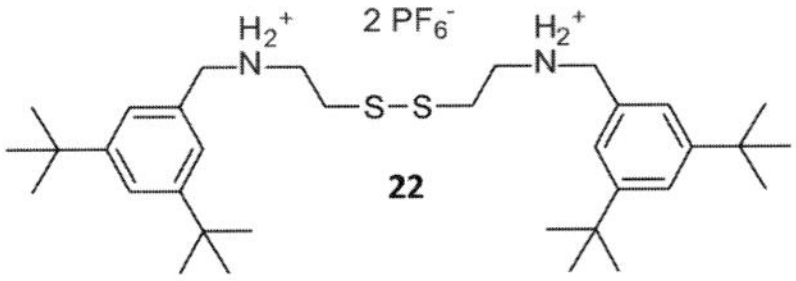

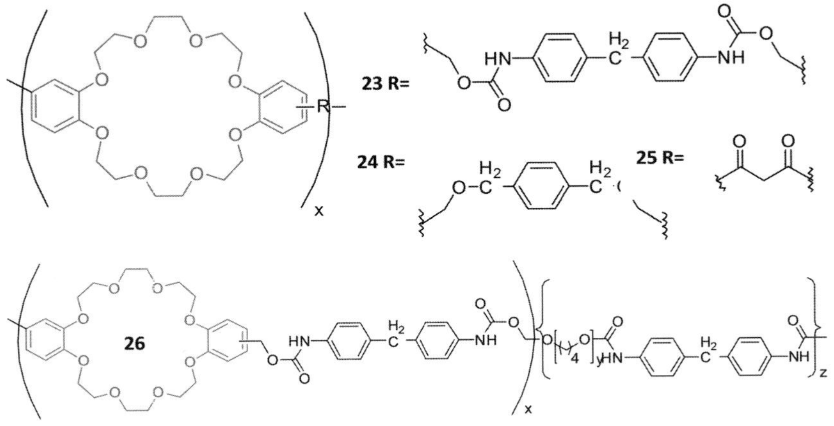

7.7
Side Chain Rotaxanes Based on Pendent Crowns

Formation of a rotaxane through the attachment of pendent groups to a polymer offers some benefits over main chain polyrotaxanes. A host or guest moiety can be deliberately designed with a functional group for incorporation into a polymer. Postpolymerization attachment also adds the ability for only a certain fraction of the polymer to be functionalized by a host or guest moiety. Alternatively, the other approach to synthesizing a side chain polyrotaxane is to form a pseudorotaxane or rotaxane monomer which will be used for polymerization to form the desired side chain polyrotaxane.

Side Chain Rotaxane

In an attempt to form a more versatile type of block copolymers, Weck, Grubbs, Stoddart, and coworkers have reported the next generation of what they call 'universal polymer backbones' (UPBs) [21]. The central idea behind the generation of UPBs is to develop a generic polymer that can be easily functionalized by noncovalent bonds to form many different types of block copolymers. To functionalize the UPBs, complexation chemistry was chosen due to the speeds at which complexation can occur and because complexed species can be made to decomplex, giving the polymer reversibility. The UPBs made by the authors took advantage of the DB24C8 and dialkylammonium complexation motif for the formation of a pseudorotaxane and sulfur–carbon–sulfur (SCS) pincer complexes for the formation of metal complexes with pyridine. Initial work began with the synthesis of homopolymers **27**, **28**, and **29** to provide a foundation for the production of a block copolymer with multiple complexing moieties. The goal was a block copolymer in which a single section of the polymer could be selectively complexed, resulting in selective functionalization at a desired location without the formation of a covalent bond.

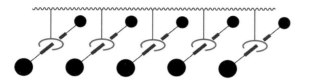

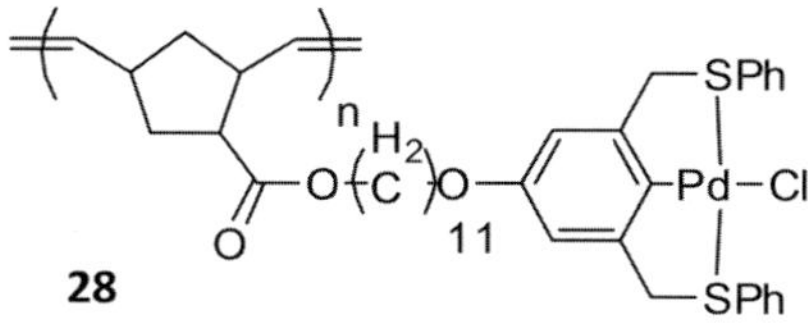

7.7 Side Chain Rotaxanes Based on Pendent Crowns

28

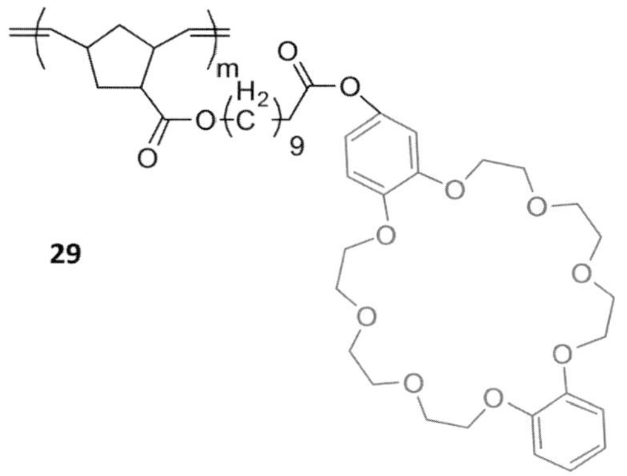

29

From the corresponding norbornene monomers two UPB block copolymers were synthesized using ROMP: first a polymer containing the SCS pincer with the dialkylammonium salt moiety (**30**) and secondly the SCS pincer with the DB24C8 moiety (**31**). Additionally homopolymers of the norbornene SCS pincer **28**, norbornene dialkylammonium **27**, and norbornene DB24C8 **29** monomers were synthesized as well. Properties of the polymers were reported as follows. Molecular weight results for polymers containing the dialkylammonium salt were reported using the protected amine *tert*-butyl carbamate analog: **30** [monomer/initiator 50/1, $M_n = 66.0$ kDa, PDI 1.14], **31** [50/1, $M_n = 29.5$ kDa, PDI 1.21]; **27** as well as **29** were each synthesized at monomer/initiator ratios of 10/1, 20/1, 50/1, 80/1 and 100/1.

Interestingly, association constants for the UPBs were obtained through ITC using chloroform as the solvent; due to chloroform's low polarity, very high association constants were reported for these systems. Association constants for the homopolymers were reported as follows: **29** with a dialkylammonium tetrakis[3,5-bis(trifluoromethyl)phenyl]borate (BAr_F^-) salt gave an impressive association constant of 2.86×10^6, and **27** with DB24C8 gave an association constant of 10×10^4. Additionally, block copolymers were examined: **30** with DB24C8: $K_a = 9 \times 10^4$ (5×10^4 when the SCS pincer was complexed) and **31** with a dialkylammonium BAr_F^- salt gave $K_a = 43 \times 10^4$ (and 54×10^4 when the SCS pincer was complexed). These studies revealed that the complexation with **30** and **31** was selective and that the complexation of one group did not affect the other appreciably. This indicated that polymers **30** and **31** could be selectively complexed with the complementary moieties to form polymers containing a single or multiple type of complex(es), making this generation of UPBs especially versatile.

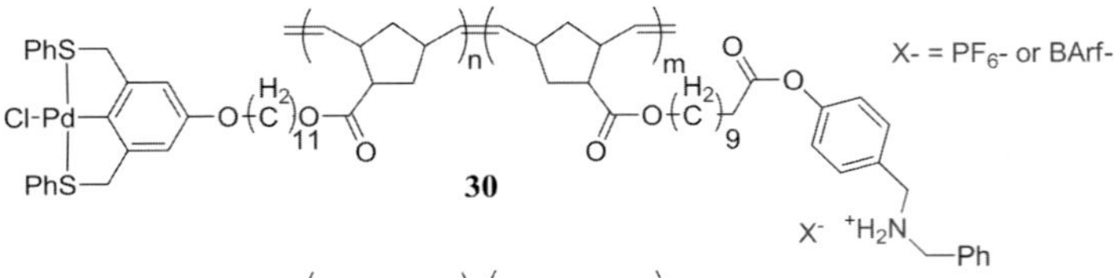

168 | 7 Crown Ether-Based Polymeric Rotaxanes

30

X⁻ = PF$_6^-$ or BArf⁻

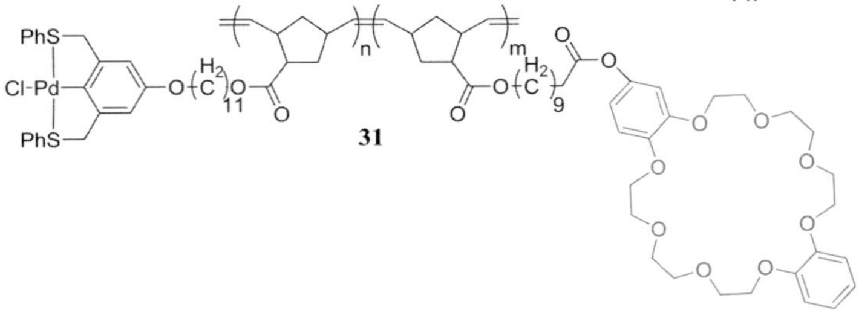

31

Taking the concept of multiple component complexes a step further Stoddart, Weck, and coworkers added a third type of complex to their block copolymer system [22]. In addition to using the DB24C8 host with a dialkylammonium guest to form a pseudorotaxane and sulfur–carbon–sulfur pincer to complex pyridine, the block copolymer incorporated a hydrogen bonding moiety via a diaminopyridine which complexes N-butylthymine.

ROMP was once again employed for polymerization; however, this time polymerization was carried out on all three monomers at once to form a random copolymer. To accurately assess complexation among the three groups, several copolymers of differing feed ratios were synthesized. Compounds and results for polymer **32** of various compositions are listed in Table 7.1.

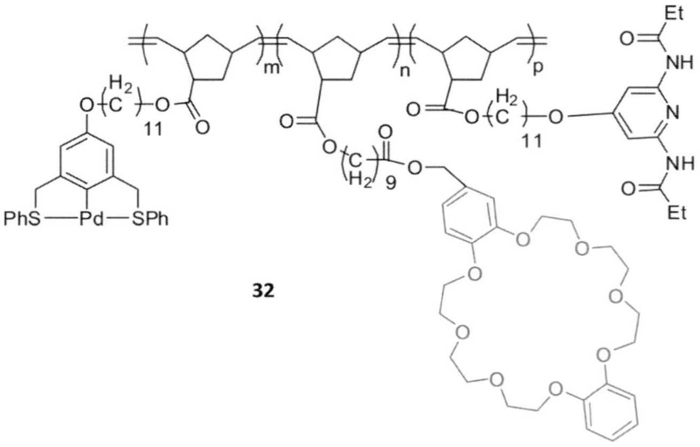

32

Table 7.1 Results for polymer **32**.

	Percent monomer			GPC Results			Association Constants	
Str	Pincer	Crown	DAP	M_n (kDa)	M_w (kDa)	PDI	$10^{-3} K_a$ (M^{-1}) crown	$10^{-3} K_a$ (M^{-1}) DAP
32a	33	33	33	42.5	56.0	1.32	231	1.60
32b	25	25	50	32.4	44.7	1.38	423	1.52
32c	25	50	25	30.6	43.6	1.42	345	1.76
32d	50	25	25	31.4	44.4	1.41	174	1.65
32e	45	45	10	21.2	30.0	1.42	459	1.44
32f	45	10	45	30.7	44.6	1.45	399	1.26
32g	10	45	45	30.0	41.5	1.38	325	1.40
				32a pincer complexed			310	
				32a with crown and pincer complexed				1.57

Most interesting of the results for **32** are the association constants; using the BAr$_F$ counterion has aided in providing high association constants for the complexation of the DB24C8 and dialkylammonium moieties through the use of low-polarity solvents such as chloroform and dichloromethane. The use of the large soft BAr$_F$ ion has simplified the system, since the dialkylammonium salt is weakly associated, and in dilute solutions ion pair formation is negligible. ITC and ^{1}H NMR data also proved that complexations between the three motifs were independent of one another.

As with the main chain polyrotaxanes, problems are often encountered during the synthesis of side chain polyrotaxanes that result in obtaining lower-than-desired molecular weights. The problem of obtaining high molecular weight polymers for side chain polyrotaxanes is often attributed to the bulkiness of the pendent groups, which can sterically hinder the reaction. Takata and coworkers, desiring a high molecular weight side chain poly[2]rotaxane, assessed the polymerization of acetylenes which have been noted for their ability to yield high-molecular-weight polymers from bulky monomers [23].

Scheme 7.3 depicts the polymerization of monomer **33** containing DB24C8 to yield **35**, whose molecular weight could not be determined by SEC due to its ionic structure and lack of solubility. Due to problems in characterizing **35**, the polymer was converted to **36a** so the molecular weight could be evaluated. SEC results for **36a** (in chloroform, polystyrene standards) revealed $M_n = 465$ kDa, PDI $= 5.9$. The researchers then converted **33** to **34** to produce a monomer that could be polymerized directly to give **36b**. SEC was attempted for **36b**, but was found to be impossible due to the polymer exceeding the mass limit of the columns (2000 kDa). Polymers **36a** and **36b**, together, provide convincing evidence that acetylene polymerizations can be used to synthesize high-molecular-weight side chain poly[2]rotaxanes.

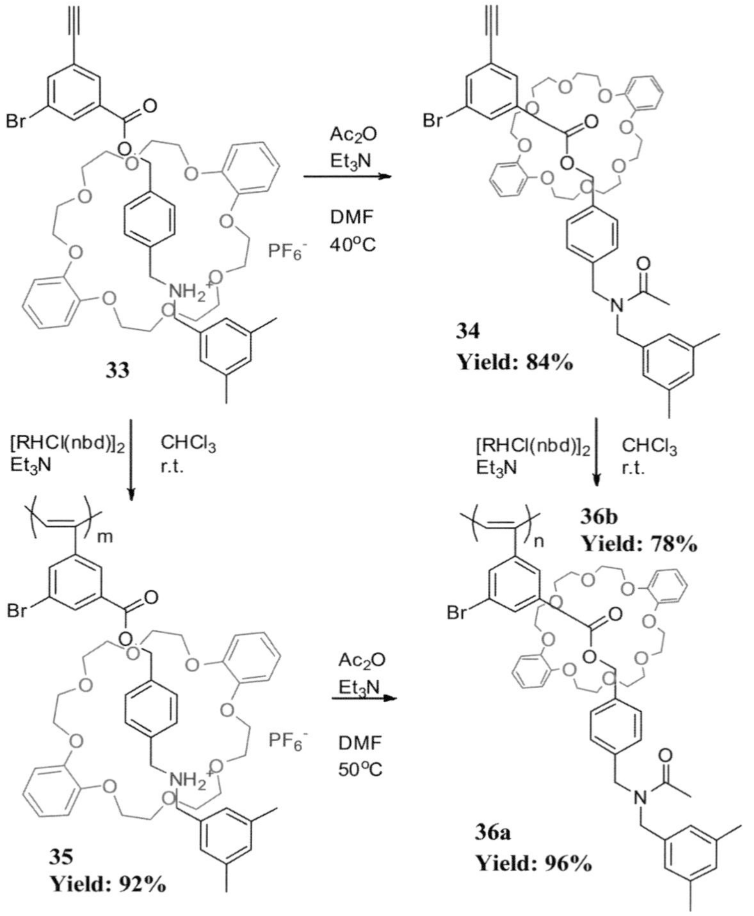

Scheme 7.3 Polymerizations utilizing the acetylene functional group to obtain high-molecular-weight polymers, developed by Takata and coworkers [23].

7.8
Poly[2]rotaxanes

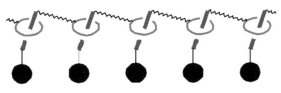

Poly[2]rotaxane

Poly[2]rotaxanes are polymers either formed from or containing a [2]rotaxane in their backbone repeat unit. It should also be noted that daisy chains are a type of main chain poly[2]rotaxane and for the purpose of clarity are grouped in a separate section.

A general technique that is often used for the synthesis of main chain poly[2]-rotaxanes involves allowing the host and guest molecules to associate to form a pseudorotaxane and then reacting the pseudorotaxane with a blocking group that both forms the rotaxane and polymer simultaneously. Seeking to improve upon the methods used in synthesizing poly[2]rotaxanes, Takata and coworkers have taken the approach of forming the rotaxane as a monomer, neutralizing it, and then polymerizing it [24]. The driving force for a change in the methodology is the difficulty associated with obtaining a high degree of polymerization. The system developed by Takata and coworkers uses the DB24C8 moiety (functionalized with an alkyne) to associate with a dialkylammonium PF_6 salt to form the desired rotaxane. The dialkylammonium rotaxane is then converted to an amine (containing an iodoaryl group), providing an AB monomer for Sonogashira head-to-tail coupling. The mechanical bonds of the new monomer, however, remain intact. Polymerization of the monomer resulted in a polymer with a $M_n = 13.1$ kDa by ^{1}H NMR and a $M_w/M_n = 1.59$ by GPC (polystyrene standards). The molecular weights obtained from GPC were considered to be inaccurate. Additionally, TGA gave a T_{d5} of 302 °C, showing the thermal stability of the poly[2]rotaxane.

A crown ether macrocycle found in a catenane developed by Mayer and coworkers is noteworthy due to its functionality as a monomer (Scheme 7.4). The system provides the first example of a catenane undergoing ROMP to form a polypseudorotaxane, which was accomplished by using the [2]catenane **37**, forming the polypseudorotaxane **38** [25]. Despite the large size of the ring system in **37**, successful polymerization was achieved by means of an entropy driven ring opening metathesis polymerization (ED-ROMP) using the second-generation ruthenium

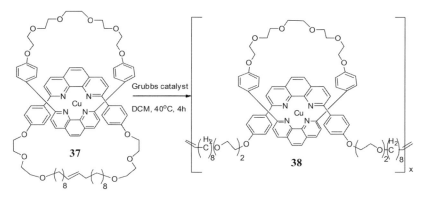

Scheme 7.4 First example of a catenane undergoing ROMP to form a polypseudorotaxane, developed by Mayer and coworkers [25].

Grubbs' catalyst. The resulting polymer **38** was reported to have $M_w = 93.0$ kDa, PDI = 1.9 by GPC.

In an attempt to find easier and alternative methods for the synthesis of polyrotaxanes, Takata and coworkers have taken advantage of [2 + 3] nitrile N-oxide/acetylene cycloadditions to produce poly[2]rotaxanes and poly[3]rotaxanes (the latter being discussed in the next section) [26]. While assessing possible cycloaddition reactions, the nitrile N-oxide/acetylene reaction was chosen over the more popular Huisgen type click chemistry due to the explosive nature of azides and the difficulty of removal of the copper catalyst required in the latter method.

Poly[2]rotaxanes were synthesized from the DB24C8/dialkylammonium PF_6 salt motif using the methodology of forming a rotaxane monomer, followed by polymerization of that monomer to form the desired poly[2]rotaxane. A rotaxane was synthesized from a difunctional acetylenic DB24C8 and a dialkylammonium PF_6 salt (which was subsequently acetylated) to form monomer **41**. The two polymers **42** and **43** were then produced from the [2 + 3] nitrile N-oxide/acetylene cycloaddition of **41** with **39**, and **41** with **40**, respectively. These syntheses were carried out without the aid of a catalyst, **42** in DMF at 80 °C with molecular sieves (4 Å) and **43** in chloroform at reflux. Molecular weights were obtained via GPC using chloroform with polystyrene standards. **42** synthesized at a monomer concentration of 0.2 M (79% yield) had $M_n = 4.3$ kDa, PDI = 2.8, while a monomer concentration of 0.5 M gave an increased yield of 96% and $M_n = 6.3$ kDa, PDI = 2.6. Additionally, **43** was synthesized at a monomer concentration of 1.0 M, providing a yield of 87%, $M_n = 9.3$ kDa, PDI = 2.1. **42** and **43** together provided justification that [2 + 3] nitrile N-oxide/acetylene cycloaddition can be used to synthesize poly[2]rotaxanes from stable and unstable nitrile N-oxides.

7.9
Poly[3]rotaxanes

The structure of a main chain poly[3]rotaxane is composed of a repetition of interlocked rotaxanes formed from combining a ditopic host with a ditopic guest to self assemble into a polymeric structure. Self assembly occurs in the same fashion as a step growth reaction, and, depending upon how the poly[3]rotaxane is synthesized, either the rotaxane can be formed directly or from the equivalent pseudorotaxane. Although the poly[3]rotaxane cartoon shows the association between a host and guest, it is not crucial that association be kept intact, only the locked mechanical interaction for the nomenclature to remain unchanged.

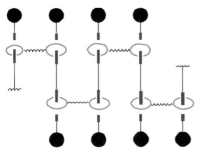

Poly[3]rotaxane

Two possibilities exist for synthesizing poly[3]rotaxanes: the first (rotaxanation then polymerization) uses a ditopic guest and two host molecules which are allowed to associate to form a [3]pseudorotaxane. Then a reaction is carried out to form the rotaxane; once the rotaxanes have been formed, polymerization is carried out using functional groups on the hosts. The second method for forming poly[3]rotaxanes (coupling then rotaxanation) uses a ditopic host and a ditopic guest, which are allowed to associate and form a pseudorotaxane. The pseudorotaxane is then reacted further to give the rotaxane; once the conversion of all pseudorotaxanes is complete, the poly [3]rotaxane is formed. As suspected, the degree of polymerization for the first method is dependent only on the efficiency of the polymerization reaction, while the degree of polymerization for the second is dependent on the association constant of the host–guest complexation and the efficiency of the stoppering process. Similarly, this leads to the conclusion that the obtainable molecular weight of the first method is dependent upon the functional groups and reaction conditions, while in the second method it is critically dependent upon the association constant of the system at hand.

Poly[3]rotaxanes reported by Takata and coworkers were synthesized using the first method described above, the host being a DB24C8 motif and the ditopic guest being a dialkylammonium PF_6 salt [27]. After blocking and acetylation of the salt, monomer 44 was formed and polymerization was carried out using the Mizoroki-Heck coupling reaction with both 1,4-diiodobenzene and 4,4′-diiodo-1,1′-biphenyl to yield polymers 45 and 46. When polymerization was carried out at 90 °C, 45 yielded a poly [3]rotaxane with $M_n = 18.0$ kDa, PDI = 2.5; likewise, polymerization of 46 at 90 °C resulted in a poly[3]rotaxane with $M_n = 14.0$ kDa, PDI = 3.1. These results demon-

7.9 Poly[3]rotaxanes

strated that main chain poly[3]rotaxanes can be synthesized by rotaxanation followed by polymerization using Mizoroki–Heck coupling.

In regard to the previously discussed [2 + 3] nitrile N-oxide/acetylene cycloaddition by Takata and coworkers, at this point the reported poly[3]rotaxanes **49** and **50** will be discussed [26]. For the synthesis of poly[2]rotaxanes the DB24C8 center (seen in **41**) was made to be difunctional with an acetylenic group and complexed with a monofunctional dialkylammonium PF_6 salt; however, for the synthesis of a poly[3]rotaxane the DB24C8 center was made to be monofunctional with the acetylenic group again and complexed with a ditopic dialkylammonium PF_6 salt that was acetylated after complexation. This brought about the formation of **47**, a [3]rotaxane monomer which was polymerized with **39** and **40** by [2 + 3] nitrile N-oxide/acetylene cycloaddition to give the poly[3]rotaxanes **49** and **50**, respectively. Reaction conditions were similar to those used for the poly[2]rotaxanes: with no catalyst, **49** synthesized in DMF at 80 °C with molecular sieves (4 Å) and **50** in chloroform at reflux. This provided the following results: **49** synthesized at a 0.5 M monomer concentration gave $M_n = 4.6$ kDa, PDI $= 1.4$ (84% yield), while **50** at a 1.0 M concentration gave $M_n = 6.0$ kDa, PDI $= 1.8$ (98% yield). Additionally, a block copolymer was synthesized from **47** and **48** to yield the poly[3]rotaxane **51**, showing the utility of the [2 + 3] nitrile N-oxide/acetylenic cycloaddition reaction; **51** synthesized at a 0.5 M monomer concentration gave $M_n = 6.4$ kDa, PDI $= 1.2$ in a 98% yield.

7.10
Polymeric End Group Pseudorotaxanes

In addition to the single host–guest complexations used to form rotaxanes, pseudorotaxanes, and supramolecular structures, different motifs may be used to produce systems with unique properties. Gibson et al. have proposed the possibility of forming such a system to produce an ABC triblock polymer, formed from the association of bis (m-phenylene)-32-crown-10 (BMP32C10) and DB24C8 with paraquat and ammonium salts, respectively [28]. However, before this or any large complex networks are made using host–guest chemistry, one must correctly assess selectivity and complexation of the individual motifs. For this reason a study evaluated the selectivity of BMP32C10- and DB24C8-terminated polymers; BMP32C10-terminated polystyrene with di(p-t-butylbenzyl) paraquat PF_6 gave an average association constant of 751 M^{-1}, (in 1:1 chloroform:acetone) and although this value is less than that for BMP32C10 with dimethyl paraquat PF_6 (which is attributed to steric, i.e., entropic, effects of the polymer), it is still considered to be efficient. However, BMP32C10-terminated polystyrene with dibenzylammonium PF_6 gave no evidence of complexation. In much the reverse trend, DB24C8-terminated polystyrene cannot complex the bulky di(p-t-butyl) paraquat and displayed an association constant 61 M^{-1} with dibenzylammonium PF_6 (in 3:2 chloroform:acetonitrile). Together this indicates that a selective system could be made from BMP32C10, DB24C8, paraquat PF_6, and dibenzylammonium PF_6 in which the BMP32C10 would associate preferentially with paraquat PF_6, and DB24C8 with dibenzylammonium PF_6. Also, a system such as this could provide switchable heierarchical structures that would provide controllable (i.e., stimuli-responsive) physical properties.

7.11
Chain Extension and Block Copolymers from End Groups

Several parameters exist for increasing or decreasing association constant between the crown ethers and paraquats. The easiest way to increase the association constant is to change the solvent. Since part of the association between crown ethers and paraquats is due to hydrogen bonding, solvents such as protic and polar aprotic solvents (which can disrupt or compete with hydrogen bonding) will lower the overall association constant of the system. Likewise the use of a nonpolar solvent can promote the complexation of paraquat and crown ether and increase the overall association constant. The problem in adjusting the solvent parameter is that the

7.11 Chain Extension and Block Copolymers from End Groups

crown ether/paraquat system is a two-component system and each component must be soluble in the desired solvent; paraquat salts are less soluble as the solvents become more nonpolar. For this reason association constants for crown ether/paraquat systems are normally reported in either acetone or acetonitrile.

Recently, however, Gibson and coworkers have reported for the first time an association constant for a crown ether/paraquat system in chloroform [29]. (Additionally, a very similar feat was accomplished by Stoddart and Weck, in which an association constant was reported via ITC using a dialkylammonium BAr_F salt with a DB24C8 compound in chloroform [21].) The study was made possible by attaching the motifs onto either the end or the center of a polymer chain; polymers for the study included paraquat-terminated poly(methyl methacrylate) **52**, paraquat-terminated polystyrene **53**, crown ether-terminated polystyrene **54**, and a crown ether-centered polystyrene **55**. Most notable of the reported results are the association constants for the two paraquat-terminated polymers. Through the use of ITC an association constant of $(2.97 \pm 0.09) \times 10^3$ M^{-1} ($\Delta G = -4.73$ kcal mol^{-1}) was found for **53** with bis(m-phenylene)-32-crown10 and a value of $(3.63 \pm 0.10) \times 10^2$ M^{-1} ($\Delta G = -3.49$ kcal mol^{-1}) was found for **52** with bis(m-phenylene)-32-crown-10. These results are very interesting in comparison to results for bis(m-phenylene)-32-crown10 with paraquat in acetone: $K_a = 760$ M^{-1} ($\Delta G = -3.93$ kcal mol^{-1}) [30, 31]. It should also be noted that **52** and **53** during complexation in chloroform gave very small changes in ΔS as a result of their polymer structures. Paraquat polymer **53** is nearly four times better in terms of K (0.80 kcal mol^{-1} in terms of ΔG) than the small-molecule analog. As for the **52** having an association constant less than its small-molecule counterpart, this is attributed to the ester moieties of polymethacylate providing a chemical environment for paraquat that is very similar to acetone, thus reducing the K value. Justification for this claim is provided by the similar chemical shifts of the paraquat protons in the NMR spectrum of the paraquat polymethacrylate in chloroform versus those of paraquat monomer in acetone.

Recently, a chain extension was reported by Gibson et al. which displayed cooperativity [32]. The chain extension system employed a DB24C8-terminated polystyrene **56** with the guest **57** to bring about the formation of a pseudorotaxane. Additionally, an attempt was made to chain extend **58** with the same bis(dialkylammonium) PF_6 salt; however, the interaction between the BMP32C10 unit of **58** and the dialkylammonium salt **57** was found to be weak. ^{1}H NMR in 1 : 1 chloroform:acetonitrile with equal concentrations (5.00 mM) of **57** and **58** revealed only 51.3% complexation. **56** with **57** was determined by ITC to have an average association constant of $1.4 \times 10^2\,M^{-1}$.

The cooperative effect for **56** was demonstrated by ^{1}H NMR analysis with monotopic, ditopic, and tritopic alkylammonium PF_6 salts. It was observed that when equal concentrations of alkylammonium PF_6 functional units were introduced to **56**, average association constants and degrees of complexation increased from monotopic to ditopic to tritopic salt. Additionally, **56** was found to undergo cooperative interaction with the paraquat salts **59** and for the production of **62** (see below);

the cooperative processes were demonstrated in each case with Scatchard and Hill plots.

Moreover, DB24C8-terminated polystyrene **56** with guest **59** also resulted in incomplete complexation at 1:1 stoichiometry. ^{1}H NMR of this complex in a chloroform:acetone mixture (3:10) provided proof that only a small amount of complexation took place, resulting in chain extension. It was speculated that chain extension was incomplete due to the nature of the difunctional guest **59**, which could allow it to conformationally fold, forming a U-shape and sandwich itself within and around the DB24C8.

7.12
Star Polymers from Crown Functionalized Polymers

As the popularity and abundance of host–guest chemistry grows, new uses and applications become available. One of the more recent uses of host–guest chemistry targets the production of star polymers. When an adequate association constant exists between the two components, one component may be developed into a multitopic core and the other part fashioned into monotopic polymers; these will combine to produce supramolecular star polymers. One such system that has seen success in this area is the crown ether/paraquat pair (see Scheme 7.5). The bis(*m*-phenylene)-32-crown-10/paraquat motif was used by Gibson and coworkers to synthesize a supramolecular triarm star polymer, which contained a crown ether core that formed pseudorotaxane units with paraquat-terminated polystyrene arms [33]. The system was reached by synthesizing paraquat-functionalized polystyrene **60** by stable free radical polymerization (SFRP) with $M_n = 32$ kDa and adding it to a tritopic 32-crown-10 compound. Solution viscosity gave evidence that the star polymer was formed; an association constant of 687 M^{-1} was obtained for the system by NMR. The model system has an association constant of approximately 1550 M^{-1}. The lower value for the polymer was again attributed to steric effects of the polystyrene, that is, entropic penalties due to conformational effects of the chain.

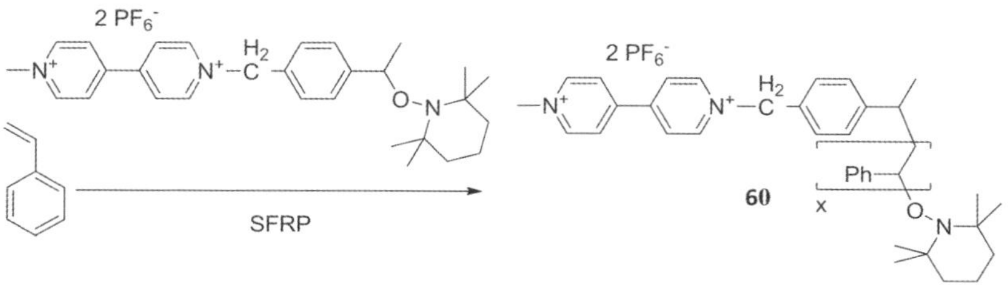

Scheme 7.5 Synthesis of paraquat-functionalized polystyrene, developed by Gibson and coworkers [33].

Further, Gibson et al. reported two tri-armed star polymers from the DB24C8-terminated polystyrene **56** (previously used for the formation of a pseudorotaxane with **57**) [32]. The first reported tri-armed pseudorotaxane takes advantage of a trialkylammonium PF$_6$ salt to complex with three molecules of **56** to form **61**. For the complexation forming **61**, ^{1}H NMR titrations gave an average association constant of 25×10^2 M^{-1} using the solvent chloroform:acetonitrile 3:10, and indicated that the system underwent cooperative complexation. The second reported that a tri-armed pseudorotaxane was formed by complexation of a tetrafunctionalized paraquat PF_6 salt with **56** [32]. Although the core has a functionality of four, only three additions of **56** were observed, resulting in the complex **62** being formed. Interestingly, the addition of **56** to the core is cooperative; however, only three of the four available sites undergo complexation, despite a small-molecule model of the system undergoing an addition of four molecules of DB24C8. The failure of the last site to undergo complexation has been attributed to either a steric clash of the arms or **56** adding between two paraquat moieties, taking up two complexing sites. Thus, both tri-armed star polymers were found to undergo cooperative complexation. Attempts were made to form a star polymer from BMP32C10 polystyrene **58** and the trialkylammonium used for the formation of **61**; however, formation of the star polymer was reported to be unsuccessful due to the low association constant previously noted for this system. Taken together, these observations indicate that selective star polymer formation is possible by proper selection of host and guest moieties in mixed systems.

References

1. Huang, F. and Gibson, H.W. (2005) *Progr. Polym. Sci.*, **30**, 982.
2. Takata, T. (2006) *Polym. J.*, **38**, 1.
3. Takata, T., Kihara, N., and Furusho, Y. (2004) *Polymer Synthesis*, vol. **171**, Springer, Berlin/Heidelberg, p. 1.
4. Farcas, A.S. and Simionescu, C.I. (2006) *Rev. Roum. Chim.*, **51**, 1153.
5. Fang, L., Olson, M.A., Benitez, D., Tkatchouk, E., Goddard, W.A. III, and Stoddart, J.F. (2010) *Chem. Soc. Rev.*, **39**, 17.
6. Söntjens, S.H.M., Sijbesma, R.P., van Genderen, M.H.P., and Meijer, E.W. (2001) *Macromolecules*, **34**, 3815.
7. Ashton, P.R., Baxter, I., Cantrill, S.J., Fyfe, M.C.T., Glink, P.T., Stoddart, J.F., White, A.J.P., and Williams, D.J. (1998) *Angew. Chem. Int. Ed.*, **37**, 1294.
8. Fang, L., Hmadeh, M., Wu, J., Olson, M.A., Spruell, J.M., Trabolsi, A., Yang, Y.-W., Elhabiri, M., Albrecht-Gary, A.-M., and Stoddart, J.F. (2009) *J. Am. Chem. Soc.*, **131**, 7126.
9. Gibson, H.W., Yamaguchi, N., Niu, Z., Jones, J.W., Slebodnick, C., Rheingold, A.L., and Zakharov, L.N. (2010) *J. Polym. Sci. Part A: Polym. Chem.*, **48**, 975.
10. Kolchinski, A.G., Busch, D.H., and Alcock, N.W. (1995) *J. Chem. Soc., Chem. Commun.*, 1289.
11. Hmadeh, M., Fang, L., Trabolsi, A., Elhabiri, M., Albrecht-Gary, A.-M., and Stoddart, J.F. (2010) *J. Mater. Chem.*, **20**, 3422.
12. Wang, F., Zheng, B., Zhu, K., Zhou, Q., Zhai, C., Li, S., Li, N., and Huang, F. (2009) *Chem. Commun.*, 4375.
13. Leung, K.C.F., Aricó, F., Cantrill, S.J., and Stoddart, J.F. (2007) *Macromolecules*, **40**, 3951.
14. Leung, K.C.F., Aricó, F., Cantrill, S.J., and Stoddart, J.F. (2005) *J. Am. Chem. Soc.*, **127**, 5808.
15. Leung, K.C.F., Mendes, P.M., Magonov, S.N., Northrop, B.H., Kim, S., Patel, K., Flood, A.H., Tseng, H.-R., and Stoddart, J.F. (2006) *J. Am. Chem. Soc.*, **128**, 10707.
16. Gibson, H.W., Nagvekar, D.S., Yamaguchi, N., Bhattacharjee, S., Wang, H., Vergne, M.J., and Hercules, D.M. (2004) *Macromolecules*, **37**, 7514.
17. Gong, C. and Gibson, H.W. (1997) *J. Am. Chem. Soc.*, **119**, 8585.

18 Kohsaka, Y., Konishi, G.-i., and Takata, T. (2007) *Polym. J.*, **39**, 861.
19 Oku, T., Furusho, Y., and Takata, T. (2004) *Angew. Chem. Int. Ed.*, **43**, 966.
20 Bilig, T., Oku, T., Furusho, Y., Koyama, Y., Asai, S., and Takata, T. (2008) *Macromolecules*, **41**, 8496.
21 South, C.R., Higley, M.N., Leung, K.C.F., Lanari, D., Nelson, A., Grubbs, R.H., Stoddart, J.F., and Weck, M. (2006) *Chem. Eur. J.*, **12**, 3789.
22 South, C.R., Leung, K.C.F., Lanari, D., Stoddart, J.F., and Weck, M. (2006) *Macromolecules*, **39**, 3738.
23 Nakazono, K., Fukasawa, K., Sato, T., Koyama, Y., and Takata, T. (2010) *Polym. J.*, **42**, 208.
24 Sasabe, H., Inomoto, N., Kihara, N., Suzuki, Y., Ogawa, A., and Takata, T. (2007) *J. Polym. Sci. Part A: Polym. Chem.*, **45**, 4154.
25 Kang, S., Berkshire, B.M., Xue, Z., Gupta, M., Layode, C., May, P.A., and Mayer, M.F. (2008) *J. Am. Chem. Soc.*, **130**, 15246.
26 Lee, Y.-G., Koyama, Y., Yonekawa, M., and Takata, T. (2010) *Macromolecules*, **43**, 4070.
27 Sato, T. and Takata, T. (2009) *Polym. J.*, **41**, 470.
28 Gibson, H.W., Ge, Z., Huang, F., Jones, J.W., Lefebvre, H., Vergne, M.J., and Hercules, D.M. (2005) *Macromolecules*, **38**, 2626.
29 Lee, M., Schoonover, D.V., Gies, A.P., Hercules, D.M., and Gibson, H.W. (2009) *Macromolecules*, **42**, 6483.
30 Allwood, B.L., Shahriari-Zavareh, H., Stoddart, J.F., and Williams, D.J. (1987) *J. Chem. Soc., Chem. Commun*, 1058.
31 Allwood, B.L., Spencer, N., Shahriari-Zavareh, H., Stoddart, J.F., and Williams, D.J. (1987) *J. Chem. Soc., Chem. Commun*, 1064.
32 Gibson, H.W., Farcas, A., Jones, J.W., Ge, Z., Huang, F., Vergne, M., and Hercules, D.M. (2009) *J. Polym. Sci. Part A: Polym. Chem.*, **47**, 3518.
33 Huang, F., Nagvekar, D.S., Slebodnick, C., and Gibson, H.W. (2004) *J. Am. Chem. Soc.*, **127**, 484.

Part Three
Properties and Functions

8
Processive Rotaxane Catalysts

Johannes A.A.W. Elemans, Alan E. Rowan, and Roeland J.M. Nolte

8.1
Introduction

For crucial chemical processes like the replication of DNA Nature has developed special types of catalysts that reduce the chance of introducing errors. These catalysts, for example, DNA polymerase III, remain in contact with the DNA chain by means of a cyclic protein which encircles and binds to the DNA, and moves along with the catalyst [1, 2]. Such a catalytic system that is linked to its substrate is called processive, since it remains bound to the substrate; an X-ray structure of the DNA polymerase III and its clamp is shown in Figure 8.1 [3].

Inspired by Nature's elegant design we asked ourselves the question whether it would be possible to construct a completely synthetic catalytic system that operates in a similar fashion, that is, binds to a synthetic polymer and moves along it while performing a reaction. We have been able prepare such a catalyst (**Mn1**), and the system is shown in Figure 8.2. Here the catalytic properties of **Mn1** in the oxidation of alkene-containing polymers and detailed studies on the mechanism of threading of the macrocyclic compound onto polymeric substrates are described.

8.2
Results and Discussion

8.2.1
Catalysis

The synthetic catalyst **Mn1** incorporates the following features: a cage compound derived from glycoluril to which a catalytic manganese(III) porphyrin complex is attached as a roof (Figure 8.2 left). The manganese complex can epoxidize alkenes, including polyalkenes, using sodium hypochlorite or iodosylbenzene as the oxidant. By adding a bulky axial ligand, for example, 4-*tert*-butylpyridine (**tbpy**), which can only coordinate to the manganese center at the outside of the porphyrin cavity, the catalyst is activated and at the same time the oxidation reaction is forced to take place on the

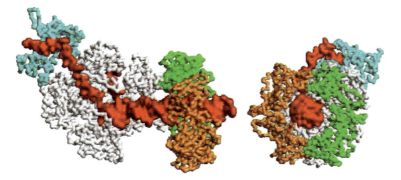

Figure 8.1 X-ray structure of the DNA polymerase III processive catalyst. Left: side view showing the cyclic protein (right) and the catalyst (left); right: end view showing the cyclic protein.

inside of the cage. Initial experiments revealed that with this set-up, low-molecular-weight alkenes such as styrene or *cis*- and *trans*-stilbene are indeed preferentially oxidized on the inside of the cage catalyst, as was concluded from kinetic studies (Figure 8.2, right). In the presence of a small ligand like pyridine (**py**), which binds preferentially inside the cage, the reaction was found to proceed on the outside of the catalyst [4].

To investigate whether **Mn1** could mimic the action of processive enzymes, the complex was tested as a catalyst in the epoxidation of polybutadiene (>98% cis) using iodosylbenzene as oxidant and a 500-fold excess of **tbpy** [6]. For comparison the same experiment was performed with meso-tetrakis(2-methoxyphenyl) porphyrin (**MnTMPP**), which is electronically related to **Mn1** but does not possess a substrate binding cage. **Mn1** turned out to be a slower catalyst than **MnTMPP**, which is in line with expectation since the reaction with polybutadiene is forced to take place inside the cavity of the former catalyst. The reaction could be blocked for only **Mn1** by

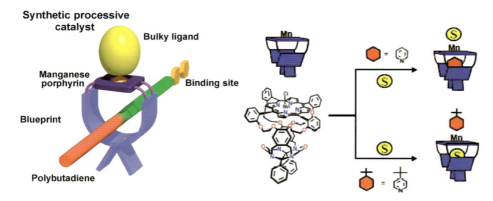

Figure 8.2 Left: blueprint of a synthetic processive catalyst. Right: cage-containing manganese(III) porphyrin catalyst (**Mn1**), which reacts with a low-molecular-weight alkene substrate (S) either on the inside or on the outside depending on the type of axial ligand that is applied to activate the catalyst.

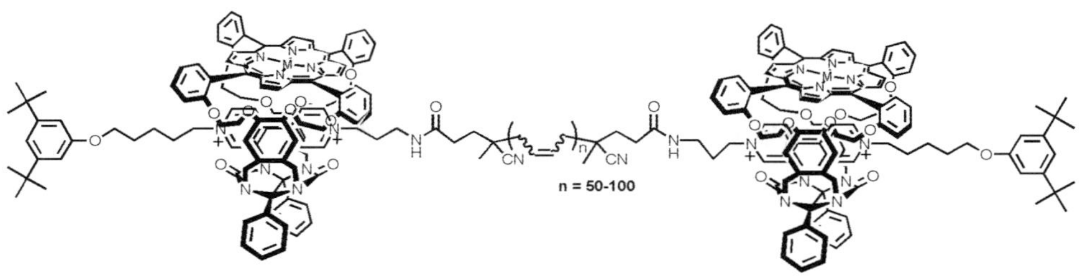

Figure 8.3 Polymer-porphyrin rotaxane catalytic system.

the addition of N,N'-dimethyl viologen dihexafluorophosphate ($K_{ass} > 10^5 \, M^{-1}$), which acts as an inhibitor. Further proof of the fact that the reaction took place on the inside of **Mn1** came from the cis-trans ratio of the product of the reaction, namely polybutadiene-epoxide. While the toroidal catalyst **Mn1** produced 80% *trans*- and 20% *cis*-epoxide polymer from polybutadiene, **MnTMPP** gave predominantly the cis product (78% cis, 22% trans). As a model compound, also a polymer, porphyrin rotaxane catalyst was synthesized (Figure 8.3). In this compound the rotaxane architecture with the manganese(III) porphyrin threaded onto the polybutadiene substrate is enforced by capping the polymer with stoppers.

Catalytic experiments with this catalyst revealed the same high trans-cis ratios as that obtained for **Mn1** and free polybutadiene, confirming that the catalysis took place on the inside of the **Mn1** (Figure 8.4).

8.2.2
Threading

The above studies provide only indirect evidence that the **Mn1** complex threads onto the polymer and moves along it during catalysis. Furthermore, the precise relationships between motion and catalysis and whether the catalysis proceeds sequentially processively (i.e. stepwise) or via a hopping mechanism remained unknown. It was decided, therefore, to study in more detail the threading of polymeric substrates by macrocycles of type **1**.

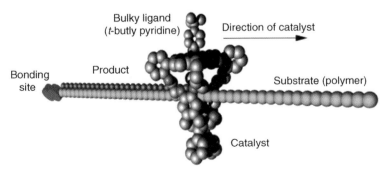

Figure 8.4 **Mn1** complex which epoxidizes the double bonds of polybutadiene via a sliding process.

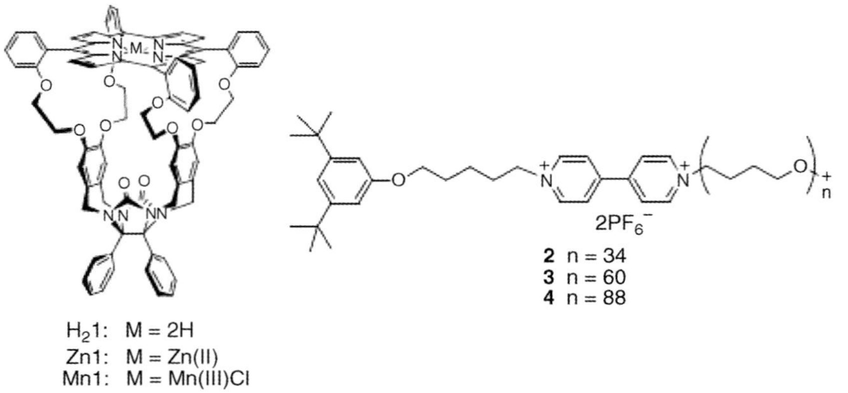

Figure 8.5 Molecular structures of several derivatives of macrocyclic host **1** (left) and polytetrahydrofurane polymers **2–4** (right).

To this end, a series of three well-defined polytetrahydrofurane polymers **2–4** were synthesized (Figure 8.5), which contain at one side an open end and at the other side a viologen (N,N′-dialkyl-4,4′-bipyridinium) trap as well as a 3,5-di(*tert*-butyl)phenyl blocking group [7]. Macrocycles of type **1** are unable to slip over the blocking group and have to traverse the whole polymer chain before they can reach the viologen trap (Figure 8.6). Upon binding of the viologen in the cavity of **1**, the fluorescence originating from the porphyrin moiety becomes quenched, allowing the process to be followed by fluorescence spectroscopy.

First, the threading process involving the process of complexation of the free base derivative of the macrocycle $H_2\mathbf{1}$ to polymers **2–4** was studied by mixing equimolar amounts of $H_2\mathbf{1}$ and a viologen-appended polymer (conc. ≈ 1 mM) in $CHCl_3$/CH_3CN 1 : 1 (v/v) and monitoring the decrease in fluorescence intensity as a function of time (Figure 8.7a). The kinetics of threading followed a second order process, as plots of $1/[H_2\mathbf{1}]$ versus time gave straight lines (Figure 8.7b). From the slopes of these lines, the rate constants for the threading process k_{on} and $\Delta G^{\ddagger}_{on}$ (the free energy

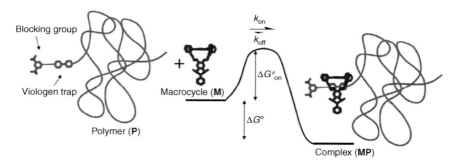

Figure 8.6 Schematic representation of the threading of macrocycle **M** onto polymer **P**, in which the host has to traverse the whole polymer chain before it can reach the viologen trap and form the complex **MP**.

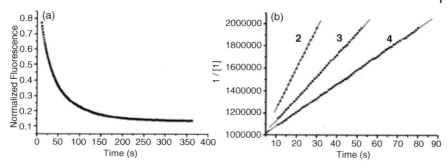

Figure 8.7 (a) Observed change in porphyrin fluorescence emission upon threading macrocycle H_21 onto polymer **2**. (b) Second order plots for the complexation of H_21 to polymers **2–4**.

difference between the ground state of the uncomplexed components and the transition state) could be determined (Table 8.1). The rate of threading became slower as the polymer length increased.

The reverse process, that is the de-threading of the polymers, was investigated by preparing a solution of millimolar concentration of a 1:1 complex of H_21 and a polymer. It can be calculated that at that concentration >99% of the host molecules are residing on the viologen trap ($K_{ass} \approx 10^7\,M^{-1}$). When this solution was rapidly diluted 1000 times, the complex started to dissociate, a process that could be followed by monitoring the increase in porphyrin fluorescence versus time. It appeared that for each of the combinations of H_21 and **2–4** the de-threading process followed first-order kinetics, unlike in the case of the threading reaction. The parameters k_{off} and $\Delta G^\ddagger_{off}$ are presented in Table 8.1.

From the threading and de-threading experiments it became clear that a length-dependent barrier, which increases with 61 J nm^{-1}, has to be surmounted by H_21 in order to reach the viologen trap. The polymer architecture plays an essential role in this process since polybutadienes thread much more slowly than polytetrahydrofuranes, apparently due to the greater bulkiness of the former polymers. Additional fluorescence studies revealed that H_21 has no problem in statistically finding the open end of the polymer chain. A sufficiently stretched-out polymer chain is required for the threading to fully occur. The threading mechanism is thought to be similar to the one proposed by Muthukumar for the translocation of a polymer chain through a hole, resembling the transportation of DNA through the opening in a virus particle [8]. After finding the hole, the macrocycle has to thread the polymer over a certain

Table 8.1 Kinetic data for the threading and de-threading of H_21 onto polymers **2–4**, measured by fluorescence spectroscopy in $CHCl_3/CH_3CN$, 1:1 (v/v) at 296 K.

Polymer	k_{on} (M^{-1} s^{-1})	$\Delta G^\ddagger_{on}$ (J mol^{-1})	k_{off} (s^{-1})	$\Delta G^\ddagger_{off}$ (J mol^{-1})
2	3.6×10^4	4.7×10^4	1.3×10^{-3}	4.7×10^4
3	1.9×10^4	4.8×10^4	6.5×10^{-4}	9.1×10^4
4	1.2×10^4	4.9×10^4	4.0×10^{-4}	9.2×10^4

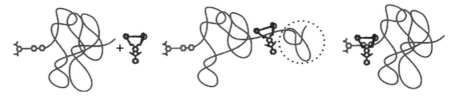

Figure 8.8 Proposed mechanism for the threading of $H_2 1$ onto a polymer, where first an entropic barrier has to be overcome in which a minimum length of the macromolecule has to be threaded before the process continues.

critical length before the process can continue (Figure 8.8). According to this mechanism, the barrier to overcome is of entropic origin and should depend on the length of the polymer chain. To investigate this phenomenon, threading experiments were carried out at different temperatures to determine the parameters $\Delta H^{\ddagger}_{on}$, $\Delta S^{\ddagger}_{on}$, and E_a. For all polymers, $\Delta H^{\ddagger}_{on}$ was positive and the same within experimental error, whereas the values of $\Delta S^{\ddagger}_{on}$ were strongly negative and were found to increase in absolute value when the polymer length became larger (Table 8.2). These results are supportive of the nucleation mechanism proposed by Muthukumar and the presence of an entropic barrier, related to the stretching of the end of the polymer chain, which has to be overcome. From the threading and dethreading experiments it became clear that a length-dependent barrier has to be taken.

To get more insight into the initial steps of polymer threading and to understand the role of the cavity of the macrocycle, a zinc ion was inserted into the porphyrin of $H_2 1$ to give **Zn1** (Figure 8.5) [9]. **Zn1** was found to thread onto polymers **2–4** much more slowly than $H_2 1$ (Figure 8.9), an observation which was intuitively attributed to coordination of the zinc ion of **Zn1** to the oxygen atoms in the polytetrahydrofuran chains during threading, since zinc porphyrins prefer to bind a fifth ligand. When the $\Delta G^{\ddagger}_{on}$-values for the threading of $H_2 1$ and **Zn1** were plotted versus polymer length, lines with similar slopes were obtained, the only difference being that in the case of **Zn1** the initial $\Delta G^{\ddagger}_{on}$-values were approximately 6 kJ mol^{-1} higher. These results indicate that the energy penalty to move along the polymer chain is similar for both macrocycles, but that the initial threading onto the polymer requires more energy in the case of **Zn1**. It was suspected that this energy penalty was caused by a blocking of the cavity of **Zn1** by a coordinating molecule, and a control experiment indeed suggested that a CH_3CN solvent molecule coordinated to **Zn1** at the inside of the cavity, albeit weakly. To investigate this hypothesis further, a 6300-fold excess of **tbpy** was added to **Zn1** before the threading experiment. From the known association

Table 8.2 Thermodynamic parameters for the threading of $H_2 1$ onto polymers **2–4** in $CHCl_3$/CH_3CN, 1 : 1 (v/v).

Polymer	$\Delta H^{\ddagger}_{on}$ (J mol^{-1})	$\Delta S^{\ddagger}_{on}$ (J K^{-1} mol^{-1})	E_a (J mol^{-1})
2	2.0×10^4	-88	2.3×10^4
3	2.0×10^4	-94	2.3×10^4
4	2.1×10^4	-97	2.3×10^4

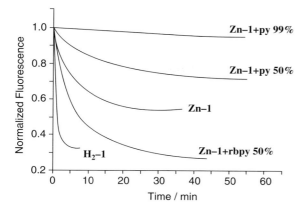

Figure 8.9 Normalized fluorescence intensity of $H_2 1$ and of $Zn1$ in the absence and presence of different axial ligands measured versus time after the addition of circa 1 equiv. of polymer **2**.

constant of this ligand ($K_{ass} = 250 \, M^{-1}$ in $CHCl_3/CH_3CN$, 1 : 1 (v/v)) and the fact that it exclusively binds to the outside of the cavity, it can be calculated that, at the concentration used, ~50% of the molecules of **Zn1** are bound to this ligand. Consequently, any more weakly binding solvent molecule bound on the inside of the macrocycle is released as a result of the pentacoordinate nature of the zinc ion in zinc porphyrins. Upon the addition of each of the polymers **2–4** to the solution containing the **Zn1-tbpy** mixture, a faster fluorescence quenching was observed than in the absence of axial ligand (Figure 8.9), indicating that the rate-limiting step for the threading of **Zn1** is indeed the initial threading of the cavity onto the polymer. To obtain additional evidence for this blocking process, the cavity of **Zn1** was intentionally blocked by adding increasing amounts of **py**, an axial ligand that is known to bind strongly and exclusively in the cavity of **Zn1** ($K_{ass} = 1.1 \times 10^5 \, M^{-1}$ in $CHCl_3$) [10]. From the fluorescence threading curves it can be clearly seen that on increasing the **py** concentration a dramatic decrease in the rate of threading takes place until a nearly complete blocking is achieved (Figure 8.9) [9].

In the above-described studies on the catalytic oxidation of polybutadiene by **Mn1** we were unable to establish whether the reaction followed a processive (i.e., stepwise) or a random sliding mechanism (Figure 8.10). From the kinetic data of the threading

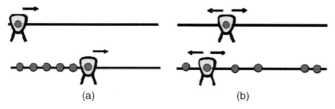

Figure 8.10 Schematic representation of the oxidation of a polybutadiene chain by **Mn1**. (a) Stepwise processive mechanism. (b) Random sliding mechanism. The dots indicate the oxidized double bonds of the polymer.

process we can calculate a lower limit for the movement of the catalyst along the polymer thread, which for the combination of $H_2 1$ and polymer 2 is 750 pm s^{-1} and for the combination of $H_2 1$ and the polybutadiene derivative 14 pm s^{-1}. These speeds are considerably higher that the translocation velocity of about 1 pm s^{-1} that can be calculated from the rate of epoxidation of polybutadiene, assuming a sequentially processive process. This suggests that the catalytic oxidation of polybutadiene by **Mn1** most likely occurs in a random sliding fashion, in which the catalyst hops from side to side, and may even leave the polymer chain from time to time.

8.3
Conclusion

We have developed a synthetic catalytic system comprising a cage compound and an attached manganese porphyrin complex that is capable of oxidizing the double bonds of polybutadiene in an unprecedented fashion. Threading studies with a metal-free manganese porphyrin cage complex and a zinc derivative of it indicate that the macrocyclic compound can thread onto the polymer and glide on its chain. The combined kinetic and thermodynamic data suggest that for this threading process an entropic barrier has to be overcome. This barrier may be related to the stretching and unfolding of the polymer chain, although further studies are required to confirm this. Comparison of the speed of movement of the porphyrin cage compounds along the polymer chain and the rate of oxidation of the polymer double bonds by the manganese porphyrin catalyst suggests that the latter oxidation has the character of a random sliding process and not of a stepwise processive process as occurs in nature, for example, during the replication of DNA by the toroidal DNA polymerase III catalytic system. Further work will be focused on the design of synthetic processive catalysts for which the sliding movement and the catalytic reaction are better matched.

Acknowledgments

The Dutch National Research School for Combination Catalysis (NRSC-C) and the Council for the Chemical Sciences of The Netherlands Organization for Scientific Research (CW-NWO) are acknowledged for financial support to J.A.A.W.E. (Veni grant), A.E.R. (Vidi grant) and R.J.M.N. (Top grant).

References

1 Benkovic, S.J., Valentine, A.M., and Salinas, S. (2001) *Annu. Rev. Biochem.*, **70**, 181–208.

2 Trakselis, M.A., Alley, S.C., Abel-Santos, E., and Benkovic, S.J. (2001) *Proc. Natl. Acad. Sci. USA*, **98**, 8368–8375.

3 Wang, J., Sattar, A., Wang, C.C., Karam, J.D., Konigsberg, W.H., and Steitz, T.A. (1997) *Cell*, **89**, 1087–1099.
4 Elemans, J.A.A.W., Bijsterveld, E.J.A., Rowan, A.E., and Nolte, R.J.M. (2000) *Chem. Commun.*, 2443–2444.
5 Elemans, J.A.A.W., Bijsterveld, E.J.A., Rowan, A.E., and Nolte, R.J.M. (2007) *Eur. J. Org. Chem.*, **5**, 751–757.
6 Thordarson, P., Bijsterveld, E.J.A., Rowan, A.E., and Nolte, R.J.M. (2003) *Nature*, **424**, 915–918.
7 Coumans, R.G.E., Elemans, J.A.A.W., Nolte, R.J.M., and Rowan, A.E. (2006) *Proc. Natl. Acad. Sci. USA*, **103**, 19647–19651.
8 Muthukumar, M. (2001) *Phys. Rev. Lett.*, **82**, 3188–3191.
9 Hidalgo Ramos, P., Coumans, R.G.E., Deutman, A.B.C., Smits, J.M.M., de Gelder, R., Elemans, J.A.A.W., Nolte, R.J.M., and Rowan, A.E. (2007) *J. Am. Chem. Soc.*, **129**, 5699–5702.
10 Elemans, J.A.A.W., Aarts, P.P.M., Claase, M.B., Rowan, A.E., Schenning, A.P.H.J., and Nolte, R.J.M. (1999) *J. Org. Chem.*, **64**, 7009–7016.

9
Emerging Biomedical Functions through 'Mobile' Polyrotaxanes
Nobuhiko Yui

9.1
Introduction

Biological systems including cells and tissues in living bodies are complex, hierarchical, and dynamic, and they always inspire us to design materials for possible applications. These structures are basically constructed from the appropriate building blocks via various intermolecular forces such as van der Waals interactions, hydrogen bonds, electrostatic interactions, and sometimes hydrophobic effects in water, and are directly adapted to performing a variety of functions such as intercellular signal transduction through plasma membranes, intracellular metabolism triggered by cytoplasmic calcium increase, and cellular proliferations [1, 2]. In the last quarter of a century, many scientists have studied interfacial phenomena between these biological systems and surfaces of artificial materials in order to design functional biomaterials for medical uses. Throughout these researches, it has been well recognized that biological responses to these surfaces include complicated acute and chronic reactions, eventually leading to cellular and tissue rejection in living bodies. In order to solve these problems, one must understand and appreciate any differences with respect to the structures and their functions between natural tissues and artificial materials. From this point of view, it should be stated that one of the dominant differences would be the mobility of molecules constructing these materials, and a quite new approach is urgently required to design biomaterials which can perform new functions in future advancing medicines.

With this in mind, we initiated a study of cyclodextrin (CD)-based polyrotaxanes as novel biomaterials, because CD molecules would be expected to move along the polymeric chain. One of the characteristics seen in polyrotaxanes is the mobility of cyclic compounds, and they can be freely rotational and sliding if any intermolecular forces with the linear chain and the neighboring cyclic compound are eliminated (Figure 9.1). In particular, polyrotaxanes consisting of α-CD molecules and a poly (ethylene glycol) (PEG) [3, 4] chain are possible biomaterials in the structural components, as α-CD and PEG have been approved by the FDA for use in foods and drugs. We have prepared biological ligand-conjugated polyrotaxanes and

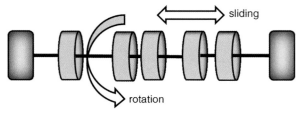

Figure 9.1 Supramolecular characteristics of 'mobile' polyrotaxanes.

examined their interaction with the receptor proteins and/or cells. Also, cytocleavable polyrotaxanes were prepared as a gene vector, and their potential in gene delivery was evaluated. Our recent achievements in the basic study of biomedical polyrotaxanes are reviewed in this chapter.

9.2
Multivalent Interaction using Ligand-Conjugated Polyrotaxanes

A variety of functional groups introduced to CD molecules in polyrotaxanes enables us to link any biologically active agents or ligands, which are designed to bind with receptor sites of plasma proteins in cell surfaces. It has been well recognized that controlling the binding of biological ligands with receptor sites of the proteins on plasma membranes of cells is a crucial event in modulating receptor-mediated cellular metabolism as well as endocytosis for drug targeting. Here, the way in which the effectiveness, and specifically the binding, on membrane proteins using very limited amounts of the agents or ligands can be enhanced is one of the most important subjects. The term 'multivalency' is defined as binding multiple copies of ligands with the receptor sites simultaneously, and it is believed to be highly effective in enhancing the binding constant between ligands and receptors [5].

In the last few decades, a variety of multifunctional polymers have been studied for multivalent interactions, and these include synthetic water-soluble polymers such as poly(acrylic acid), poly(amino acids) such as poly(aspartic acid), polysaccharides such as dextran, and dendritic polymers [6] However, an increase in the binding constant using such multifunctional polymers is not as striking as was expected. This is mainly due to thermodynamically unfavorable situation in the interaction: increasing the number of ligands in the polymer eventually causes excessively increased density of the ligands. It can easily be appreciated that many ligands introduced to the polymeric chain can contribute to increasing the enthalpic gain, whereas their binding to the polymeric chain leads to entropic loss at the same time.

We therefore postulated that the mobility of ligands linked to α-CD molecules in polyrotaxanes contributes to preventing such an entropic loss, leading to an increase in the multivalent interaction with protein receptors. We believed that polyrotaxanes are advantageous, giving thermodynamic benefits by enhancing the multivalent interaction with biological systems: freely mobile ligands linked to CD molecules in polyrotaxanes can effectively bind receptor proteins in a multivalent manner

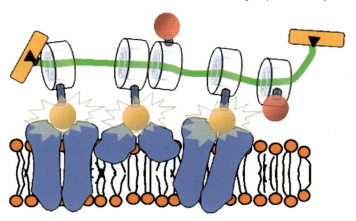

Figure 9.2 Multivalent binding of ligand-conjugated polyrotaxanes with receptor sites of plasma proteins.

(Figure 9.2). One of our representative studies using polyrotaxanes in multivalent interaction is to clarify the effect of maltose-polyrotaxane conjugates on interaction with lectin, Concanavalin A (Con A) [7–10]. The maltose-conjugated polyrotaxanes were prepared by condensation reaction between β-maltosylamine and carboxypropanoyl-α-CD molecules in the polyrotaxanes consisting of α-CD molecules and a PEG chain ($M_n = 20\,000$).

The extent of the multivalent interaction was examined by a Con A-induced hemagglutination test using erythrocyte suspension. In fact, the introduction of maltose molecules into the polyrotaxanes enhanced the binding with Con A, giving up to 3000 times better binding than maltose itself. Special attention should be paid to the fact that the extent of the enhancement was dependent upon the number of α-CD molecules in the polyrotaxanes as well as the number of maltose molecules. Pulse NMR experiments revealed that the mobility of maltose molecules linked with α-CD molecules estimated from spin-spin relaxation time is much influenced by the number of α-CD molecules in the polyrotaxanes. Finally, it was found that the extent of multivalent binding between the maltose and Con A molecules in the polyrotaxanes was closely proportional to the mobility of maltose molecules (Figure 9.3). This finding strongly suggests that the mobility of ligands in polymers can dominate the binding with receptor proteins and that polyrotaxane is a suitable tool which can reduce spatial mismatches of the binding due to their mobility along the polymeric chain.

9.3
The Formation of Polyrotaxane Loops as a Dynamic Interface

Nature always inspires us by showing how effectively and quickly the hierarchical architectures modulate a variety of functions in cellular and/or tissue levels. Spatial and temporal control of such sophisticated systems in living bodies can be precisely achieved by adjusting noncovalent interactions of the constituent molecules as well

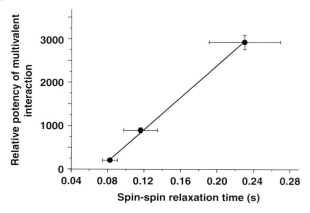

Figure 9.3 Enhanced multivalent interaction between Con A and maltose-conjugated polyrotaxanes in relation to the mobility of maltose molecules (spin-spin relaxation time of maltose C1 proton).

as their dynamism. In particular, the dynamism of proteins and lipids in cells, such as the lateral diffusion of the plasma proteins and the fluidity of the lipid membranes, is directly related to signal transduction and amplification across the plasma membrane in the course of intracellular metabolism. It is thus accepted that modulating cellular and tissue metabolisms via their dynamic interfaces with biomaterials is one of the crucial goals in the development of advanced medicine and therapy.

In the last few decades, we have emphasized the importance of molecular mobility in biomaterials design. For instance, we clarified that the mobility of biological ligands in polymers can dominate the enhancement of the binding with their receptor proteins in a multivalent manner, as mentioned above [7–10]. Also, we have demonstrated that supramolecular polyrotaxanes are a promising tool which exhibits such mobile properties. Polyrotaxane is a family of molecular assemblies in which many cyclic compounds are threaded onto a linear polymeric chain capped at both terminals with bulky end-groups [3, 4]. It is easily imagined that the cyclic compounds can be freely sliding and/or rotational along the chain in the polyrotaxane if the intermolecular forces acting between the cyclic compounds and the chain are eliminated. Such mobile properties seen in polyrotaxanes will be one of the most representative and fascinating characteristics of supramolecular-structured polymers [2, 10].

In order to create surfaces exhibiting the mobile properties of polyrotaxanes as dynamic interfaces, it is indispensable to consider the following two issues: how to fix the polyrotaxane while maintaining the mobility of cyclic compounds, and how to face the polyrotaxane structure toward cells and tissues. The grafting of polyrotaxanes onto a solid substrate may be one of the approaches to create a polyrotaxane surface. However, the dynamic nature of the polyrotaxane is not so well made use of in this approach, because increasing the graft density causes the alignment of polyrotaxane terminals at the outermost surface. A rather more effective idea is to fix both terminals of polyrotaxane onto a solid substrate so as to form polyrotaxane loops on the surface, such that cyclic compounds can be freely mobile along the

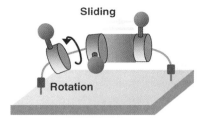

Figure 9.4 Polyrotaxane loop as a dynamic interface with biological systems.

polymeric chain on the solid substrate and be accessible as an interface with cells and tissues.

Quite recently, we prepared a water-soluble polyrotaxane consisting of α-CD molecules and a PEG ($M_n = 3000$) chain using a thiol (SH)-containing terminal stopper, and then fixed it onto a gold substrate so as to form polyrotaxane loops (Figure 9.4) [10, 11]. The formation of the polyrotaxane loops on the gold substrate was characterized by quartz cell microbalance (QCM) and X-ray photoelectron spectroscopy (XPS) measurements. It was confirmed from analyzing the ratio of SH and to SAu bonds in the XPS measurement that the majority of the polyrotaxanes immobilized on the Au surface formed a loop structure when applying the solution concentration of less than 1×10^{-5} M, while higher solution concentrations resulted in the formation of a grafted polyrotaxane structure. Also, the formation of about 4–5 polyrotaxane loops was estimated per the area of 100 nm^2 from QCM measurement. This result indicates that gold surface is almost fully covered with the polyrotaxane loops, assuming the approximate size of the polyrotaxane applied (less than 2 nm wide and about 20–30 nm long). Characterization of the surface properties of polyrotaxane loops as biomaterials is now in progress. However, it is certain that the success in forming polyrotaxane loops on a solid substrate will initiate the exploitation of an interdisciplinary study of dynamic interfaces, which have much potential to regulate biological responses via supramolecular mobility.

9.4
Cytocleavable Polyrotaxanes for Gene Delivery

Inventing nonviral gene vectors is one of the important challenges for advanced medicines, since viral vectors are not acceptable in the living body for reasons of safety [12]. Gene vectors are generally required to form a polyplex (polyion complex) with DNA which should be stable during circulation within the system and before cellular uptake via endocytosis, escape from endosom/lysosome to cytoplasm (preventing lysosomal digestion) after the endocytosis, and release of DNA at the nucleus. Cationic polymers such as poly(L-lysine) and poly(ethyleneimine) have recently been studied as a candidates for nonviral gene vectors, however, the problem of their low transfection efficiency and high cytotoxicity has not been solved.

Generally, stable polyplex formation of cationic polymers with anionic DNA requires an excessive amount of cationic groups, which causes a difficulty in the

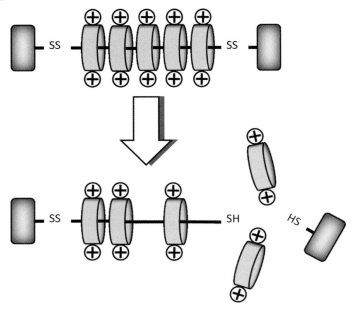

Figure 9.5 Supramolecular dissociation of cytocleavable polyrotaxanes for gene delivery.

DNA release and an increase in the cytotoxicity. In order to solve this problem, we have proposed the use of cytocleavable polyrotaxanes as gene vectors; the cytocleavable polyrotaxanes consist of cationic α-CD molecules and a PEG chain capped with tyrosine via disulfide (SS) linkage (Figure 9.5) [13–15] Cationic groups of α-CD molecules should participate in polyplex formation with anionic DNA, and this polyplex can be dissociated to release DNA by the intracellular cleavage of SS linkage in response to a high cytoplasmic glutathione concentration.

The cytocleavable polyrotaxanes were prepared by the following steps: (1) the capping of PEG ($M_n = 4000$) with cystamine to obtain a SS-introduced PEG-bisamine (SS–PEG), (2) the formation of an inclusion complex between α-CD molecules and the SS–PEG chain, (3) the capping reaction of z-protected L-tyrosine to both terminals of the inclusion complex, and (4) the introduction of dimethylaminoethylcarbamoyl (DMAEC) groups into α-CDs in the polyrotaxane. The polyrotaxanes were found to form a polyplex with plasmid DNA (pDNA) at relatively low N/P ratio, which may be due to the mobility of CD molecules along a PEG chain in the polyrotaxane through preventing spatial mismatch between DMAEC groups and phosphate groups in pDNA. Furthermore, it was confirmed that this polyplex was dissociated under reducing conditions to release pDNA via the SS cleavage and intermolecular ionic exchange with anionic polymers.

In the *in vitro* experiments using cultured NIH3T3 cells, the cytocleavable polyrotaxane/pDNA polyplex was found to exhibit much greater transfection activity than the reference polyplex using a noncytocleavable polyrotaxane, and the magnitude reached over 500 times that of the reference (Figure 9.6). This indicates that the SS

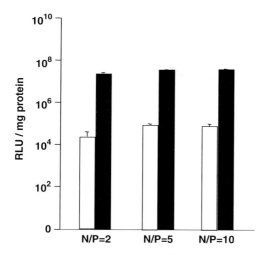

Figure 9.6 pDNA transfection at different N/P ratios by cytocleavable polyrotaxane (white bar) and noncytocleavable polyrotaxane (black bar).

cleavage of cytocleavable polyrotaxanes plays an important role in intracellular pDNA release for gene expression. Also, quite an interesting feature which can be seen from Figure 9.6 is that the pDNA transfection ability of the cytocleavable polyrotaxane was independent of the N/P ratio, although the pDNA transfection ability of the reference polyplex slightly increased with the N/P ratio. This result indicates that the mechanism of nuclear pDNA transfection using the cytocleavable polyrotaxane can be different from that of conventional cationic polymers, whose transfection ability has been generally reported to be highly dependent upon the N/P ratio.

The cytocleavable polyrotaxane/pDNA polyplex showed little cytotoxicity, although the cytotoxicity of the reference polyplex increased with the N/P ratio (Figure 9.7). It is considered that the dissociation of the polyrotaxane into building-blocks contributes to the prevention of toxicity generally seen in cationic polymers. These results obviously demonstrate the efficiency of the cytocleavable polyrotaxanes as a gene vector, and confirm the feasibility of our design concept, which can cope with two critical issues in gene delivery: the catching and release of pDNA at appropriate biological environments and the prevention of cytotoxicity observed for conventional cationic polymers.

9.5
Conclusion

The emergence of biomedical functions utilizing the mobility of polyrotaxane structures is illustrated by our recent achievements. The key to achieving the biomedical functions is to ensure high mobility of cyclic compounds along a polymeric chain, which is one of the structural characteristics seen in polyrotaxanes. For instance, multivalent interaction between ligands and receptor sites of proteins was surprisingly enhanced by increasing the mobility of ligands, including the rotational

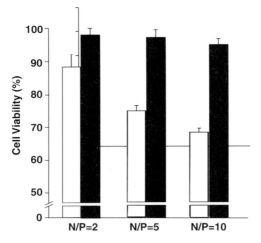

Figure 9.7 Cellular viability test by MTT assay at different N/P ratios for = cytocleavable polyrotaxane (white bar) and noncytocleavable polyrotaxane (black bar).

and sliding motion of CD molecules conjugated into polyrotaxanes. In order to further display such a 'mobile' characteristic as a dynamic interface with biological systems, we have recently succeeded in forming polyrotaxane loops on a solid substrate. The surface properties of the polyrotaxane loops are still under investigation. However, we are convinced that the 'mobile' nature of polyrotaxane surfaces can effectively modulate cellular and tissue responses, exploiting the new paradigm of supramolecular materials in the field of biomedical applications. Another important issue is the ability of polyrotaxanes to dissociate the supramolecular structures into pieces if one of both terminal stoppers is cleaved. This characteristic enables us to modulate binding with biological substrates, and in fact exploits outstanding properties in the design of nonviral DNA vectors. In order to expand the concept mentioned above, one of our next concerns is to regulate the mobility of cyclic compounds along a polymeric chain in polyrotaxanes. The key point in this issue will be how to control intermolecular forces between the cyclic compound and the linear polymeric chain. It is strongly expected that direct switching of the mobility in polyrotaxanes may be useful not only for presenting direct evidence of the mobility effect on multivalent binding but also for regulating the clustering of plasma membrane proteins via multivalent binding. We thus believe that the ON-OFF-like switching of ligand mobility is a very promising property of advanced biomaterials to modulate cellular functions, including cellular adhesion and intracellular metabolism.

Acknowledgments

The authors are grateful to Prof. Hideyoshi Harashima, Assoc. Prof. Kentaro Kogure (present address: Kyoto Pharmaceutical University), Assis. Prof. Hidetaka Akita, Hokkaido University, Prof. Atsushi Maruyama, Assis. Prof. Arihiro Kano, Kyushu

University, Prof. Goro Mizutani, Assoc. Prof. Tooru Ooya (present address: Kobe University), Assoc. Prof. Haruyuki Sano (present address: Ishikawa National College of Technology), Assoc. Prof. Yoshiko Miura, Assis. Prof. Ryo Katoono, Dr. Masaru Eguchi, Dr. Hak Soo Choi, Dr. Atsushi Yamashita, and Dr. Jun Yamaguchi, JAIST, for their collaboration. A part of this study received financial support from Special Coordination Funds for Promoting Science and Technology and a Grant-in-Aid for Scientific Research (B) of the Ministry of Education, Culture, Sports, Science and Technology, the Japanese Government.

9.6
Appendix

In my opinion, the structure of polyrotaxanes has been potentially familiar to Asian people, because we have created our corroborative culture of using polyrotaxane structures for about 2000 years, for example, preserving and carrying coins threaded onto a string, and making a Chinese 'suan pan' (abacus) for calculation and mathematics. This may indicate that we Asian people congenitally have learned tacit knowledge of naturally imagining and utilizing polyrotaxane structures in our daily lives. Figure 9.8 shows a hundred of Kan-Ei-Tsuho coins threaded onto a linen string, which is one of the definite examples from my antique collections. An image of releasing Kan-Ei-Tsuho coins from this polyrotaxane-structured bundle immediately will remind Japanese people of Heiji Zenigata, a famous hero in Japanese period dramas, who catches criminals by throwing a coin.

Of course, it is obvious that Prof. Akira Harada's scientific achievement of preparing a pseudopolyrotaxane consisting of α-CD molecules and a PEG chain was outstanding and strongly inspired me to initiate new researches into supramolecular biomaterials using polyrotaxane structures about 15 years ago when I moved

Figure 9.8 A hundred Kan-Ei-Tsuho coins threaded onto a linen string. The Kan-Ei-Tsuho was first manufactured in 1636, and has been well known as one of the representative Japanese coins commercially available throughout the Edo era.

to JAIST to set up my own laboratory. Also, I have learnt from and been much inspired by nature when designing more ideal biomaterials such as 'mobile' polyrotaxanes. However, at the same time, it is true that the image of biodegradable polyrotaxanes in my mind actually arose from the Kan-Ei-Tsuho bundle as shown in Figure 9.8 in order to tacitly judge the validity as biomaterials. If I did not have such cultural perspectives as an Asian, I would have not initiated the study of biodegradable polyrotaxanes. Thus, it can be honestly stated that some ideas of biomedical-oriented polyrotaxanes during my career were derived from the interplay of an intuitive and cultural mind with a purposive and scientific point of view.

References

1. (a) Yui, N. (ed.) (2002) *Supramolecular Design for Biological Applications*, CRC Press, Boca Raton, USA; (b) Yui, N., Mrsny, R.J., and Park, K. (eds) (2004) *Reflexive Polymers and Hydrogels: Understanding and Designing Fast Responsive Polymeric Systems*, CRC Press, Boca Raton, USA.
2. (a) Ooya, T. and Yui, N. (1999) *Crit. Rev. Ther. Drug Carrier Syst.*, **16**, 289–330; (b) Yui, N. and Ooya, T. (2006) *Chem. Eur. J.*, **12**, 6730–6737.
3. (a) Harada, A. and Kamachi, M. (1990) *Macromolecules*, **23**, 2821–2823; (b) Harada, A. and Kamachi, M. (1992) *Nature*, **356**, 325–327.
4. (a) Gibson, H.W. and Marand, H. (1993) *Adv. Mater.*, **5**, 11–21; (b) Wenz, G. (1994) *Angew. Chem. Int. Ed.*, **33**, 803–822.
5. Mammen, M., Choi, S.K., and Whiteside, G.M. (1998) *Angew. Chem. Int. Ed.*, **37**, 2754–2794.
6. (a) Cairo, C.W., Gestwicki, J.E., Kanai, M., and Kiessling, L.L. (2002) *J. Am. Chem. Soc.*, **124**, 1615–1619; (b) Gestwicki, J.E., Cairo, C.W., Strong, L.E., Oetjen, K.A., and Kiessling, L.L. (2002) *J. Am. Chem. Soc.*, **124**, 14922–14933.
7. Ooya, T., Eguchi, M., and Yui, N. (2003) *J. Am. Chem. Soc.*, **125**, 13016–13017.
8. (a) Hirose, H., Sano, H., Mizutani, G., Eguchi, M., Ooya, T., and Yui, N. (2004) *Langmuir*, **20**, 2852–2854; (b) Hirose, H., Sano, H., Mizutani, G., Eguchi, M., Ooya, T., and Yui, N. (2006) *Polym. J.*, **38**, 1093–1097.
9. Ooya, T., Utsunomiya, H., Eguchi, M., and Yui, N. (2005) *Bioconj. Chem.*, **16**, 62–69.
10. Yui, N., Katoono, R., and Yamashita, A. (2009) *Adv. Polym. Sci.*, **222**, 55–77.
11. Yang, D.H., Katoono, R., Yamaguchi, J., Miura, Y., and Yui, N. (2009) *Polym. J.*, **41**, 952–953.
12. (a) Wolff, J.A. (2002) *Nature Biotechnol.*, **20**, 768–769; (b) Nori, A. and Kopecek, J. (2005) *Advanced Drug Delivery Reviews*, **57**, 609–636; (c) Nishiyama, N., Iriyama, A., Jang, W., Miyata, K., Itaka, K., Inoue, Y., Takahashi, H., Yanagi, Y., Tamaki, Y., Koyama, H., and Kataoka, K. (2005) *Nature Mater.*, **4**, 934–941.
13. Ooya, T., Choi, H.S., Yamashita, A., Yui, N., Sugaya, Y., Kano, A., Maruyama, A., Akita, H., Ito, R., Kogure, K., and Harashima, H. (2006) *J. Am. Chem. Soc.*, **128**, 3852–3853.
14. Yamashita, A., Yui, N., Ooya, T., Kano, A., Maruyama, A., Akita, H., Kogure, K., and Harashima, H. (2006) *Nature Protocols*, **1**, 2861–2869.
15. Yamashita, A., Kanda, D., Katoono, R., Yui, N., Ooya, T., Maruyama, A., Akita, H., Kogure, K., and Harashima, H. (2008) *J. Controlled Release*, **131**, 137–144.

10
Slide-Ring Materials Using Polyrotaxane

Kazuaki Kato and Kohzo Ito

10.1
Introduction

Gels are divided into two categories: chemical gels and physical gels [1]. Physical gels have noncovalent cross-linking junctions that are formed by ionic interaction, hydrophobic interaction, hydrogen bonding, microcrystal formation, helix formation, and so on. In general, the noncovalent cross-links in chemical gels are not as strong as the covalent ones, and physical gels show a sol–gel transition in response to changes in temperature, pH, and solvents. The mechanical behavior of physical gels is complicated because recombination of cross-linking points occurs upon deformation; thus, it is not clear whether affine deformation occurs in physical gels. This recombination causes hysteresis in the stress–strain curve of a physical gel, and this implies that it cannot regain its original shape from the deformed one as quickly as a chemical gel.

When cross-linked polymeric materials are immersed in a good solvent, they absorb the liquid solvent until the swelling force associated with the mixing entropy between the chains and the solvent balances the elastic force of the chains between the junction points. These cross-linked polymeric systems containing the solvent are called chemical gels [1]. The swelling behavior of chemical gels has been explained in detail by Flory and Rehner [2]. In addition, Tanaka discovered volume phase transition in chemical gels, wherein their swelling and shrinking behaviors exhibit discontinuous profiles with hysteresis [2, 3]. This novel discovery regarding cross-linked polymeric materials has attracted considerable interest from researchers in the field of polymer science. As a result, some interesting aspects of chemical gels have been uncovered, for example, the kinetics of the volume phase transition by Tanaka and Fillmore [4], the frozen or fixed inhomogeneous structure of chemical gels by Shibayama [5], and the abnormally small friction behavior of gels by Gong and Osada [6, 7]. In addition, various new types of gels have been developed. For example, Gong and Osada developed a double network gel having a high modulus up to the sub-megapascal range with a failure compressive stress as high as 20 MPa [8, 9]. This double network gel comprises both soft and hard components that prevent

Supramolecular Polymer Chemistry, First Edition. Edited by Akira Harada.
© 2012 Wiley-VCH Verlag GmbH & Co. KGaA. Published 2012 by Wiley-VCH Verlag GmbH & Co. KGaA.

fracture, like that observed in some biomaterials. Furthermore, Yoshida incorporated a dissipative system into gels to realize a self-oscillating gel device [10] and also developed comb-type grafted hydrogels that exhibit a rapid deswelling response to temperature changes [11].

Recently, polymer network structures have been developed using polyrotaxane in the supramolecular chemistry. Supramolecules and their topological characteristics have attracted considerable interest [12], for example, rotaxanes, whose cyclic molecules are threaded on a single polymer chain and trapped by capping the chain with bulky end groups [13–17]. Ogino prepared the first rotaxane using cyclodextrin (CD) as the ring molecule [18]. CDs are cyclic oligosaccharides comprising six, seven, or eight glucose units with inner diameters of 0.44, 0.58, or 0.74 nm, respectively; these oligosaccharides are called α-, β-, and γ-CDs, in that order [19–21]. As compared to other cyclic molecules, CDs are readily available in both high purities and large quantities and can be modified using various functional groups. The most important feature of CDs is their amphiphilic property; that is, they are hydrophobic on the inside and hydrophilic on the outside. Therefore, water-soluble CDs tend to include small hydrophobic molecules in their cavities; this phenomenon is referred to as inclusion complex formation. In 1990, Harada and Kamachi reported the first synthesis of pseudo-polyrotaxane, in which many α-CD molecules were threaded on a single polymer chain of polyethylene glycol (PEG) [22]: CDs mixed with PEG in water were threaded onto a PEG self-assembly. Subsequently, in 1992, both ends of the pseudo-polyrotaxane were capped with bulky groups to form polyrotaxane [23]. In recent years, this novel architecture in supramolecular chemistry has attracted great attention as the basis of a new technique for developing functional polymeric materials [13–17, 24].

The first report of a physical gel based on the polyrotaxane architecture was presented by Harada et al. [25]. They discovered that when α-CDs were mixed with long PEG chains at a high concentration in water, a sol–gel transition occurred because of hydrogen bonding between the α-CDs threaded on the PEG chains in different pseudo-polyrotaxanes. In addition, Yui and coworkers prepared some hydrogels using biodegradable CD polyrotaxane [26] as the cross-linker for use in regenerative medicine [27]. The biodegradable polyrotaxane has a hydrolysis part, namely, an ester bond between a bulky end group and the polymer axis. As a result, the erosion time of a biodegradable hydrogel strongly depends on its polyrotaxane content. Further, on the basis of the concept of dynamic covalent bond chemistry, Takata et al. synthesized recyclable cross-linked polyrotaxane gels using topologically networked polyrotaxane that is capable of undergoing reversible assembly and disassembly [13, 28]. They cross-linked poly(crown ether)s with dumbbell-shaped axle molecules, which showed a reversible cleavage of the disulfide bond. As a result, a novel reversible cross-linking/decross-linking system that could recycle networked polymeric materials was realized.

We have recently developed another cross-linking structure based on the polyrotaxane architecture [29]. We prepared polyrotaxane containing a small amount of α-CD and then cross-linked α-CDs on different polyrotaxanes, as shown in Figure 10.1. As a result, the cross-linking junctions in a figure-eight shape are not

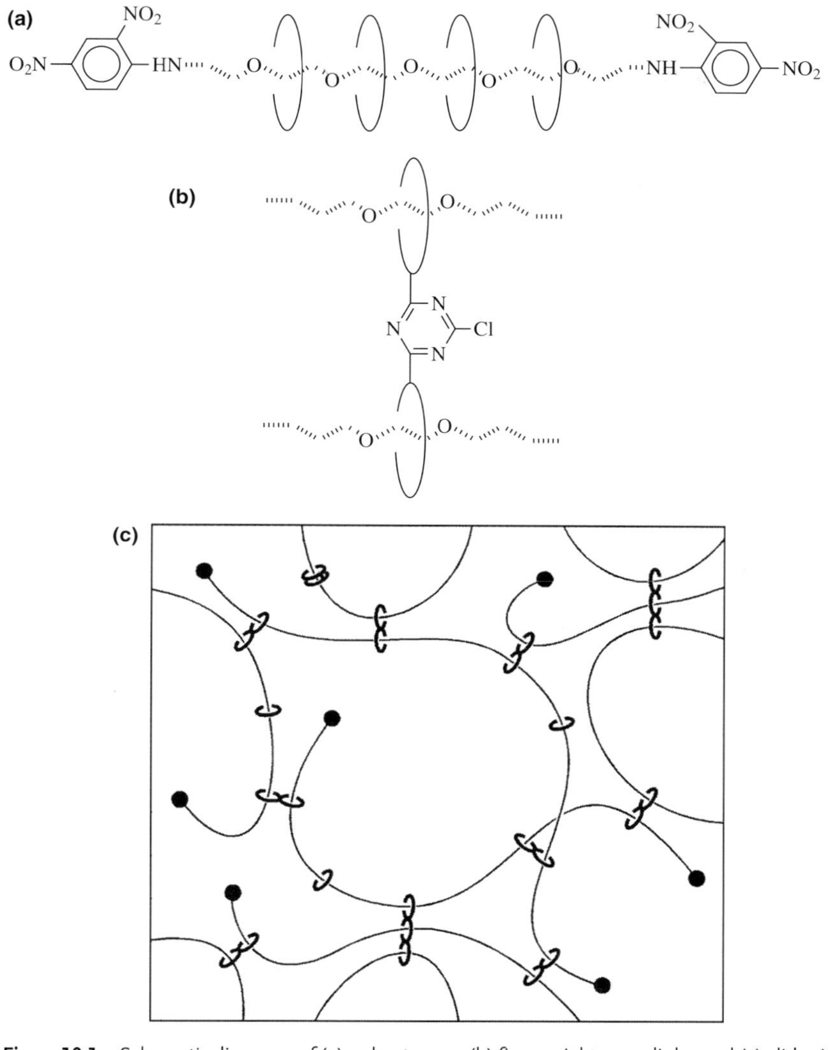

Figure 10.1 Schematic diagrams of (a) polyrotaxane, (b) figure-eight cross-links, and (c) slide-ring or topological gel. Reprinted with permission from Wiley-VCH 2001, [29].

fixed to the PEG chains and can move freely in the polymer network. We refer to this new cross-linked network polymer as a *slide-ring material*. Such a polymeric material with freely movable cross-links was theoretically considered as a sliding gel by de Gennes in 1999 [30]. In addition, the historical significance of slide-ring materials or gels was reviewed by comparing them with the slip-link model of Granick and Rubinstein [31]. Here, we provide an overview of the general features, synthesis, structure, and mechanical properties of slide-ring gels. It is noteworthy that the concept of freely movable cross-links is not limited to slide-ring gels containing solvents, but it can also be applied to slide-ring polymeric materials prepared without solvents. As mentioned later, it is clear that the freely movable cross-links drastically change the mechanical properties of polymeric materials, leading to remarkable scratch-resistant properties. This may bring about the first major paradigm shift in cross-linked polymeric materials since the discovery of chemical cross-linking by Goodyear.

10.2
Pulley Effect of Slide-Ring Materials

It is well known that a chemical gel generally has a large inhomogeneous structure due to the gelation process, which considerably reduces its mechanical strength, as schematically shown in Figure 10.2. In a chemical gel, long polymer chains are divided into shorter pieces of different lengths by fixed cross-links. As a result, the tensile stress is concentrated in the shorter chains, thereby breaking down the chemical gel easily. To resolve this problem, we have recently developed a slide-ring gel having freely movable cross-links based on the supramolecular architecture with topological characteristics [29]. In slide-ring gels, the polymer chains with bulky end groups exhibit neither covalent cross-links as in chemical gels nor attractive interactions as in physical gels, but are topologically interlocked by figure-eight-shaped cross-links. These cross-links can therefore freely pass along the polymer chains in order to equalize the tension of the threading polymer chains in a pulley-like manner, in what we refer to as the *pulley effect* [29].

Figure 10.3 schematically shows a comparison between chemical and slide-ring gels upon tensile deformation [29]. The polymer chains in the chemical gel are progressively broken due to the length of the heterogeneous polymer between the fixed cross-links. The polymer chains in the slide-ring gel, on the other hand, can pass through the figure-eight-shaped cross-links that act as pulleys. It is noteworthy that the equalization of tension can take place not only in a single polymer chain but also among adjacent polymers that are interlocked by figure-eight-shaped cross-links. As a result, a slide-ring gel exhibits a high degree of stretchability, up to 24 times its length, and a large change in volume of up to 24 000 times its

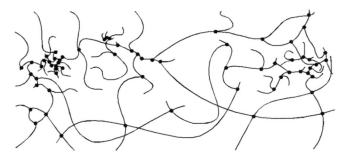

Figure 10.2 Schematic diagram of chemical gel with fixed network junctions.

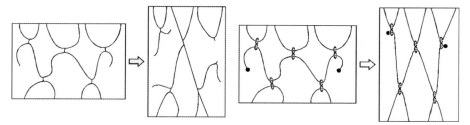

Figure 10.3 Schematic comparison between chemical (left) and slide-ring gels (right). In the slide-ring gel, the pulley effect automatically disperses the tension in the polymer chains upon tensile deformation, while in the chemical gel, the tensile stress is concentrated in the short chains. Reprinted with permission from Wiley-VCH 2001, [29].

original weight. Furthermore, we expect that the pulley effect will be able to produce various unique properties in a slide-ring gel, these being different from those in a chemical gel, as will be discussed later.

10.3
Synthesis of Slide-Ring Materials

The first slide-ring gel was synthesized by Okumura and Ito [29]. We used diamino-terminated PEG (PEG-DAT) with a molecular weight of 20 000 as the axis of the polyrotaxane, because PEGs having larger molecular weights form sparse inclusion complexes with α-CD. The average molecular weight of the polyrotaxane was determined to be 82 000 by ^{1}H NMR and ultraviolet spectroscopy, which indicated that 64 α-CD units were captured in the polymer chain. The polymer chain can hold a maximum of about 230 α-CD units if the α-CDs are densely packed; the inclusion ratio was determined to be 28%. There are more than a thousand hydroxyl groups on the α-CDs contained in polyrotaxane, while the axis of bis(2,4-dinitrophenyl)-PEG (PEG-DNB2) has no hydroxyl groups. By intermolecular cross-linking of the α-CDs contained in the polyrotaxanes, a transparent gel was obtained. Such a gel was not formed by cross-linking of the mixture of PEG-DNB2 and α-CD of the same composition as that of the polyrotaxane under the same conditions. Furthermore, we treated the slide-ring gel with a strong base (1 N NaOH) at 50 °C so as to remove the bulky stoppers (dinitrophenyl groups) at each end of the polyrotaxanes, with a rapid liquefaction of the gel observed after 7 h [29]. This indicates the first evidence of the movable cross-links; thus, the polymer network in the gel is maintained by the topological restrictions, not by the chemical cross-links.

Araki and Ito have recently developed a new method for the synthesis of sparse polyrotaxane using a minimum number of reagents, and an easy, one-step, high-yield method of preparation for PEG-COOH of any molecular weight with a high degree of modification, by using 2,2,6,6-tetramethylpiperidine-1-oxyl radical (TEMPO) mediated oxidation, as shown in Figure 10.4 [32]. TEMPO oxidation yields a carboxyl content that is higher than 95% in PEG-COOH through the conversion of the

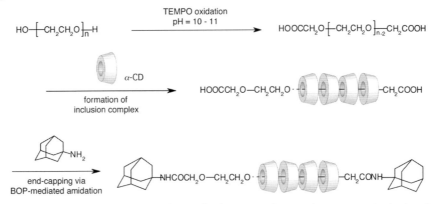

Figure 10.4 High-yield synthesis scheme of polyrotaxane of PEG and α-CD capped at both ends with adamantane. Reprinted with permission from American Chemical Society 2005, [32].

terminal hydroxyl groups. The preparation of the polyrotaxane from PEG-COOH was further investigated by comparing it to that from PEG-DAT. The PEG-based yield of the combination of PEG-COOH and 1-adamantanamine used as a capping agent ranged from 91% to 98%, which is much higher than that of PEG-DAT. These novel improvements result in a high-yield production of polyrotaxanes at low cost that can potentially be applied to the large-scale manufacture of polyrotaxanes for use in prospective applications.

Yui et al. have controlled the amount of α-CDs in polyrotaxanes that contain PEG by using a mixture of dimethyl sulfoxide (DMSO) and water during the formation of the inclusion complex [33, 34]. The solvent has a significant influence on the inclusion ratio of α-CDs in the polyrotaxanes. Recently, Hadziioannou et al. changed the time, temperature, and initial ratio of α-CD to PEG (having a molecular weight of 20 000) and indicated that the inclusion number of α-CDs varied from 3 to 125 [35]. It was found that very sparse pseudo-polyrotaxanes were formed at a high temperature, while the threaded α-CD acted like a nucleus with a favorable driving force for the final inclusion complex at a lower temperature.

Polyrotaxane modification is very important for the preparation of slide-ring materials [17]. The modifications of polyrotaxane were analogous to those performed upon other oligosaccharides or polysaccharides, including cellulose, chitin, starch, and cyclodextrin. The modification of polyrotaxane, however, brings a problem that is peculiar to the supramolecular structure of polyrotaxane, namely the degradation or decomposition of molecules. It is essential, then, that a solution to this problem be found. Whereas molecular degradation during the modification of other polymers is also problematic, in these other cases the resulting damage could be limited to a slight decrease in the molecular weight of the obtained derivatives without affecting their overall properties. However, the fragmentation of any moiety in a polyrotaxane molecule always causes significant changes in the polyrotaxane structure, in that all the scissions that take place in cyclic molecules, linear axes, bulky end groups, and their binding moieties, induce the dissociation of the cyclic molecules, which results in a loss of the polyrotaxane structure. Therefore, any decomposition that includes hydrolytic, oxidative, thermal, or biological degradation must be strictly avoided during the modification of polyrotaxane. Typical modifications of polyrotaxane have been reviewed elsewhere [17].

Thus, we can see progress on a number of fronts in the synthesis and modification of a polyrotaxane consisting of PEG and α-CD that promotes the application of slide-ring materials obtained from the polyrotaxane. Since the first polyrotaxane was synthesized from PEG and α-CD, a great many have studies concentrated on PEG-based polyrotaxane. Thanks to the knowledge that was accumulated, a decade later the first slide-ring material was prepared from a PEG-based polyrotaxane. As described below, all the fundamental studies to date have been performed with the first generation of slide-ring material. Recently, we have developed a new scheme for the synthesis of polyrotaxane that can be easily and generally applied to other polymer chains and CDs. This enables us to produce the next generation of slide-ring materials, which are addressed in detail in Chapter 7.

10.4
Scattering Studies of Slide-Ring Gels

A slide-ring gel has remarkable physical properties, such as high extensibility. In order to elucidate its physical appearance and properties, slide-ring gels and pregel solutions were investigated using small-angle neutron scattering SANS, SAXS, and quasi-elastic light scattering (QELS).

Shibayama et al. measured the SANS of sparse polyrotaxanes and investigated the sol–gel transition using solvents of deuterated DMSO (d-DMSO) or deuterated sodium hydroxide (NaODaq), and revealed the following results [36]. (1) The polyrotaxane chains show rod-like conformations in d-DMSO and behave like Gaussian chains in NaODaq. (2) The degree of inhomogeneity of a slide-ring gel in NaODaq is at a minimum around the sol–gel transition, while the gel in d-DMSO increases monotonically with the concentration of cross-linkers. (3) The scattering intensity of the slide-ring gel in NaODaq can be described by a Lorentz function, while that in d-DMSO is given by the sum of a squared Lorentz function and the scattering function for a rod. These differences were ascribed to the stacking behavior of α-CDs in polyrotaxanes. Hadziioannou et al. also demonstrated the rod-like conformation of polyrotaxanes in d-DMSO [37]. The polymer conformation in a pregel solution is considered to be primarily responsible for the structure and properties of slide-ring gels. Mayumi et al. investigated the concentration-induced conformational change in hydroxypropylated polyrotaxane (H-PR) composed of polyethylene glycol (PEG) and hydroxypropylated α-cyclodextrins (CDs) at various concentrations from the overlap concentration c^* in the semidilute regime by SANS [38]. It was found that the persistence length of H-PR was dependent on the polymer concentration c_p, while those of PEG remained unchanged in the same molar concentration regime. This unusual concentration-dependence of the polymer conformation for H-PR may indicate that in H-PR, the CDs could slide freely and rapidly over the whole range of PEG chains in the neighborhood of c^*, but their mobility was suppressed as c_p increased due to the molecular interaction among CDs, which, for example, gave rise to hydrogen-bond formation.

Shibayama et al. also revealed a significant difference between the behavior of chemical gels and slide-ring gels by subjecting them to SANS upon uniaxial deformation [39]. It is well known that chemical polymer gels have inherently large inhomogeneous structures due to the non-random distribution of their cross-links [5]. These inhomogeneities usually increase with swelling or deformation because the cross-linking junctions are permanently fixed on the polymer chains, and the gel cannot adjust its cross-link distribution or polymer length in the network when its environment changes [5]. As a result, the usual chemical gel exhibits an abnormal butterfly pattern [40–46], that is, prolate patterns that are *parallel* to the stretching direction. This behavior has been qualitatively explained by Onuki [47] and by Panykov and Rabin [48, 49]. A slide-ring gel, however, shows a normal butterfly pattern, that is, prolate patterns that are *perpendicular* to the stretching direction, as shown in Figure 10.5 [39]. This is the first study to report that cross-linked polymer gels exhibit a normal butterfly pattern. The normal butterfly pattern was also observed in homogeneous polymeric materials such as polymer films and solutions, due to the orientation of the polymer chains along the elongation or flow direction [50]. This indicates that the slide-ring gel displays a homogeneous structure when it is elongated. It is considered that this distinct difference between the slide-ring and chemical gels is a result of the pulley effect.

Amemiya et al. performed SAXS of slide-ring gels, focusing on the structure of the movable cross-links [51]. The main advantage of SAXS over SANS is that the exposure time is shorter due to the high brilliance of synchrotron radiation X-rays. This short exposure time makes it possible to study the structure of gels in various types of evaporable solvents. The SAXS results showed that in a poor solvent, the sliding cross-links form aggregates that prevent the pulley effect, while in a good solvent, the

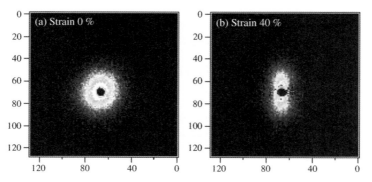

Figure 10.5 SANS patterns of slide-ring gel before (left) and after (right) uniaxial deformation of 1.4 times in length along horizontal axis. A normal butterfly pattern perpendicular to the deformed direction can be clearly observed, and the scattering intensity decreases with increasing extension ratio. Reprinted with permission from American Chemical Society 2005, [39].

polymer chains freely pass through the cross-links, acting as pulleys. A vertically elliptic pattern was observed in the two-dimensional SAXS profiles of covalently bonded chemical gels in a good solvent under uniaxial horizontal deformation, while an isotropic profile was observed for the slide-ring gels in a good solvent under deformation. This difference between the deformation mechanisms of the slide-ring and chemical gels confirmed the pulley effect of the slide-ring gels.

Using QELS, we investigated the sliding modes of cyclic molecules in polyrotaxanes and those of cross-linking junctions in the slide-ring gel [52]. It was found that a sparse polyrotaxane that included α-CDs exhibited the sliding mode in a solution in addition to the self- and cooperative-diffusion modes, while the sliding mode was not observed in a dense polyrotaxane that included α-CDs. After gelation of the sparse polyrotaxane, the self-diffusion mode of the polyrotaxane disappeared, but the sliding mode could still be observed. This indicates that the figure-eight-shaped cross-links in a slide-ring gel slide in the polymer network, passing through the polymer chains. The diffusion constant of α-CD in the sliding mode was two orders of magnitude smaller than that in free diffusion, which may have been caused by the trans–gauche transformation of the PEG chain.

10.5
Mechanical Properties of Slide-Ring Gels

The mechanical properties of slide-ring gels are quite different from those of conventional physical and chemical gels [53, 54]. A physical gel shows a J-shaped stress–strain curve with large hysteresis. This large hysteresis is caused by the recombination among noncovalent cross-links in a polymer network upon deformation. A chemical gel, on the other hand, shows no hysteresis because it has stable covalent cross-links that do not exhibit recombination upon deformation. In addition, a chemical gel shows an S-shaped stress–strain curve similar to that of cross-linked natural rubber, which is well explained by the three-chain model or the fixed-junction model. In the fixed-junction model, if we assume the affine deformation of the junction points and the additivity of the individual conformational entropies of the three Gaussian chains, we can obtain the dependence of the normal stress σ on the extension ratio λ for a uniaxial deformation, as expressed by the following well-known expression [55, 56]:

$$\sigma = \nu kT(\lambda - \lambda^{-2}), \tag{10.1}$$

where ν is the cross-linking density, k is the Boltzmann constant, and T is the temperature. The stress–strain curve of a chemical gel shows a concave-down profile in the low extension region, which is well fitted by Eq. (10.1), while it shows a concave-up profile that deviates from Eq. (10.1) in the high extension region due to non-Gaussian behavior or the stretching effect. As a result, a chemical gel shows an S-shaped stress–strain curve.

The slide-ring gel exhibited a J-shaped curve that is different from that of a chemical gel. Furthermore, it showed no hysteresis loop, in contrast to the physical

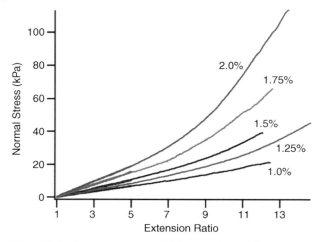

Figure 10.6 Stress–strain curve of slide-ring gel with different concentrations of cyanuric chloride as cross-linker. The slide-ring gel shows a J-shaped stress–strain curve without hysteresis.

gel, as shown in Figure 10.6 [53, 54]. This behavior can be explained in a qualitative manner by the modified three-chain model or the free-junction model that corresponds to the pulley effect, in which it is assumed that three Gaussian chains are able to slide toward one another in the following manner [54].

Figure 10.7 shows a comparison between chemical and slide-ring gels with respect to the stress–strain curve in a low extension region. The chemical gel shows a concave-down stress–strain curve, as mentioned earlier, while the slide-ring gel exhibits a concave-up curve, which agrees qualitatively with the J-shape in experimental results, as shown in Figure 10.6. A difference can also be observed in the compression region [54]. The free junction model indicates that the slide-ring

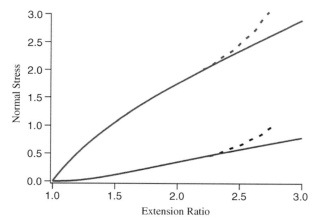

Figure 10.7 Comparison between fixed (upper) and free (lower) junction models with regard to stress–strain curves on uniaxial stretched deformation. The dotted curves reflect the deviation from the Gaussian chains.

material shows a significantly small amount of normal stress at a low compression ratio, and that the normal stress increases drastically when the compression ratio exceeds $\lambda \approx 0.4$. On the other hand, the fixed-junction model shows that the normal stress increases gradually with compression. It is noted that this difference is not limited only to within the gel materials, but it could be observed in the cross-linked polymeric materials without solvents. Consequently, the inherent pulley effect that arises from freely movable cross-links provides mechanical properties that are significantly different from those of the fixed cross-linking junctions.

Many biomaterials such as mammalian skin, vessels, and tissues show J-shaped stress–strain curves, which usually provide toughness without elastic instability, among other advantages [57–59]. The J-shaped stress–strain curve represents the toughness because its low shear modulus drastically reduces the energy released in a fracture, which would drive crack propagation. In addition, the material becomes stiffer as the extension ratio approaches the fracture point. Slide-ring materials exhibit J-shaped stress–strain curves that are similar to the curves exhibited by biomaterials such as mammalian skin, vessels, and tissue. This indicates that slide-ring materials are suitable for use as substitutes for various types of biomaterials. If artificial arteries were made of slide-ring materials, they might be better adapted for use with native arteries than fixed cross-linked materials.

In their recent study on the viscoelastic properties of slide-ring gels, Hadziioannou et al. discovered a relaxation mode that was ascribed to the sliding of α-CD rings on the PEG chain [60]. Then, based on the viscoelastic behavior of slide-ring gels in DMSO or water, they indicated that a slide-ring gel in water contained highly cross-linked chemical and physical domains with short flexible chains, while the network in the DMSO had long rigid chains due to low cross-link density and the presence of long α-CDs transient tubes. This finding is consistent with the experimental results of SANS and SAXS mentioned above. The long α-CDs transient tubes in DMSO also caused a difference in the relaxation time. Furthermore, Hadziioannou et al. collected systematic data to verify the implications suggested by SANS and viscoelastic measurements [37].

Koga and Tanaka performed a molecular simulation of Brownian dynamics by using a simple model of polymer networks with tri-functional sliding junctions in order to study the elastic properties of a slide-ring gel as compared to those of conventional chemical gels [61]. A J-shaped stress–strain curve was obtained, and the mobility and distribution of the sliding junctions along the polymer chains changed drastically upon deformation. They also observed that the uniaxial deformation formed aggregations of the sliding junctions with a decrease in the number of elastically effective chains.

10.6
Sliding Graft Copolymers

We have recently developed other slide-ring materials from polyrotaxane structures by grafting a number of functional groups, mesogen, and polymer chains onto cyclodextrins, to form what are called sliding graft copolymers. Kataoka et al. found thermoreversible gelation and microphase formation of aqueous solutions of a methylated polyrotaxane (MePR) [62, 63]. The aqueous solutions of MePR show a lower critical solution temperature (LCST) and form an elastic gel with increasing temperature. The sol–gel transition of the MePR solutions was induced by the formation and deformation of aggregates of methylated α-CDs of MePR that were due to hydrophobic dehydration and hydration, respectively. Wide-angle X-ray diffraction (WAXD) investigation revealed the localization and highly ordered arrangement of methylated α-CDs along the PEG chain in the gel. This arrangement of CDs was also reflected in changes in elasticity and the long relaxation behavior of the solution around the sol–gel transition. The quasiequilibrium shear modulus of MePR solutions showed these critical phenomena against temperature. The scaling exponents measured at two different concentrations were almost equal to the values predicted by a gel percolation theory [64]. Therefore, the heat-induced gelation of aqueous MePR solutions is well explained by a model in which clusters assembled with methylated α-CDs are gradually connected to the network as the temperature increases.

Kidowaki et al. developed a novel side-chain liquid crystalline polyrotaxane (LCPR) whose mesogenic side chains can slide and rotate on the linear backbone. To obtain LCPR, 6-(4′-cyanobiphenyl-4″-yloxy)-hexanoyl chloride was prepared and reacted with a polyrotaxane consisting of α-CD, poly(ethylene glycol) ($M_w = 35\,000$), and adamantane end groups [65]. The molecular characterization of LCPR was performed with FT-IR, ^{1}H NMR, and gel permeation chromatography analysis. Its phase structures and transitions were investigated by differential scanning calorimetry, WAXD, and polarized light microscopy experiments. The experimental results show that LCPR has a glass transition at 70 °C and a mesomorphic isotropic transition at 129 °C. Araki et al. developed a novel sliding graft copolymer in which many linear poly-ε-caprolactone (PCL) side chains are bound to the cyclodextrin rings of a polyrotaxane, which was prepared by ring-opening polymerization of ε-caprolactone initiated by the hydroxyl groups of the polyrotaxane [66]. An amorphous, flexible, and sufficiently tough elastomer film that possessed remarkable scratch-resistant properties was prepared by cross-linking the obtained SGC – a supramolecule possessing a number of mobile side chains – with hexamethylene diisocyanate (HMDI). The resulting slide-ring elastomer shows scratch-resistant properties that can be applied to coating materials for automobiles, cell phones, mobile computers, fishing rods, golf clubs, and so on.

10.7
Recent Trends of Slide-Ring Materials

10.7.1
Introduction: Diversification of the Main Chain Polymer

As we have described thus far, the fundamental properties of slide-ring materials prepared from a polyrotaxane consisting of PEG and α-CD have been intensively investigated. The development of procedures for the synthesis of polyrotaxane and of various modifications to the α-CD threads has enabled us to handle these materials easily and to realize their practical use. Through these intensive studies and applications, PEG as a general-purpose polymer has grown to become a base polymer of slide-ring materials that have peculiar properties. Thus, the diversity of main chain polymers creates the expectation that there will explosive growth in the use of slide-ring materials. Replacing one main chain polymer with another, however, is not at all a straightforward process.

The general procedure for the synthesis of slide-ring material is shown in Figure 10.8. Slide-ring materials are synthesized by the cross-linking of polyrotaxanes, which are obtained by end-capping reactions of the corresponding pseudo-polyrotaxanes. Since pseudo-polyrotaxane formations depend on self-organizing

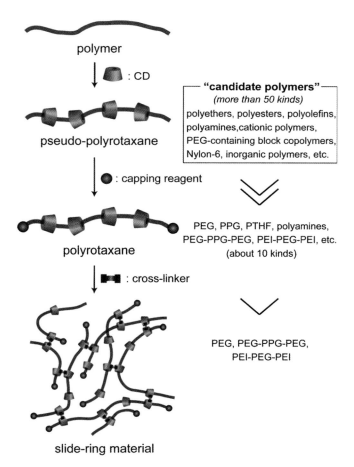

Figure 10.8 General synthesis procedure of slide-ring materials and varieties of main chain polymers. Many types of candidate polymers that spontaneously form inclusion complexes with CDs have been found so far.

processes between a polymer and CD, it is not easy to predict whether or not the polymer and CD will form an inclusion complex. Harada and coworkers have systematically studied such complex formations with a wide variety of polymers and CDs [67]. They found that various polymers, for example, polyethers, polyesters, polyolefins, and inorganic polymers form pseudo-polyrotaxanes with CDs, depending on the size of the inner cavities of the CDs and the thickness of the polymers. Wenz and coworkers investigated these complex formations with various polyamines and cationic polymers [15]. Tonelli and coworkers found that Nylon 6-based composites form a pseudo-polyrotaxane with α- and β-CD [68]. Block copolymers of PEG and PPG were also found to form pseudo-polyrotaxanes by Yui and coworkers [69] and by Li and coworkers [70]. As the result of their enormous body of research work, we now know various candidates for the main chain polymers of slide-ring materials.

The current crucial problem is the subsequent end-capping reactions needed to obtain polyrotaxanes. Since the first report on PEG-based polyrotaxane in 1992 by Harada et al. [22], most studies of polyrotaxanes have concentrated on a single polyrotaxane, despite the existence of a number of other candidate polymers. Although the achievement of the first capping reaction came only two years after the first report on the corresponding pseudo-polyrotaxane, most of the other candidate pseudo-polyrotaxanes have yet to be converted into polyrotaxanes. The reason for this is that the capping reactions must meet the diverse requirements of the candidate polymers. Unlike water-soluble PEG, hydrophobic polymers require inhomogeneous complexations with CDs and specific techniques to achieve the subsequent end-cappings. Takata and coworkers achieved the synthesis of a polyrotaxane consisting of poly(tetrahydrofuran) and methylated α-CD via solvent-free complexation and end-capping [71]. Unlike PEG, which is thin enough to be accommodated with α-CD, thick polymers require larger CDs such as β-CD and γ-CD, so that larger end-capping groups are needed. At the same time, such bulky capping reagents must have enough reactivity and solubility to prevent the unraveling of the CD threads. In the case of α-CD-based polyrotaxanes, for example, the 2,4-dinitrophenyl and adamantyl groups are bulky enough for capping while the 4-nitrophenyl group is not sufficiently bulky. For β-CD-based polyrotaxanes, the 2,4-dinitrophenyl and adamantyl groups are already too small, so that trityl and 9-anthracenyl groups, for example, are needed. On the other hand, in the case of γ-CD-based polyrotaxane, it remains unclear at what point the bulk size becomes critical. For example, while some reports suggest that the trityl group is large enough, others do not support this. In fact, there have been only a few reports on polyrotaxanes with γ-CD.

A number of creative ways of obtaining polyrotaxanes with γ-CD have been reported. Herrmann et al. demonstrated the first polyrotaxane with a mixture of β-CD and γ-CD, which was realized through photochemical modification of the main chain polymer [72]. Since the resultant joints of the polymer are not bulky enough for γ-CD but are for β-CD, the β-CDs captured by the joints can block γ-CDs. Okada et al. synthesized a polyrotaxane consisting of PPG and γ-CD through the photochemical coupling reaction of the 9-anthracenyl group at both ends of the pseudo-polyrotaxane [73]. Although the 9-anthracenyl group is small enough to form pseudo-polyrotaxane with γ-CD, the dimerized group is large enough to prevent the unraveling of γ-CD. In terms of the application of these polyrotaxanes to slide-ring materials, it is not advisable for these polyrotaxanes to have many bulky joints on the main chain polymers, as these disable the sliding of γ-CDs. Recently, Kato et al. succeeded in synthesizing a polyrotaxane consisting of poly(dimethylsiloxane) (PDMS) and γ-CD by a capping reaction with bulk reagents prepared from p-methoxytritylamine [74]. Takahashi et al. synthesized a polyrotaxane consisting of PEG and γ-CD capped with a monoamino β-CD derivative.

There has recently been an increase in the number of studies of γ-CD-based polyrotaxanes along these lines. This accumulated knowledge on the capping reaction to obtain γ-CD-based polyrotaxanes will enable us to design more diverse polyrotaxanes. Indeed, many of the candidate polymers form pseudo-polyrotaxanes with γ-CD selectively. Therefore, we await a breakthrough in the synthesis of γ-CD-based polyrotaxanes that will herald the development of next-generation slide-ring materials. Our focus above has been on the following two types of candidates for main chain polymers of slide-ring materials: (1) inorganic polymers and (2) reactive main chain polymers. The former is quite attractive as a new class of organic–inorganic materials. The latter has the potential to control the dynamics of the cross-linking points of the materials. We describe the details of these in later chapters, based on our recent breakthrough syntheses of γ-CD-based polyrotaxanes.

10.7.2
Organic–Inorganic Hybrid Slide-Ring Materials

In the application of slide-ring materials, it is necessary to consider not only the characteristic properties that arise from the slidable cross-linking points, but other properties as well, such as the chemical, thermal, and optical properties that arise mainly from the main chain polymers. PEG, which has thus far been the main axis polymer of slide-ring materials, offers such advantages as hydrophilicity and biocompatibility. Thus, the slide-ring materials prepared from PEG are also hydrophilic enough to yield hydrogels, and are nontoxic to human cells and muscle. In the same way, it would be possible to endow the materials with various properties and functions by changing the main chain.

Silicone is a general-purpose inorganic polymer that has various properties such as thermostability, chemical stability, and gas permeability, and that also has good electrical-insulation properties. Because of their distinctive properties as an inorganic material, hybrid materials with organic compounds, the so-called *organic–inorganic hybrid materials*, have been investigated intensively. For example, the hybrid of silicone and a hydrophilic polyolefin yields a flexible and hydrophilic material with high oxygen permeability that is currently widely used in soft contact lenses. Since inorganic and organic materials are usually immiscible, their mere mixture causes macroscopic phase separation. To avoid this separation, there are three main types of methods for creating organic–inorganic hybrid materials, as shown in Figure 10.9. The first involves a covalently bound network of organic and inorganic polymers. This type of network is generally prepared by polymerization of an organic monomer

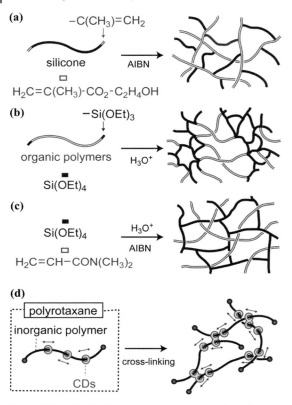

Figure 10.9 (a)–(c) Three main preparation methods for organic–inorganic hybrid materials and (d) newly proposed method based on synthesis of inorganic polymer-based polyrotaxane and subsequent cross-linking.

in the presence of a telechelic silicone, which is caught up in the growing organic chains covalently. The second method employs a network that functions quite differently from the first one. Telechelic organic polymers modified with trialkoxysilyl moieties are bound to polysiloxane networks by so-called sol–gel processing. The third method involves an entangled network that does not have covalent bonds or strong interactions between organic and inorganic polymers: the so-called interpenetrating polymer network (IPN). This kind of network is formed, for example, by the kinetically-controlled simultaneous polymerization of an organic monomer and the formation of an inorganic network by sol–gel processing.

We propose here a new class of organic–inorganic hybrid materials based on rotaxanation between the inorganic polymers and organic cyclic molecules, and on their subsequent cross-linking to obtain slide-ring materials, as shown in Figure 10.9d. Since the cyclic component is bound topologically, not covalently, to the inorganic polymer network, it would be impossible for macroscopic phase separation to occur. It is notable that the relative positions of both components can

10.7 Recent Trends of Slide-Ring Materials

change because of the sliding of the organic components. The key to this kind of hybrid is the synthesis of a polyrotaxane that has inorganic main chain polymers.

Okumura et al. systematically studied pseudo-polyrotaxane formations between CDs and PDMS, which is a representative silicone [75, 76]. They reported that PDMS forms densely packed pseudo-polyrotaxanes with γ-CD at MW < 3300, while no complexes are obtained from α- or β-CD. They also reported on selective complexations between poly(dimethylsilane) and γ-CD [77, 78]. These crucial reports on the first inorganic polymer-based pseudo-polyrotaxanes, which have been frequently cited, have aroused great expectations of new materials. Porbeni et al. made a detailed study of the thermal stability of PDMS-based pseudo-polyrotaxanes depending on their molecular weights and supported their selective complexation with γ-CD [79]. On the other hand, Sukhanova et al. and Marangoci et al. reported the high-yield production of pseudo-polyrotaxanes between PDMS and β-CD [80, 81].

Recently, we have finally succeeded in synthesizing the first polyrotaxane and slide-ring material consisting of PDMS and γ-CD, using the scheme of synthesis shown in Figure 10.10 [74]. The key to the synthesis of polyrotaxane is the capping reaction of the pseudo-polyrotaxane with bulky capping regents in controlled conditions. First, the bulkiness of these capping regents is crucial. As mentioned above, it remains controversial whether an unmodified trityl group is bulky enough to block γ-CD or not. On the other hand, our result shows that a p-methoxy substituent on a trityl group is effective. Through this study, the point at which bulkiness produces the blocking of

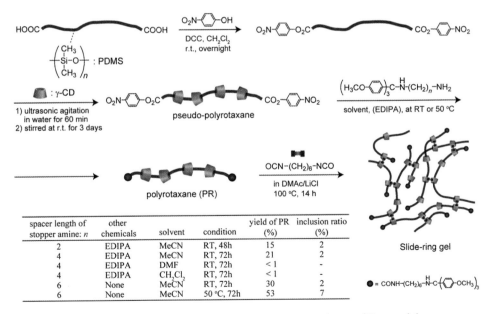

Figure 10.10 Synthesis scheme of polyrotaxane and slide-ring gel consisting of PDMS as main chain and γ-CD as cyclic component. The embedded table shows the various capping reaction conditions and the production yields of polyrotaxanes on a polymer basis. Reprinted with permission from American Chemical Society 2009, [74].

γ-CD became much clearer than before. Second, the activated polymer ends of pseudo-polyrotaxanes are very effective in the capping process. In the case of PEG-α-CD polyrotaxane, the carboxyl groups on both ends of the pseudo-polyrotaxane could bind with the bulky amine that is directly mediated by the BOP reagent. In the case of PDMS-γ-CD, however, this condensation reaction did not occur. The origin of this absence of reaction could be that the BOP is disabled by its encapsulation with γ-CD, since the reagent is thought to form an active cyclic complex with carboxylic acid and amines in the first stage of condensation. We therefore first activated the carboxyl groups as a p-nitrophenyl ester, and the active ester then reacted with the bulky amines without the help of any condensation agent. Third, the length of the methylene spacer affected the production yield. This result clearly shows that the higher accessibility of the amino group to the polymer ends with a longer spacer is advantageous to the reaction.

In all cases, appropriate solvents for the capping reaction are necessary. Acetonitrile and acetone, poor solvents for both PDMS and γ-CD, both showed a good performance. On the other hand, good solvents for either PDMS or γ-CD, such as dichloromethane or DMF, resulted in failure. We found that PDMS was extracted from the pseudo-polyrotaxane with dichloromethane. This indicates that the pseudo-polyrotaxane of PDMS and γ-CD tends to dissociate immediately, unlike that of PEG and α-CD. In the case of the pseudo-polyrotaxane of PEG and α-CD, DMF is generally used for the end-capping reaction. Since this is a good solvent for both components, the solid or gel of pseudo-polyrotaxane gradually dissolved during the capping process. When the capping reaction is faster than the dissociation, PEG-α-CD polyrotaxane is obtained. Thus, the slow dissociation is advantageous to the end-capping. No such convenient solvent was found for the pseudo-polyrotaxane of PDMS and γ-CD, and it would also be difficult to find such solvents for other pseudo-polyrotaxanes. Therefore, the end-capping carried out in poor solvents for both components is a solution to the end-capping reactions for such readily dissociative pseudo-polyrotaxanes.

The solubility of PDMS-γ-CD polyrotaxane is quite unique. The polyrotaxane is soluble in DMF and DMAc only in the presence of lithium chloride. This result obviously shows that the interruption of hydrogen bonds among γ-CD is essential to the dissolution. The most surprising result, however, is that polyrotaxane is dissolved by solvents that are poor solvents for PDMS, in spite of the low inclusion ratio (7%). Since the main chain polymer is not miscible in such hydrophilic solvents, the hydrophilic γ-CD would have to somehow cover the hydrophilic main chain. Quite recently, various scattering studies and direct observation by electron microscopes have revealed a unique micelle formation in solution. Polyrotaxanes consisting of hydrophilic polymers and CDs have sometimes been synthesized with the aim of realizing their solubilization in hydrophilic solvents. In such cases, densely packed polyrotaxanes were required and prepared. On the other hand, the sparsely covered PDMS-γ-CD polyrotaxane is a typical amphiphilic polyrotaxane; to the best of our knowledge, there is no good solvent for both components. Since the hydrophilic parts can slide along the hydrophilic chain, this kind of polyrotaxane can behave like an amphiphilic block copolymer and random copolymer, as shown in Figure 10.11.

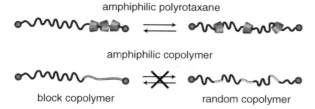

Figure 10.11 Intramolecular phase separation and dispersion of single amphiphilic polyrotaxane, suggesting structural analogy with amphiphilic block and random copolymers.

The intramolecular phase separation and dispersion can be controlled by various stimuli and environments, and various unusual properties then manifest themselves, raising the expectation of a new class of amphiphilic polymers.

Cross-linking of the obtained silicone-based polyrotaxane yields a new type of slide-ring material. The silicone-based polyrotaxane was dissolved in DMF/LiCl, to which hexamethylene diisocyanate was then added as a cross-linker to obtain an organic–inorganic hybrid slide-ring gel, as shown in Figure 10.12. The obtained gel is transparent and flexible. The transparency indicates that no macroscopic phase separation or aggregation has occurred, even during the cross-linking process. The amphiphilic polyrotaxane has no opportunity to induce macroscopic phase separation because of the topological bonds between both components. Thus, the topological network also shows no macroscopic phase separation. This process of obtaining a material produced by the synthesis of an amphiphilic polyrotaxane and the subsequent topological cross-linking gives us a new idea regarding the nanoscale mixing required for organic–inorganic hybrid materials. To begin with, polyrotaxane is a hybrid molecule in which a polymer and cyclic molecules are topologically bound, no matter how small an affinity both components have for one another. Although the synthesis of a polyrotaxane that consists of two immiscible components is not easy, such amphiphilic polyrotaxanes are promising in terms their potential contribution to new hybrid materials.

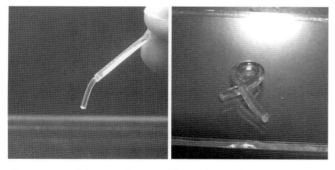

Figure 10.12 Slide-ring gel prepared from solution of polyrotaxane consisting of PDMS and γ-CD in DMF/LiCl. Reprinted with permission from American Chemical Society 2009, [74].

10.7.3
Design of Materials from Intramolecular Dynamics of Polyrotaxanes

As we described in detail in the above chapters, all the characteristic properties of slide-ring materials depend on the unusual mobility of the cross-linking points, particularly the sliding along the polymer chains. The direct correlation between microscopic mobility and macroscopic properties has been intensively studied using PEG-α-CD polyrotaxane-based materials. The results obtained convince us that a next-generation slide-ring material that can be endowed with a variety of macroscopic properties on demand will be created once we succeed in controlling the microscopic sliding. It is true that the sliding motion must be affected by factors such as the solvents, temperature, and the modified groups on the outer parts of CDs. However, attaining control of the inner friction between CDs and the main chain polymer threads offers a quite promising prospect for attaining direct control of the macroscopic properties. If we could introduce an appropriate obstacle into the main chain step-by-step to increase the friction between CDs and the chain, it would be possible to observe the stepwise changes in various macroscopic properties. For this purpose, it is necessary to prepare a polyrotaxane that has reactive moieties in the main chain polymer. The reactive moieties must be convertible under mild conditions in order to avoid side reactions and deformation of the polyrotaxane structures.

Polydienes are promising candidates in this regard, since the unsaturated carbons are susceptible to addition reactions with various reagents. By changing the bulkiness of the reagents or the degree of modification, the thickness of the main chains or the number of obstacles could be systematically changed, resulting in an increase in the friction between the CDs and the modified main chains, as depicted in Figure 10.13. Michishita et al. found that polybutadiene (PBD) selectively forms an inclusion

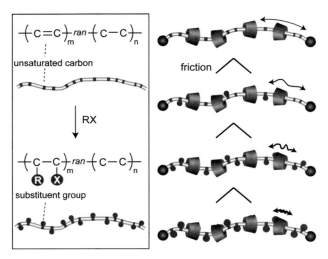

Figure 10.13 Systematic control strategy for intramolecular friction of polyrotaxanes by addition reactions to unsaturated carbons of polydienes of the main chain.

complex with γ-CD when the MW of PBD is higher than 1000 [82]. Later, Kuratomi et al. discovered that 1,4-polybutadiene can also form pseudo-polyrotaxanes with α-CD only in the absence of the bulky 1,2-structure [83]. Polyisoprene also selectively formed an inclusion complex with γ-CD [84]. In this way, the favorable complex formations with γ-CD require the complex capping reactions mentioned earlier in the case of PDMS.

We have quite recently overcome the low versatility of end-capping reactions by employing a breakthrough method, which is illustrated in Figure 10.14 [85]. Using a single method, various polyrotaxanes were successfully synthesized using PEG, PDMS, and even PBD as the main chains. The end-capping reactions were achieved by transesterification at both ends of pseudo-polyrotaxanes that had excess CDs. In general, excess CDs are necessary for accelerating pseudo-polyrotaxane formation, and the excess CDs are then washed out before and/or after the subsequent capping reactions. Instead of utilizing excess CDs, other reactive bulky reagents, so-called capping reagents, are normally used. This novel method, however, actually takes advantage of the excess CDs for the transesterification reaction. Transesterification is generally accelerated by adding an excess of substitutive alcohols (CDs in this case) and by removing the elimination alcohols (p-nitrophenol in this case). Since p-nitrophenol can efficiently form inclusion complexes with CDs, this method satisfies both the conditions needed for favoring the transesterification reaction.

This end-capping with excess CDs eliminates the need for capping reagents, which has long troubled us. Any size of CD that is used for pseudo-polyrotaxane formation is large enough for the end-capping, which means that any CD can play a dual role as both the cyclic component and the end-capping groups at the same time. In fact, this method was applied to the synthesis of the two polyrotaxanes that include α-CD and γ-CD. Moreover, the first polyrotaxane made up of γ-CD and PBD was successfully synthesized using the same method. Elimination of the need of capping reagents in

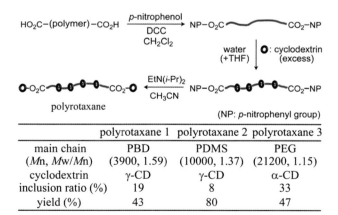

Figure 10.14 Versatile synthesis of polyrotaxanes requiring no capping regents. This single method produces three different polyrotaxanes, including a novel one consisting of PBD and γ-CD. Reprinted with permission from the American Chemical Society 2010, [85].

this way drastically increased the versatility and simplified the process of polyrotaxane synthesis. In our current work we are modifying the reactive main chain of PBD-based polyrotaxane with various groups and cross-linking these in order to elucidate the effect of this on the macroscopic properties of next-generation slide-ring materials.

10.8
Concluding Remarks

For a long time, since their discovery by Goodyear in 1839, cross-linking junctions had been considered to be fixed to the polymer chains. The elasticity of rubber at a low extension ratio is well explained by a fixed-junction model by assuming affine deformation of the network with fixed-junction points and entropy change in the Gaussian chains. We have recently discovered a new class of materials that comprise freely movable cross-linking junctions, that is, slide-ring materials. These materials are prepared by cross-linking only cyclic molecules, which are sparsely threaded on a linear polymer chain. Because the cross-linking junctions can move within the polymer network, the structure and properties of the polymeric materials are drastically different from those of the conventional cross-linked or non-cross-linked materials. This phenomenon is called the pulley effect. We observed from SANS and SAXS studies that slide-ring gels exhibit the normal butterfly pattern upon uniaxial deformation, this being different from the abnormal butterfly pattern exhibited by chemical gels with fixed junctions. The mechanical behavior of the slide-ring materials was well explained by the free junction model, taking into account the pulley effect. The slide-ring materials showed a J-shaped stress–strain curve that was different from the S-shaped curve shown by polymeric materials with fixed junctions, such as cross-linked rubbers and chemical gels. The peculiar mechanical behavior of the slide-ring materials is similar to that of biomaterials such as mammalian skin, vessels, and tissues, and may endow them with remarkable scratch-resistant properties. The slide-ring gel is a concrete example of the slip-link model, which had hitherto been considered only theoretically, and it can be regarded as a third cross-linking concept alongside the conventional chemical and physical concepts in which the polymer network is interlocked by topological restrictions. The concept of the pulley effect can be applied not only to gel materials such as soft contact lenses and polymer batteries but also to polymeric materials prepared without solvents, such as paints, textiles, and films. This concept offers a new framework for polymeric materials

The knowledge of such materials with an entirely new cross-linking concept has given us various ideas for next-generation materials. For example, replacement of the main chain with an inorganic polymer results in a new class of organic–inorganic materials in which the relative position of both phases can change within given topological restrictions. Introducing various substituents to the unsaturated carbons in reactive main chains such as polydienes enables us to control the intramolecular friction between the chains and the cross-linking junction. Increasing or decreasing the friction may affect the macroscopic mechanical behaviors directly. Such next-generation materials rely on recent developments in the synthesis of polyrotaxanes, which are the precursor macromolecules of slide-ring materials. The preparation of slide-ring materials was originally based on the supramolecular chemistry of polyrotaxane, and the subsequent development of these materials was led by polymer physicists until recently. Now that fundamental studies of these materials are beginning to reveal the correlations between macroscopic behavior and molecular structure, cooperation between polymer physicists and supramolecular chemists is now necessary for the diversification of slide-ring materials.

Acknowledgments

The authors wish to thank their colleagues Dr. Yasushi Okumura, Mr. Changming Zhao, Dr. Masatoshi Kidowaki, Dr. Jun Araki, Dr. Yasuhiro Sakai, Dr. Toshiyuki Kataoka, Dr. Aoi Inomata, and Mr. Koichi Mayumi for their continuous support during the course of this study. The authors also gratefully acknowledge the support of Dr. Takeshi Karino, Dr. Satoshi Okabe, and Dr. Mitsuhiro Shibayama in the SANS measurements and Mr. Kentaro Kayashima, Mr. Yuya Shinohara, and Dr. Yoshiyuki Amemiya in the SAXS measurements. Finally, the contribution of Dr. Akira Harada (Osaka University) in leading the authors to this new field of polymer science is sincerely acknowledged.

References

1 Osada, Y. and Kajuwara, K. (eds) (2000) *Gels Handbook*, Academic Press, Elsevier, Amsterdam.
2 Tanaka, T. (1978) *Phys. Rev. Lett.*, **40**, 820.
3 Tanaka, T., Fillmore, D.J., Sun, S.–T., Nishio, I., Swislow, G., and Shah, A. (1980) *Phys. Rev. Lett.*, **45**, 1636.
4 Tanaka, T. and Fillmore, D.J. (1979) *J. Chem. Phys.*, **70**, 1214.
5 Shibayama, M. (1998) *Macromol. Chem. Phys.*, **199**, 1.
6 Gong, J.P., Higa, M., Iwasaki, Y., Katsuyama, Y., and Osada, Y. (1997) *J. Phys. Chem.*, **101**, 5487.
7 Kagata, G., Gong, J.P., and Osada, Y. (2003) *J. Phys. Chem. B*, **107**, 10221.
8 Gong, J.P., Katsuyama, Y., Kurokawa, T., and Osada, Y. (2003) *Adv. Mater.*, **15**, 1155.
9 Tanaka, Y., Gong, J.P., and Osada, Y. (2005) *Prog. Polym. Sci.*, **30**, 1.
10 Yoshida, R., Takahashi, T., Yamaguchi, T., and Ichijo, H. (1996) *J. Am. Chem. Soc.*, **118**, 5134.
11 Yoshida, R., Uchida, K., Kaneko, Y., Sakai, K., Kikuchi, A., Sakurai, Y., and Okano, T. (1995) *Nature*, **374**, 240.
12 Lehn, J.H. (1995) *Supramlecular Chemistry: Concept and Perspectives*, VCH, Weinheim.
13 Takata, T., Kihara, N., and Furusho, Y. (2004) *Adv. Polym. Sci.*, **171**, 1.
14 Huang, F.H. and Gibson, H.W. (2005) *Prog. Polym. Sci.*, **30**, 982.
15 Wenz, G., Han, B.H., and Muller, A. (2006) *Chem. Rev.*, **106**, 782.
16 Harada, A., Hashidzume, A., and Takashima, Y. (2006) *Adv. Polym. Sci.*, **201**, 1.
17 Araki, J. and Ito, K. (2007) *Soft Matter*, **3**, 1456.
18 Ogino, H. (1981) *J. Am. Chem. Soc.*, **103**, 1303.
19 Szejtli, J. and Osa, T. (1996) *Comprehensive Supramolecular Chemistry*, vol. **3**, Cyclodextrins, Pergamon, Elsevier, Oxford.

20. Dodziuk, H. (2006) *Cyclodextrins and their Complexes: Chemistry, Analytical Methods, Applications*, Wiley-VCH, New York.
21. Wenz, G. (1994) *Angew. Chem. Int. Ed. Engl.*, **33**, 803.
22. Harada, A. and Kamachi, M. (1990) *Macromolecules*, **23**, 2821.
23. Harada, A., Li, J., and Kamachi, M. (1992) *Nature*, **356**, 325.
24. Frampton, M.J. and Anderson, H.L. (2007) *Angew. Chem. Int. Ed.*, **46**, 1028.
25. Li, J., Harada, A., and Kamachi, M. (1994) *Polymer J.*, **26**, 1019.
26. Okumura, Y., Ito, K., and Hayakawa, R. (1998) *Phys. Rev. Lett.*, **80**, 5003.
27. Watanabe, J., Ooya, T., and Yui, N. (2000) *J. Artif. Organs*, **3**, 136.
28. Kihara, N., Hinoue, K., and Takata, T. (2005) *Macromolecules*, **38**, 223.
29. Okumura, Y. and Ito, K. (2001) *Adv. Mater.*, **13**, 485.
30. de Gennes, P.G. (1999) *Physica A*, **271**, 231.
31. Granick, S. and Rubinstein, M. (2004) *Nat. Mater.*, **3**, 586.
32. Araki, J., Zhao, C., and Ito, K. (2005) *Macromolecules*, **38**, 7524.
33. Oya, T., Noguchi, M., and Yui, N. (2003) *J. Am. Chem. Soc.*, **125**, 13016.
34. Oya, T., Utsunomiya, H., Noguchi, M., and Yui, Y. (2005) *Bioconjugate Chem.*, **16**, 62.
35. Fleury, G., Brochon, C., Schlatter, G., Lapp, A., and Hadziioannou, G. (2005) *Soft Matter*, **1**, 378.
36. Karino, T., Okumura, Y., Ito, K., and Shibayama, M. (2004) *Macromolecules*, **37**, 6177.
37. Fleury, G., Schlatter, G., Brochon, C., and Hadziioannou, G. (2006) *Adv. Mater.*, **18**, 2847.
38. Mayumi, K., Osaka, N., Endo, H., Yokoyama, H., Sakai, Y., Shibayama, M., and Ito, K. (2008) *Macromolecules*, **41**, 6580.
39. Karino, T., Okumura, Y., Zhao, C., Kataoka, T., Ito, K., and Shibayama, M. (2005) *Macromolecules*, **38**, 6161.
40. Bastide, J. and Leibler, L. (1988) *Macromolecules*, **21**, 2647.
41. Bastide, J., Leibler, L., and Prost, J. (1990) *Macromolecules*, **23**, 1821.
42. Mendes, E., Lindner, P., Buzier, M., Boue, F., and Bastide, J. (1991) *Phys. Rev. Lett.*, **66**, 1595.
43. Zielinski, F., Buzier, M., Lartigue, C., and Bastide, J. (1997) *Prog. Colloid. Polym. Sci.*, **90**, 115.
44. Rouf, C., Bastide, J., Pujol, J.M., Schosseler, F., and Munch, J.P. (1994) *Phys. Rev. Lett.*, **73**, 30.
45. Ramzi, A., Zielinski, F., Bastide, J., and Boue, F. (1995) *Macromolecules*, **28**, 3570.
46. Shibayama, M., Kawakubo, K., Ikkai, F., and Imai, M. (1998) *Macromolecules*, **31**, 2586.
47. Onuki, A. (1992) *J. Phys. II*, **2**, 45.
48. Panyukov, S. and Rabin, Y. (1996) *Macromolecules*, **29**, 7960.
49. Rabin, Y. and Panyukov, S. (1997) *Macromolecules*, **30**, 301.
50. Shibayama, M., Kurokawa, H., Nomura, S., Roy, S., Stein, R.S., and Wu, W.L. (1990) *Macromolecules*, **23**, 1438.
51. Shinohara, Y., Kayashima, K., Okumura, Y., Zhao, C., Ito, K., and Amemiya, Y. (2006) *Macromolecules*, **39**, 7386.
52. Zhao, C., Domon, Y., Okumura, Y., Okabe, S., Shibayama, M., and Ito, K. (2005) *J. Phys. Cond. Matt.*, **17**, S2841.
53. Okumura, Y. and Ito, K. (2003) *Nippon Gomu Kyokaishi*, **76**, 31.
54. Ito, K. (2007) *Polym. J.*, **39**, 488.
55. Treloar, L.R.G. (1975) *The Physics of Rubber Elasticity*, 3rd edn, Oxford University Press, Oxford.
56. Mark, J.E. and Erman, B. (2007) *Rubber Elasticity: A Molecular Primer*, 2nd edn, Cambridge University Press, Cambridge.
57. DoITPoMS, University of Cambridge http//www.doitpoms.ac.uk/tlplib/bioelasticity/index.php (April 2006).
58. Vincent, J.F.V. (1990) *Structural Biomaterials*, Princeton University Press, Princeton.
59. Vogel, S. (2003) *Comparative Biomechanics: Life's Physical World*, Princeton University Press, Princeton.
60. Fleury, G., Schlatter, G., Brochon, C., and Hadziioannou, G. (2005) *Polymer*, **46**, 8494.
61. Koga, T. and Tanaka, F. (2005) *Euro. Phys. J. E*, **17**, 225.
62. Kataoka, T., Kidowaki, M., Zhao, C., Minamikawa, H., Shimizu, T., and Ito, K. (2006) *J. Phys. Chem., B*, **110** (48), 24377.

63 Kidowaki, M., Zhao, C., Kataoka, T., and Ito, K. (2006) *Chem. Commun.*, 4102.
64 De Gennes, P.G. (1979) *Scaling Concepts in Polymer Physics*, Cornell University Press, Ithaca, New York.
65 Kidowaki, M., Nakajima, T., Araki, J., Inomata, A., Ishibashi, H., and Ito, K. (2007) *Macromolecules*, **40**, 6859.
66 Araki, J., Kataoka, T., and Ito, K. (2008) *Soft Matters*, **4**, 245.
67 Harada, A., Hashidzume, A., Yamaguchi, H., and Takashima, Y. (2009) *Chem. Rev.*, **109**, 5974.
68 Huang, L., Allen, E., and Tonelli, A.E. (1999) *Polymer*, **40**, 3211.
69 Fujita, H., Ooya, T., and Yui, N. (1999) *Macromolecules*, **32**, 2534.
70 Li, J., Ni, X., and Leong, K. (2003) *Angew. Chem., Int. Ed.*, **42**, 69.
71 Liu, R., Maeda, T., Kihara, N., Harada, A., and Takata, T. (2007) *J. Polym. Sci. Part A*, **45**, 1572.
72 Herrmann, W., Schneider, M., and Wenz, G. (1997) *Angew. Chem., Int. Ed. Engl.*, **36**, 2511.
73 Okada, M., Takashima, Y., and Harada, A. (2004) *Macromolecules*, **37**, 7075.
74 Kato, K., Inoue, K., Kidowaki, M., and Ito, K. (2009) *Macromolecules*, **42**, 7129.
75 Okumura, H., Okada, M., Kawaguchi, Y., and Harada, A. (2000) *Macromolecules*, **33**, 4297.
76 Okumura, H., Kawaguchi, Y., and Harada, A. (2001) *Macromolecules*, **34**, 6338.
77 Okumura, H., Kawaguchi, Y., and Harada, A. (2002) *Macromol. Rapid Commun.*, **23**, 781.
78 Okumura, H., Kawaguchi, Y., and Harada, A. (2003) *Macromolecules*, **36**, 6422.
79 Porbeni, F.E., Edeki, E.M., Shin, I.D., and Tonelli, A.E. (2001) *J. Polymer*, **42**, 6907.
80 Sukhanova, T.E., Bronnikov, S.V., Grigor'ev, A.I., Gubanova, G.N., Perminova, M.P., Marangoci, N., Pinteala, M., Harabagiu, V., and Simionescu, B. (2007) *Russ. J. Appl. Chem.*, **80**, 1111.
81 Marangoci, N., Farcas, A., Pinteala, M., Harabagiu, V., Simionescu, B., Sukhanova, T., Bronnikov, S., Grigoryev, A., Gubanova, G., Perminova, M., and Perichaud, A. (2008) *High Perform. Polym.*, **20**, 251.
82 Michishita, T., Okada, M., and Harada, A. (2001) *Macromol. Rapid Commun.*, **22**, 763.
83 Kuratomi, Y., Osaki, M., Takashima, Y., Yamaguchi, H., and Harada, A. (2008) *Macromol. Rapid Commun.*, **29**, 910.
84 Michishita, T., Takashima, Y., and Harada, A. (2004) *Macromol. Rapid Commun.*, **25**, 1159.
85 Kato, K., Komatsu, H., and Ito, K. (2010) *Macromolecules*, **43**, 8799.

11
Stimuli-Responsive Systems

Akihito Hashidzume and Akira Harada

11.1
Introduction

Living organisms detect stimuli and responses using their 'five senses' (Figure 11.1). The responses of living organisms, including contraction–extension movement of muscles, exclusion of foreign materials by white blood cells and antibodies, enzymatic reactions (e.g., syntheses of proteins, DNA, and ATP and digestion), and transportation of molecular oxygen and nutrients, are based on a number of chemical reactions. In turn, these reactions are conducted by stimuli-responsive supramolecular polymer systems formed from biological macromolecules [1–3]. Hence, biological supramolecular polymers have inspired numerous research groups to investigate artificial stimuli-responsive supramolecular polymer systems (Figure 11.2).

Phase diagrams indicate that all substances undergo phase transitions upon changing temperature, pressure, and/or concentration. Thus, all substances are stimuli-responsive. However, typically stimuli-responsive systems are defined as functional systems whose properties are controlled by stimuli.

Using the above definition, stimuli applicable to stimuli-responsive supramolecular polymer systems and the responses of these systems are overviewed in the next section (Section 11.2). The subsequent section (Section 11.3) describes a few examples of stimuli-responsive supramolecular polymer systems. Interested readers can refer to leading and comprehensive reviews [4–28] as well as to other chapters in this book.

11.2
Stimuli and Responses

11.2.1
Stimuli

11.2.1.1 Temperature
The most common stimulus is temperature. Because all events proceed to thermally stable states, all materials are active toward heat, and thus their states depend on

Supramolecular Polymer Chemistry, First Edition. Edited by Akira Harada.
© 2012 Wiley-VCH Verlag GmbH & Co. KGaA. Published 2012 by Wiley-VCH Verlag GmbH & Co. KGaA.

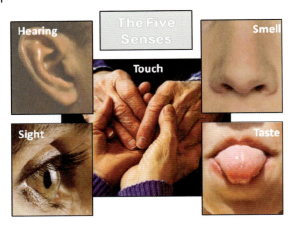

Figure 11.1 The five senses.

temperature. Moreover, noncovalent bonds are more temperature dependent because they are usually weaker than covalent bonds. Altering the temperature of a system can shift the equilibrium, and this causes a change in the degree of polymerization of a supramolecular polymer. A change in the temperature of a system can also increase or decrease the quality of the medium, leading to variations in the conformation of a supramolecular polymer.

11.2.1.2 Pressure, Force, Stress, and Ultrasound

All materials are dependent on pressure, as they are on temperature. However, fewer examples of pressure-responsive systems are known than there are of temperature-responsive systems, presumably because of experimental difficulties in varying the pressure.

Applying force or stress to some materials causes the structure or conformation to change. Because noncovalent bonds may be more dependent on force or stress than covalent bonds, supramolecular polymers may respond more strongly to applied force or stress.

Sonic and ultrasonic waves, which are oscillations of pressure, can vibrate and stress materials. In supramolecular polymer systems, vibrations and stress result in conformational and structural changes.

Figure 11.2 Conceptual illustration of stimuli-responsive supramolecular polymer systems.

11.2.1.3 pH

Another popular stimulus is pH. As the pH of a medium is varied, weak acids and bases are converted from their acidic form to their basic form or vice versa around their pK_a or pK_b, respectively. Because these conversions switch the acid and base moieties from charged to uncharged states or vice versa, their electrostatic interactions change, causing structural or conformational variations in the system. Hence, pH-responsive systems are designed using weak acid or base moieties, for example, carboxylic acid, hydroxyl, amine, and pyridine groups.

11.2.1.4 Chemicals

Another popular stimulus class for suparmolecular polymer systems is chemicals, which can be categorized as either modifiers of the medium quality or competitors for noncovalent bonding sites.

The strength of noncovalent bonds strongly depends on the quality of the medium. For example, the strengths of electrostatic interactions and hydrogen bonding depend on the dielectric constant of the medium, these interactions being stronger in less polar media. Thus, changing the solvent or adding polar or nonpolar additives may control the electrostatic interaction and hydrogen bonding. Because solvophobic interactions are apparently attractive interactions due to the exclusion of solutes from the medium, the strength of solvophobic interactions can be controlled by adding chemicals, for example, organic solvents, salts, urea, and amphiphiles, which alter the quality of the medium.

Supramolecular polymers are usually formed from compounds (monomers) possessing two or more binding sites. Thus, compounds containing a single binding site, that is, competitors, compete with the corresponding monomers to reduce the degree of polymerization in supramolecular polymers.

11.2.1.5 Electromagnetic Waves or Light

An electromagnetic wave, namely light, is another popular stimulus, which has been well documented in supramolecular polymer systems. The energy of an electromagnetic wave varies over a wide range, depending on the wavelength. Because far ultraviolet (UV) light ($\lambda = 10 - 200$ nm) has an energy greater than or equal to that of covalent bonds, it is not typically employed as a stimulus for supramolecular polymer systems. On the other hand, infrared (IR) light ($\lambda = 0.7 - 2000$ μm) or an electromagnetic wave with a wavelength longer than that of IR light does not have sufficient energy to excite electronic transitions, which are necessary to induce structural changes. (Because the energy of IR light corresponds to the transition of a vibrational quantum state, irradiation with IR light increases the temperature of a system [29].) Thus, near UV or visible light ($\lambda = 200 - 750$ nm) is more often used as a stimulus for supramolecular polymer systems.

Photo-responsive (i.e. light-responsive) systems are constructed using moieties that absorb near UV or visible light, which leads to structural changes. Photoinduced structural changes include photoisomerization, photodimerization, photoinduced decomposition, and others [30]. Residues that undergo photoisomerization include azobenzene, stilbene, diarylethene, spyropyrans, and others (Figure 11.3a), while

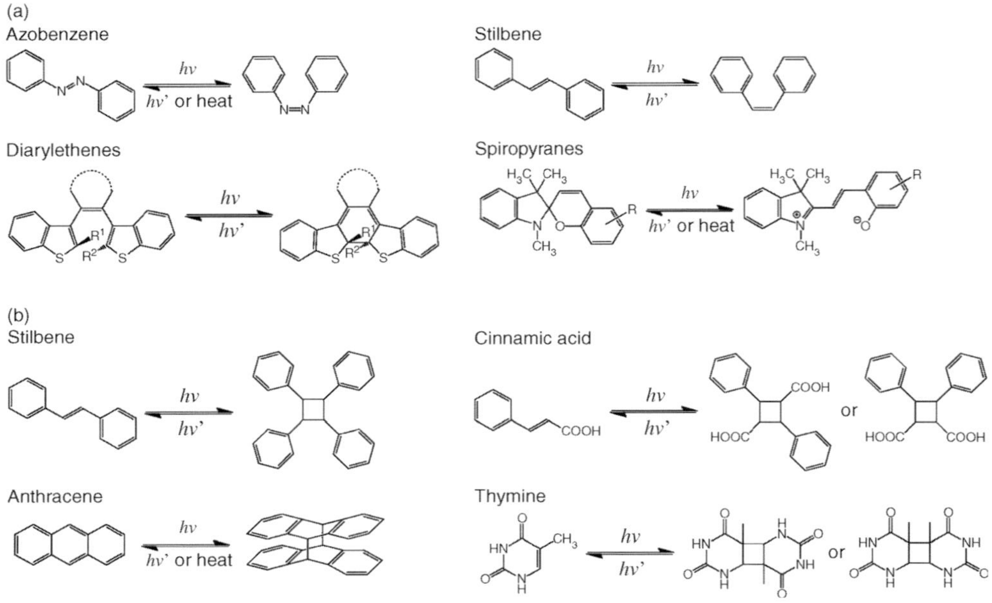

Figure 11.3 Typical examples of compounds for photoisomerization (a) and photodimerization (b).

stilbene, cinnamic acid, anthracene, thymine, and so on undergo photodimerization (Figure 11.3b). Although there are only a few examples of supramolecular polymer systems containing residues that undergo photoinduced decomposition, residues that undergo photoinduced decomposition include triphenylmethanol [31].

11.2.1.6 Redox

Redox is a promising stimulus for supramolecular polymer systems because the redox state can be controlled chemically or electrically. However, compared to temperature-, pH-, chemical-, and photo-responsive systems, there are fewer examples of redox-responsive supramolecular polymer systems. When materials undergo redox reactions, the charged states switch, altering the electrostatic interactions or cleaving and reforming bonds, which leads to structural or conformational changes. Redox-responsive residues include metal complexes (e.g., metalocenes and porphyrins), aromatic moieties, disulfides, peroxides, among others, which act as electron acceptors or donors.

11.2.2
Responses

The application of a stimulus causes structural or conformational changes in supramolecular polymers. Here, a structural or conformational change in a supramolecular polymer is based either on a structural change of the unit (i.e., the monomer) or a change in the strength of the noncovalent bonds between the monomers. Occasionally, these changes occur simultaneously. The structural or conformational changes alter the properties of a supramolecular polymer, and the changes in the properties are defined as responses.

11.2.2.1 Movement
In biological systems, movement is one of the most important responses, because living organisms must move to feed or escape dangers. Movements are based on flagellae or muscle fibers in biological systems. On the other hand, movement on the macroscopic scale remains a challenge in artificial supramolecular polymer systems.

11.2.2.2 Capture and Release of Chemicals
Capture and release of chemicals is another important response in biological systems. Structural changes of supramolecular polymers upon applying a stimulus can lead to capture or release of chemicals. This can be utilized for controlled release, and it is therefore important in pharmaceutical applications. The ability to capture and release allows one to actively transport chemicals.

11.2.2.3 Chemical Reactions
Chemical reactions controlled by stimuli-responsive supramolecular polymer systems are another important class of responses in biological systems. However, controlling chemical reactions remains a challenge in artificial supramolecular polymer systems. Catalytic systems based on supramolecular polymers can be divided into three elementary processes: binding of substances, conversion of substances to products by the action of functional residues, and release of the products. Chemical reactions can be controlled by commanding one of these three processes.

The addition of stimuli-responsive competitors or a stimuli-responsive structural change of a residue around the binding site can control substance binding. If an inhibitor, which has a structure similar to that of the substance but does not participate in the reaction, is captured by the binding site, then the reaction is terminated. However, switching the structure of the inhibitor by a stimulus causes the binding site to release the inhibitor, allowing the reaction catalyzed by the supramolecular polymer to resume. Thus, binding of the substance can be also controlled if a residue around the binding site changes its structure in response to the stimulus.

Functional residues act by different mechanisms. For example, when the nucleophilicity or electrophilicity of functional residues increases upon the formation of hydrogen-bonding networks, reactions can be controlled by breaking and reforming the hydrogen-bonding network, which may be due to alteration of the pH or the local structure. When functional residues are linked through noncovalent bonds like a coenzyme, stimuli-responsive attachment and detachment of the functional residues can control the reactions.

Furthermore, structural changes of the binding site can control the release of the product.

11.2.2.4 Change in Viscoelastic Properties, or Gel-to-Sol and Sol-to-Gel Transitions

Changes in the viscoelastic properties or sol–gel transition represent one of most popular responses in artificial supramolecular polymer systems. The viscoelastic properties of solutions strongly depend on the size of the solute molecules (i.e., the molar mass of the solute). Thus, when the degree of polymerization (or the association number) of supramolecular polymers increases or decreases upon applying a stimulus, the viscoelastic properties of the solution also change. In extreme cases, supramolecular polymer systems undergo sol-to-gel or gel-to-sol transitions. Gel-to-sol transitions are occasionally utilized for the controlled release of chemicals.

11.2.2.5 Change in Color

A stimulus-responsive color change is a useful response for sensor applications. In recent decades, sensors based on molecules have attracted increasing interest from researchers [32–37]. In supramolecular polymer systems, color changes may be based on changes in the coordination state of metal complexes, changes in the structure of aromatic compounds, or changes in the aggregation mode of aromatic compounds. However, only a few examples of drastic stimuli-responsive color changes have been reported for supramolecular polymer systems [38–41].

11.3
Examples of Stimuli-Responsive Supramolecular Polymer Systems

11.3.1
Temperature-Responsive Systems

Telechelic polymers, which have binding sites at both ends, are an important class of building blocks for supramolecular polymers. Woodward and coworkers [42] prepared telechelic polymers bearing hydrogen-bonding sites at both ends from poly(isobutylene) diol of $M_n = 5.05 \times 10^3$ and 4,4'-methylenebis(phenyl isocyanate) followed by an end-capping reaction with several alcohols and amines (Scheme 11.1).

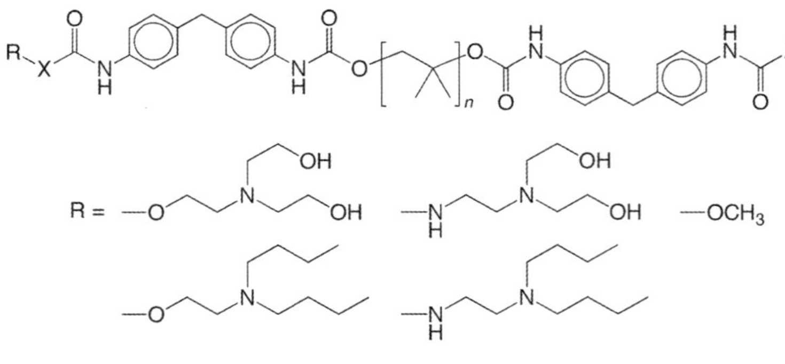

Scheme 11.1

11.3 Examples of Stimuli-Responsive Supramolecular Polymer Systems

The storage modulus decreased with increasing temperature because the hydrogen-bonding interactions became weaker at higher temperatures. Sivakova and coworkers [43] prepared telechelic polymers possessing nucleobase moieties at both the ends from poly(tetrahydrofuran) (pTHF) of $M_n \approx 1.4 \times 10^3$ (Scheme 11.2). Rheological measurements indicated that bis-N^6-anisoyladenocine-modified pTHF exhibited the behavior of a critical gel. That is, it has a storage modulus parallel to the loss modulus with respect to the frequency in the log-log coordinates in the temperature range between 100 and 120 °C, indicating the formation of an expanded network structure. On the basis of differential scanning calorimetry, IR spectroscopy, and wide-angle X-ray diffraction data, it was deduced that the adenosine moieties were segregated from pTHF segments by a π–π stacking interaction at lower temperatures, whereas thermal rearrangements at higher temperatures caused the π–π stacking interaction to weaken but hydrogen bonding to strengthen, resulting in the formation of a more extended network structure. The rheological behavior depends strongly on the structure of the nucleobase functions.

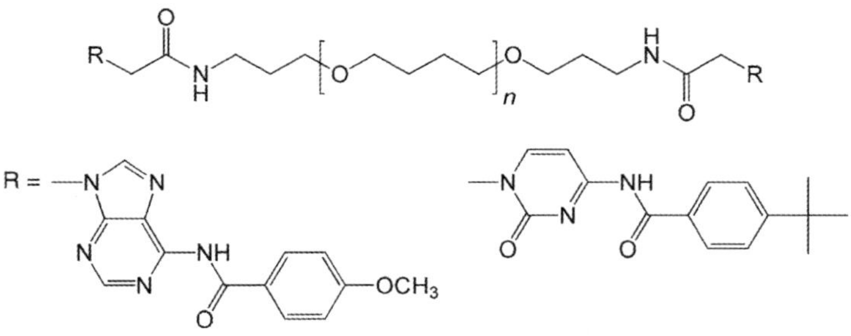

Scheme 11.2

Kuroiwa and coworkers [38] reported supramolecular coordination polymers formed from Co(II)/4-dodecyloxyethyl-1,2,4-triazole complexes, which underwent a temperature-responsive sol–gel transition (Scheme 11.3). At higher temperatures (>about 25 °C), solutions of the Co(II) complex in chloroform were blue and exhibited a gel-like behavior. As the temperature decreased, the solutions turned into a sol around 25 °C, accompanied by a color change from blue to pink. These observations indicated that the gel–sol transition is based on a change in the coordination state of the Co(II) complex from tetrahedral (at higher temperatures) to octahedral (at lower temperatures) (Figure 11.4).

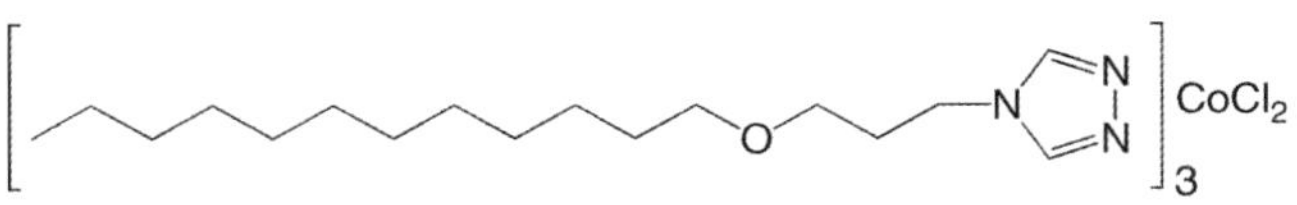

Scheme 11.3

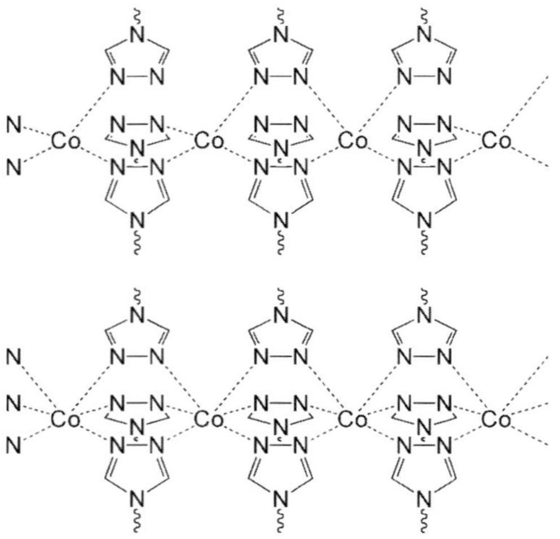

Figure 11.4 Conceptual illustration of a temperature-responsive organo-gel formed from lipophilic Co(II)/4-dodexyloxyethyl-1,2,4-triazole complexes: gel state of tetrahedral complexes (upper) and sol state of octahedral complexes (lower).

Hamachi and coworkers [44] prepared various organo- and hydro-gelators based on glycosylated amino acids and reported the formation of supramolecular polymers responsive to several stimuli. Hydrogels formed from select derivatives of N-acetyl-galactosamine-appended amino acid (Scheme 11.4) underwent a sol–gel transition accompanied by a reversible temperature-responsive volume change; the volume shrank at higher temperatures, but swelled at lower temperatures [45]. These gelators formed very large fibrous supramolecular structures by combining hydrogen-bonding and hydrophobic interactions. This caused the structures to become entangled, resulting in the formation of hydrogels. At higher temperatures, the hydrogels shrank because of dehydration. These hydrogels can be utilized for the controlled release of DNA and the removal of bisphenol A.

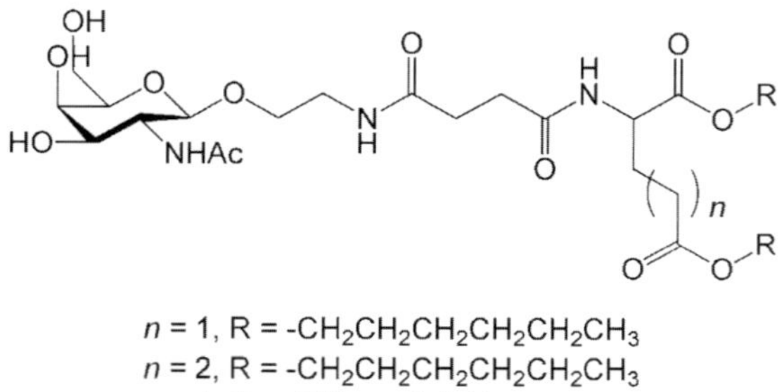

$n = 1$, R = -CH$_2$CH$_2$CH$_2$CH$_2$CH$_2$CH$_3$
$n = 2$, R = -CH$_2$CH$_2$CH$_2$CH$_2$CH$_2$CH$_3$
$n = 1$, R = -CH$_2$-cycloC$_6$H$_{11}$

Scheme 11.4

11.3.2
Pressure-, Force-, and Sonication-Responsive Systems

Craig and coworkers [46, 47] prepared supramolecular polymer network structures from poly(4-vinylpyridine) (PVP) and bifunctional metal complexes, that is, 2,3,5,6-tetrakis{(dimethylamino)methyl}phenylene-1,4-bis(palladium trifluoromethanesulfonate), 2,3,5,6-tetrakis{(diethylamino)methyl}phenylene-1,4-bis(palladium trifluoromethanesulfonate), 2,3,5,6-tetrakis{(dimethylamino)methyl}phenylene-1,4-bis-(platinum trifluoromethanesulfonate), and 2,3,5,6-tetrakis{(diethylamino)methyl}-phenylene-1,4-bis(platinum trifluoromethanesulfonate) (Scheme 11.5). As the shear

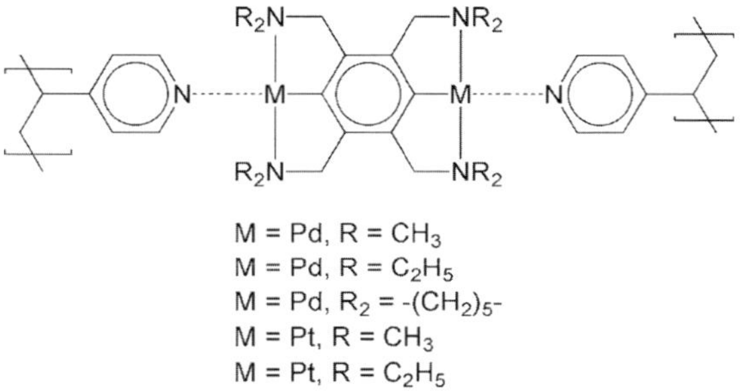
R2N
NR2
N
M
M
N
R2N
NR2
M = Pd, R = CH3
M = Pd, R = C2H5
M = Pd, R2 = -(CH2)5-
M = Pt, R = CH3
M = Pt, R = C2H5

Scheme 11.5

rate increased in steady shear viscosity measurements, the mixtures exibited shear-thickening and then shear-thinning, implying the formation of shear-responsive supramolecular networks. The shear rate at which the mixture exhibited shear-thickening and shear-thinning corresponded to the rate constant of the coordination bond dissociation. Zhao and coworkers [48] also reported shear-responsive supramolecular polymer networks formed from telechelic polymers with tridentate ligand moieties, that is, a 2,6-bis(1′-methylbenzimidazolyl)-4-pyridyl group, at both ends (Scheme 11.6) as well as transition and lanthanoid metal ions, for example, Zn(II)

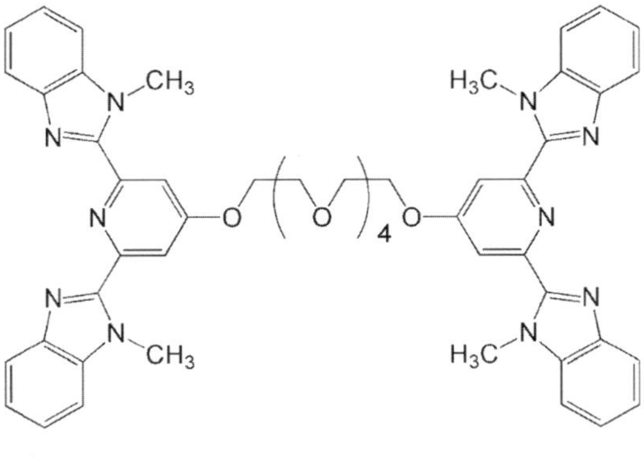
CH3
N
N
H3C
N
N
N
O
O
4
O
N
N
N
CH3
N
H3C
N

Scheme 11.6

and La(III). As the shear stress increased in storage and loss moduli measurements under oscillatory shear stress sweep, the storage and loss moduli decreased rapidly at about 200 Pa, indicating that the network structures collapsed because of coordination bonds breaking. The network structures were gradually restored to reach the initial state after about 18 min.

Sijbesma and coworkers [49, 50] prepared sonication-responsive supramolecular coordination polymers from poly(tetrahydrofuran) (pTHF) with diphenylphosphine or dicyclohexylphosphine moieties at both ends and Pd(II) metal ions (Scheme 11.7). Size exclusion chromatography analyses demonstrated that the molar mass of the supramolecular coordination polymer decreased after sonication for several hours, indicating that the coordination bonds were cleaved. The molar mass increased gradually, reaching the equilibrium state after about 24 h. Paulusse and coworkers [51] also prepared sonication-responsive supramolecular coordination polymer networks from pTHF carrying diphenylphosphine moieties and Rh(I) and Ir(I) metal ions, which formed crosslinks with three or four diphenylphosphine groups (Scheme 11.8). These mixtures underwent a gel-to-sol transition upon sonication for 3 min. After sonication ceased, the mixtures turned into gels. Although Rh(I) gels were restored quickly (about 1 min), Ir(I) gels were restored slowly (1.5 h) (Figure 11.5).

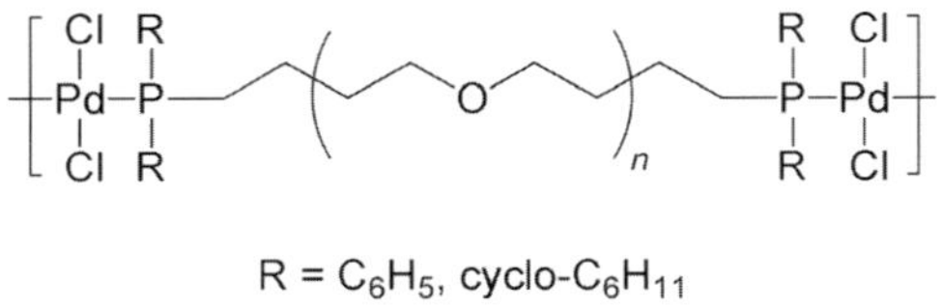

$R = C_6H_5, \text{cyclo-}C_6H_{11}$

Scheme 11.7

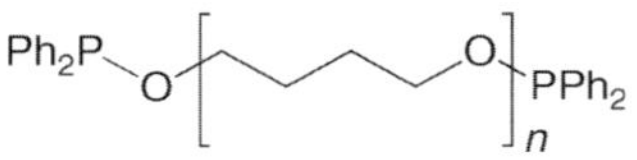

Scheme 11.8

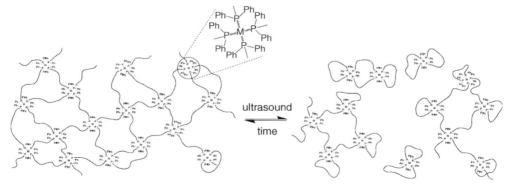

Figure 11.5 Conceptual illustration of an ultrasonic-responsive organo-gel formed from pTHF possessing diphenylphosphine moieties at both ends and Rh(I) or Ir(I) metal ions.

Naota and coworkers [52, 53] reported sonication-induced gelation of solutions with lower-molecular-weight Pd(II) complexes in organic solvents. Koori and Naota [52] prepared dinuclear Pd(II) complexes (Scheme 11.9) from Pd(II) acetate and N,N'-bis(salicylidene)-1,n-alkanediamines ($n=5$, 6, 7, and 8). Among these dinuclear Pd(II) complexes, solutions of the anti-complex of $n=5$ in various organic solvents underwent gelation after sonication for a few seconds. Warming these gelated solutions returned them to the initial sol state as these transitions were completely reversible. Thus, sonication induced an interaction of the anti-complex with the solvent molecules to make the conformation of the anti-complex appropriate for the formation of linear supramolecular polymers through π–π stacking interaction, resulting in gelation. Isozaki and coworkers [53] reported that a dipeptide Glu-Glu possessing Pd(II) complexes underwent sonication-induced gelation (Scheme 11.10). At higher temperatures, the gel turned into a sol. Scanning electron micrograms of xerogel of the dipeptide demonstrated β-sheet monolayers. On the basis of detailed NMR studies, it was deduced that sonication cleaved intramolecular hydrogen bonding, which made the conformation of the dipeptide suitable for the formation of β-sheets, resulting in gelation.

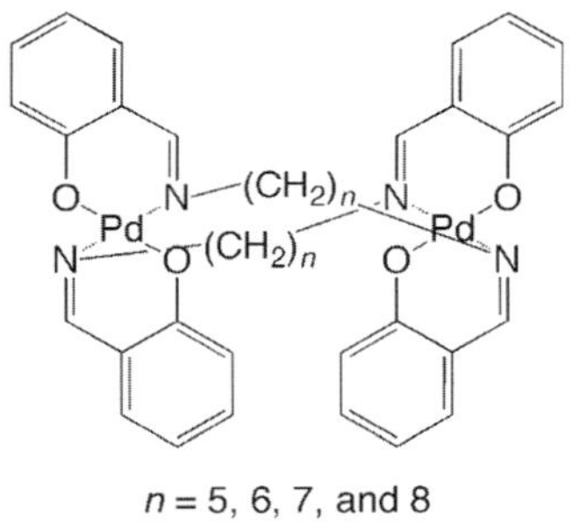

n = 5, 6, 7, and 8

Scheme 11.9

11.3.3
pH-Responsive Systems

Jones and coworkers [54] synthesized the first- and third-generation poly(propylenimine) dendrimers bearing bis(m-phenylene)-32-crown-10 periphery moieties, and studied the interaction of the dendrimers with paraquat diol via NMR (Scheme 11.11). In the neutral state, binding was anticooperative because of steric hindrance. In contrast, in the acidic state, where the tertiary amines in the interior of the dendrimers were protonated, binding was noncooperative and the average apparent binding constants were larger than those in the neutral state. This is because the crown ether moieties were more accessible and less sterically hindered for the protonated dendrimers which assumed a rigid and expanded conformation.

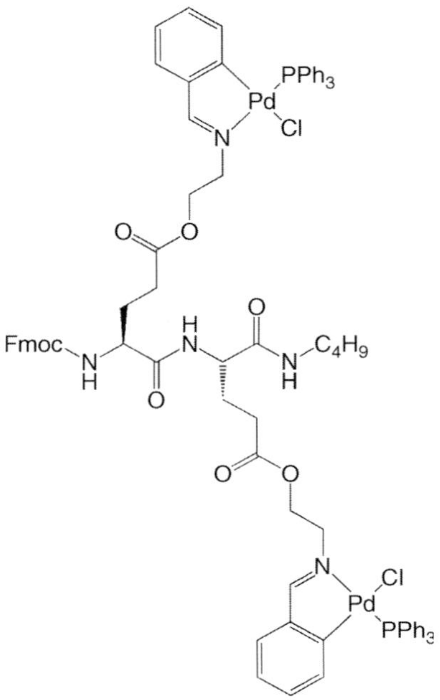

Scheme 11.10

Ghoussoub and Lehn [55] reported a pH-responsive hydrogel utilizing K^+-induced formation of guanine quartets. A mixture of telechelic oligo(ethylene glycol) with guanine moieties at both ends, 2,2,2-cryptand, and K^+ exhibited a gel-like behavior under acidic conditions because the nitrogen atoms in 2,2,2-cryptand were protonated, and free K^+ induced the formation of the guanine quartets which acted as crosslinks (Scheme 11.12). On the other hand, the mixture was a sol under basic conditions because of deprotonated 2,2,2-cryptand-captured K^+.

Martens and coworkers [56] synthesized biosynthetically A-B-A triblock protein copolymers and reported their pH-responsive gelation. The A-B-A triblock protein copolymers consisted of a pH-responsive silk-like polypeptide [(Gly-Ala)$_3$-Gly-Glu]$_{24}$ (S) and a Gln, Asn, and Ser-rich collagen-like polypeptide (C), that is, **CSSC** and **SCCS**. Both copolymers were soluble in water under neutral and basic conditions, but both formed transparent gels under acidic conditions (pH 2 – 4). Circular dichroism spectroscopic studies indicated that the **S** block assumed a β-turn structure under acidic conditions to form long fibrous supramolecular structures.

Zhou and coworkers [57] introduced pH-responsiveness to hydrogels formed from glycosylated amino acids [44] by mixing with carboxylic acid-carrying analogs (Scheme 11.13). After carefully optimizing the conditions, a one-to-one mixture of N-acetylgalactosamine glutamate ester and the carboxylic acid-carrying analog of $n = 6$ was most useful for a pH-responsive volume change of the hydrogel. A controlled release using pH-responsive volume change of hydrogel was investigated with hydrophilic (vitamins B1, B6, and B12) and hydrophobic substances (quercetin

11.3 Examples of Stimuli-Responsive Supramolecular Polymer Systems | 243

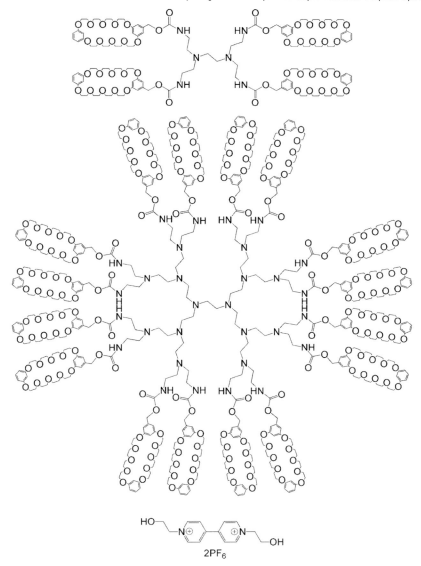

Scheme 11.11

and myricetin). The mixed hydrogel efficiently released hydrophilic substances, but not hydrophobic substances, presumably because of hydrophobic interaction with the interior of supramolecular gel fibers.

Mesoporous silica nanoparticles are a promising class of drug containers [58]. Stimulus-responsive gate opening is useful for controlled release applications. Park and coworkers [59] as well as Angelos and coworkers [60, 61] utilized the

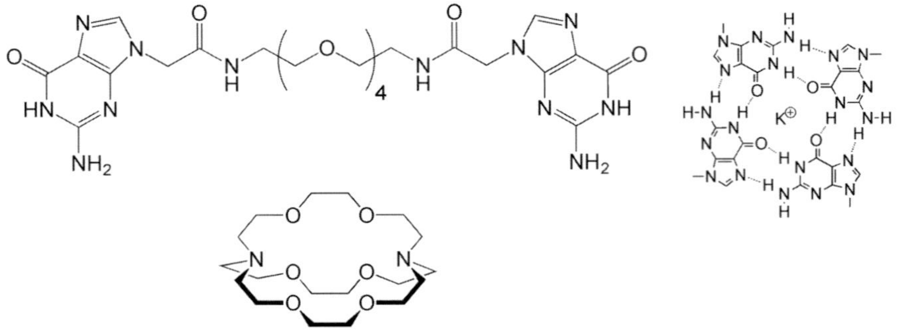

Scheme 11.12

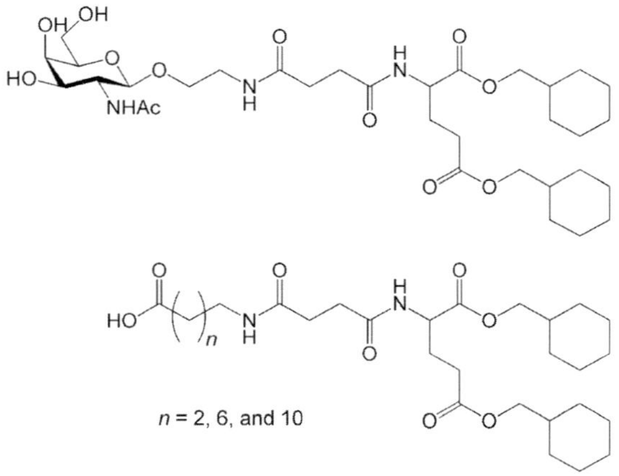

n = 2, 6, and 10

Scheme 11.13

pH-responsive formation of *pseudo*-rotaxanes as a valve for mesoporous silica particle containers. Park and coworkers [59] used poly-*pseudo*-rotaxanes of α- or γ-cyclodextrin (α-CD or γ-CD) with poly(ethyleneimine) (PEI) [62] (Scheme 11.14). A fluorescent guest, calcein, was loaded into the pores of the silica nanoparticles by soaking the particles in a phosphate-buffered saline solution of calcein followed by the formation of poly-*pseudo*-rotaxanes α-CD or γ-CD with PEI chains grafted on the silica nanoparticles. The loaded calcein molecules were quickly released by dethreading the CD molecules from the PEI chains upon adjusting the pH to 5.5. Angelos and coworkers [60, 61] used *pseudo*-rotaxanes of cucurbit[6]uril (CB[6]) with ammonium cations (Scheme 11.15). In the case where the axles bore two amino groups (Scheme 11.15a), the *pseudo*-rotaxanes were dissociated under basic conditions (pH = 10) and the valve opened, leading to release of the Rhodamine B loaded into the pores of the nanoparticles [60]. On the other hand, in the case where the axles had

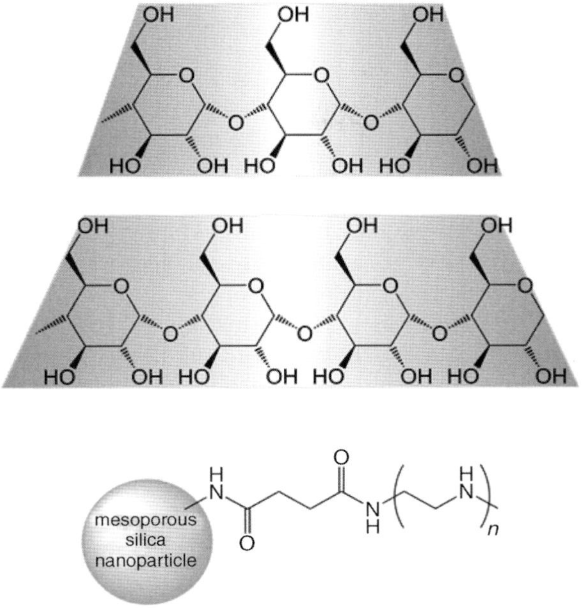

Scheme 11.14

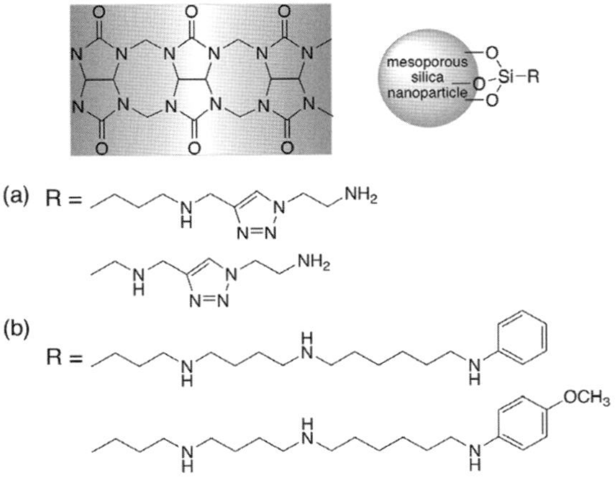

Scheme 11.15

two amino groups and one anilino group (Scheme 11.15b), the valve opened to release Rhodamine B under basic and acidic conditions. Under basic conditions, the *pseudo*-rotaxanes were dissociated, whereas, under acidic conditions, the CB[6] included the outer part of the axles, that is, anilinium and ammonium groups, and the valve opened [61].

11.3.4
Chemical-Responsive Systems

Stimuli-responsive Janus [2]rotaxanes are a prototype of molecular muscles [63]. Some research groups have reported stimuli-responsive Janus [2]rotaxanes based on CDs [64]. Tsukagoshi and coworkers [65] synthesized a CD-based Janus [2] rotaxane with a longer spacer, that is, two hexamethylene units linked with a ureido bond (Scheme 11.16). On the basis of the characterization data from ROESY NMR, circular dichroism, and pulsed-field-gradient spin-echo NMR spectroscopy, it was concluded that the inclusion site of the CD moiety was switched by changing the polarity of the medium. This resulted in a change in the hydrodynamic size of the Janus [2]rotaxane. The hydrodynamic size in dimethyl sulfoxide (DMSO) was smaller than that in a mixed solvent of DMSO/water (1/1, v/v) (Figure 11.6).

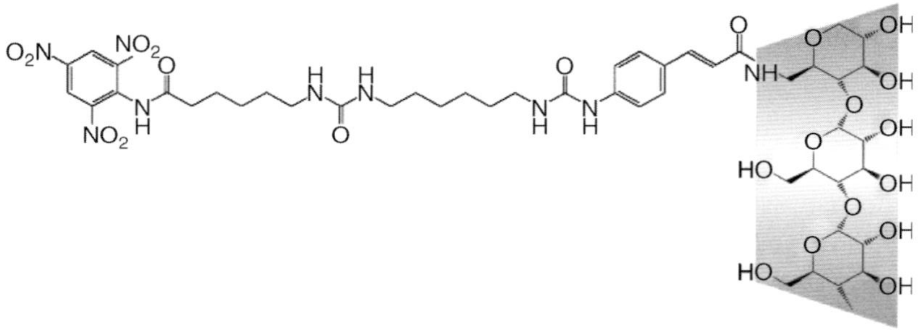

Scheme 11.16

Deng and coworkers [66, 67] reported the first chemical-responsive hydrogel based on a relatively simple CD derivative, that is, 2,4,6-trinitrophenyl-modified 6-aminocinnamoyl-β-CD (Scheme 11.17). Atomic force microscopy indicated that this β-CD derivative formed long fibrils, which assumed linear supramolecular polymers at lower concentrations and bundles and networks at higher concentrations. NMR and circular dichroism spectroscopy revealed that the formation of an inclusion complex and hydrogen bonding between the CD moieties were responsible for the formation of the bundles and networks. The hydrogel exhibited chemical responsiveness; gel-to-sol and sol-to-gel transitions occurred upon the repetitive additions of competitive guests and hosts, that is, methyl orange and α-CD. Ogoshi and coworkers [68] also reported a chemically responsive gel-to-sol transition in supramolecular single-walled carbon nanotube (SWNT) hydrogels. Hydrogels were formed from SWNTs solubilized with pyrene-modified β-CD and dodecyl-modified poly(sodium acrylate) through the inclusion complexes of β-CD moieties with dodecyl groups (Scheme 11.18 and Figure 11.7). Upon the addition of

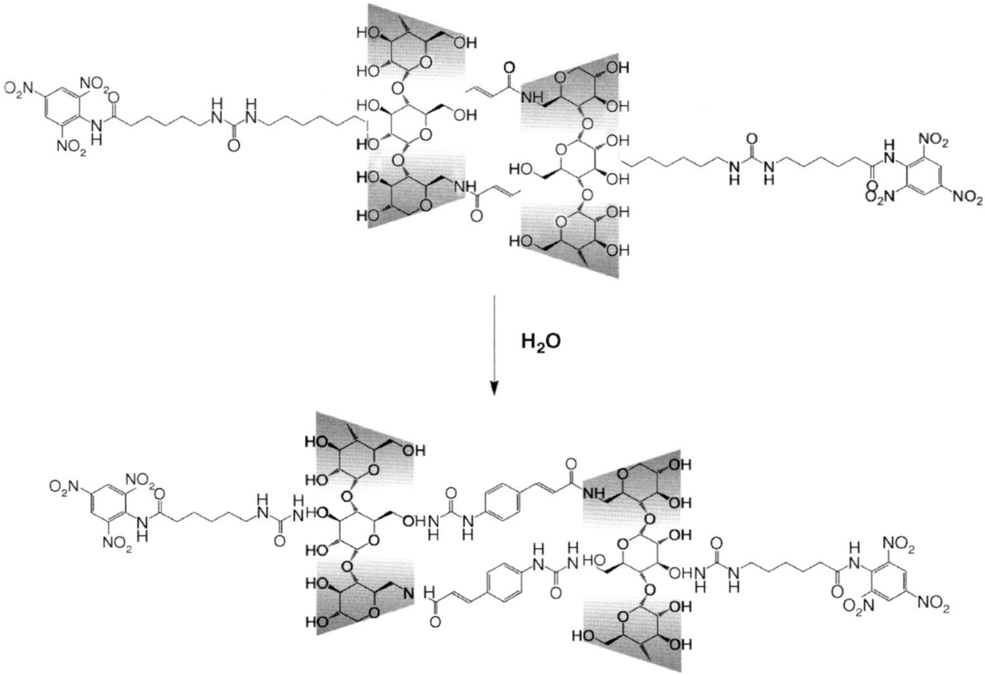

Figure 11.6 Conceptual illustration of water-induced contraction of the Janus[2] rotaxane of an α-CD derivatives.

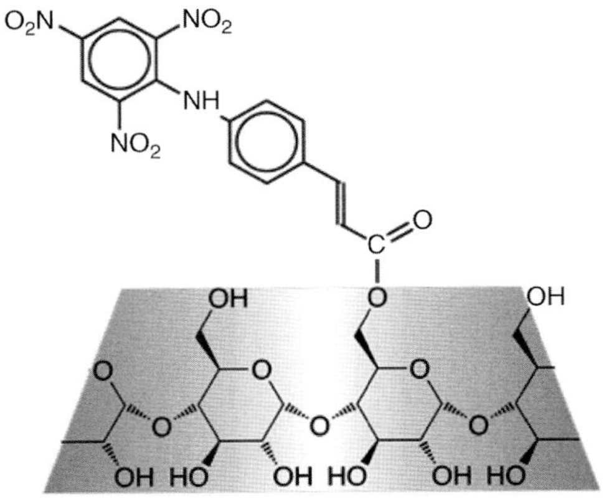

Scheme 11.17

competitive guests, that is, α-CD or sodium 1-adamantane carboxylate, the hydrogel turned into a sol.

The movement of supramolecular polymer systems controlled by external stimuli remains a challenge. Recently, Gu and coworkers [69] and Lund and coworkers [70]

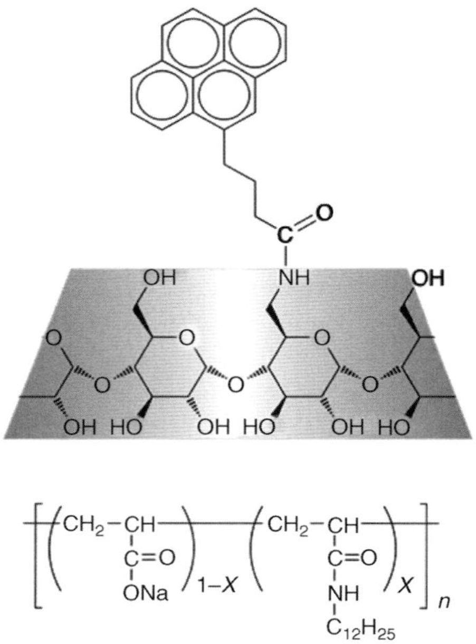

Scheme 11.18

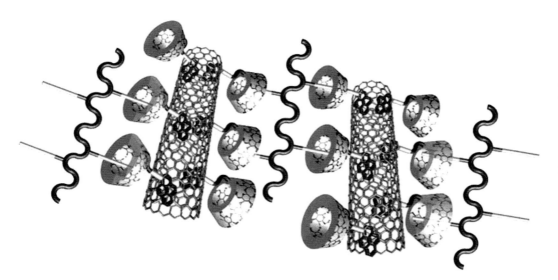

Figure 11.7 Conceptual illustration of a hydrogel formed from SWNT solubilized with a pyrene-carrying β-CD derivative and dodecyl-modified poly(sodium acrylate).

reported that chemical stimuli, that is, single-strand DNA segments, could control the movement of DNA walkers on DNA origami possessing dangling oligonucleotide chains controlled with chemical stimuli.

11.3.5
Photo-Responsive Systems

In the course of the research of Harada and coworkers on CD-based supramolecular polymers, Kuad and coworkers [71] constructed a photo-responsive alternating supramolecular copolymer system using a β-CD dimer linked with a stilbene moiety and an adamantane dimer linked with 1,3-bis(4-pyridinium) propane (Scheme 11.19). 2D ROESY NMR studies indicated the formation of supramolecular copolymers for both the β-CD dimers of *trans-* and *cis*-stilbene units. Pulsed field gradient spin echo NMR and atomic force microscopy confirmed the photo-responsive structural change. The β-CD dimer with a *trans*-stilbene unit formed supramolecular cyclic dimers with an admantane dimer, whereas the β-CD dimer with a *cis*-stilbene unit formed supramolecular copolymers.

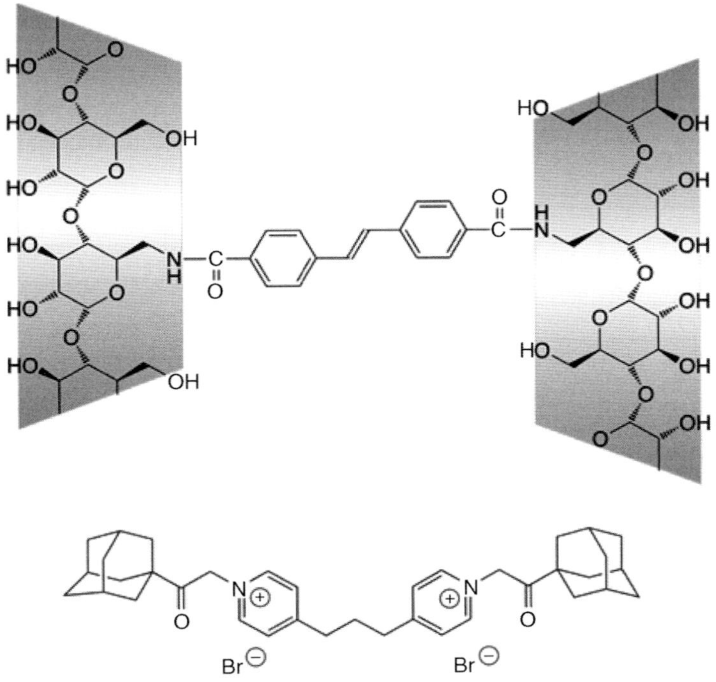

Scheme 11.19

Some publications have reported symmetrical organo- and hydro-gelators possessing photo-responsive moieties that undergo *trans-cis* isomerization at the center (Scheme 11.20) [72, 73]. These gelators formed gels in the trans form, but not in the cis form. Frkanec and coworkers [72] used maleic acid and fumaric acid as photoresponsive moieties, and Koumura and coworkers [73] employed an azobenzene moiety. Eastoe and coworkers [74] reported a gel-to-sol transition using photo-dimerization of a stilbene-carrying organo-gelator (Scheme 11.21).

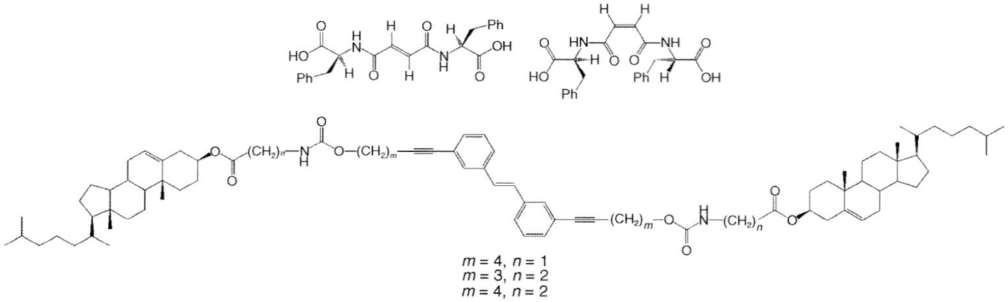

Scheme 11.20

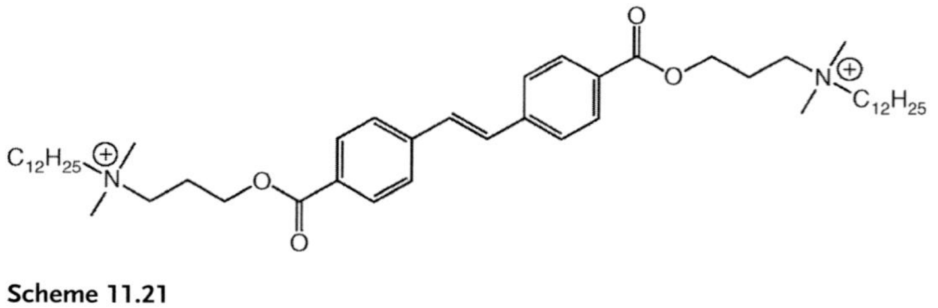

Scheme 11.21

In the course of the research of Hamachi and coworkers on organo- and hydro-gelators based on glycosylated amino acids, Matsumoto and coworkers [75] synthesized a variety of compounds based on fumaric acid diamide to obtain photo-responsive hydrogels (Scheme 11.22). Some of the compounds did form photo-responsive hydrogels: their solutions underwent a gel-to-sol transition upon *E*-to-*Z* photoisomerization and vice versa. Hydrogels were utilized for the controlled release of vitamin B12. Moreover, control of biomaterials, for example, bacteria and the F_1-ATPase motor, was realized using photo-responsive hydrogels.

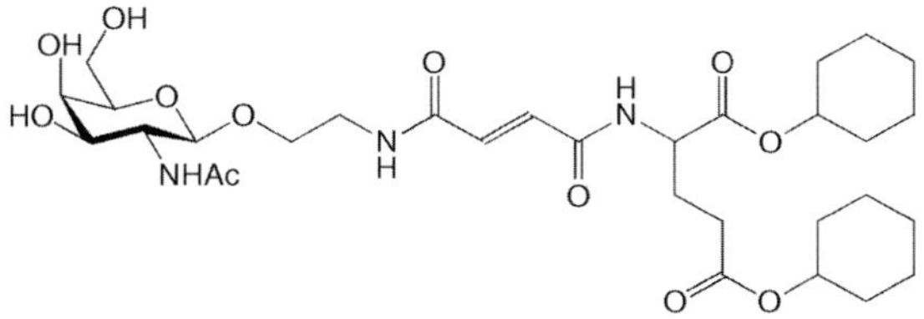

11.3 Examples of Stimuli-Responsive Supramolecular Polymer Systems

Scheme 11.22

α-CD and azobenzene (or stilbene) are a popular pair for photo-responsive molecular recognition. α-CD includes the trans isomer, but not the cis isomer. Yamauchi and coworkers [76] synthesized a stilbene-modifed α-CD (Scheme 11.23). The α-CD derivative formed a stable supramolecular dimer in its trans state, but in the cis state supramolecular assemblies were formed. Wang and coworkers [77] reported photo-responsive supramolecular assemblies using a combination of α-CD and an azobenzene-containing surfactant (Scheme 11.24). In the trans state, α-CD preferably included the azobenzene moiety in the surfactant and the complexed surfactant molecules were in the molecularly dispersed state. In the cis state, on the other hand, α-CD included the decamethylene moiety and the complexed surfactant molecules formed vesicular assemblies. Ravoo and coworkers [78] synthesized CD-based amphiphiles, which formed vesicles, and investigated the binding properties of the CD-vesicles [79, 80]. More recently, Nalluri and Ravoo [81] reported photo-responsive aggregation of CD-vesicles via combination with an azobenzene dimer (Scheme 11.25).

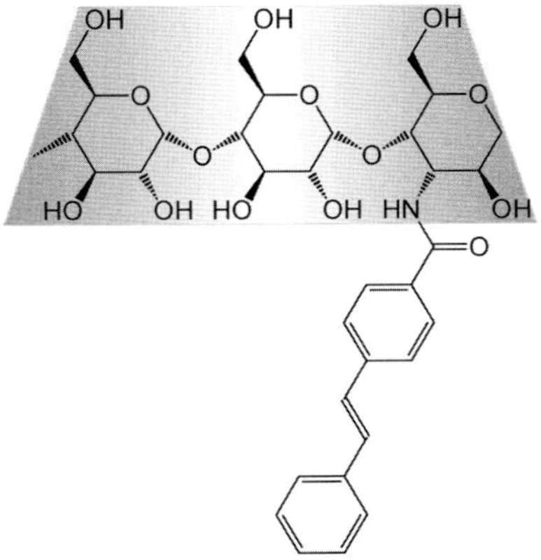

Scheme 11.23

Li and coworkers [82, 83] utilized α-CD and azobenzene to synthesize Janus [2] rotaxanes, which exhibited photo-responsive size changes. They synthesized α-CD derivatives with two recognition sites (i.e., azobenzene and heptamethy-

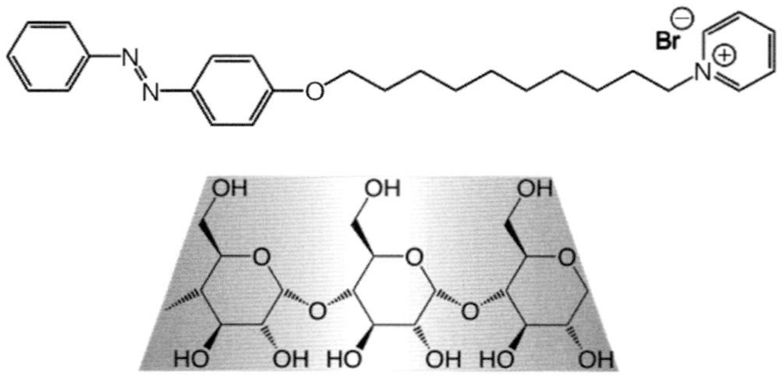

Scheme 11.24

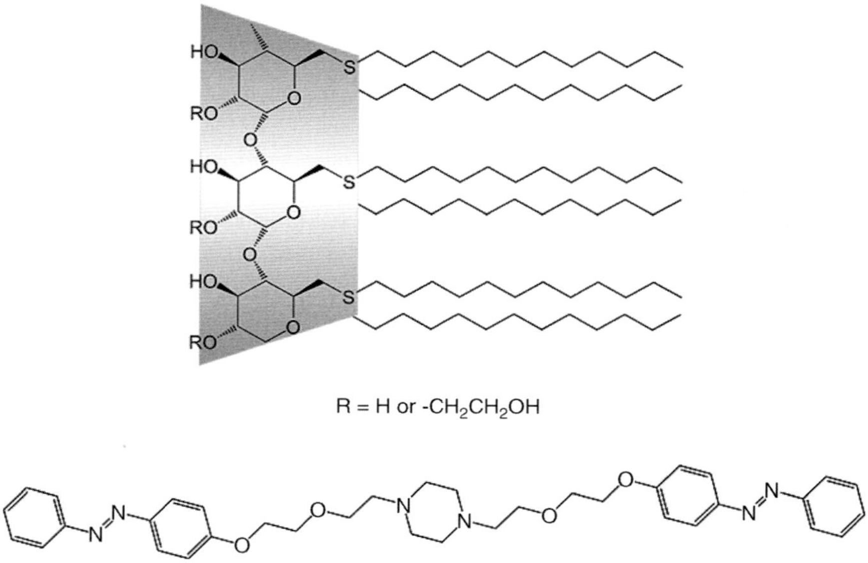

R = H or -CH$_2$CH$_2$OH

Scheme 11.25

lene) linked with long linkers (i.e., oligo(ethylene glycol)) (Scheme 11.26). These α-CD derivatives formed double-threaded dimers. For α-CD derivatives modified at their 6-position, it was possible to cap both ends with adamantane stoppers to form Janus [2]rotaxanes. Pulsed field gradient spin echo NMR studies indicated that the hydrodynamic radius of the trans form was larger than that of the cis form for both the double-threaded dimer and the Janus [2]rotaxanes (Figure 11.8).

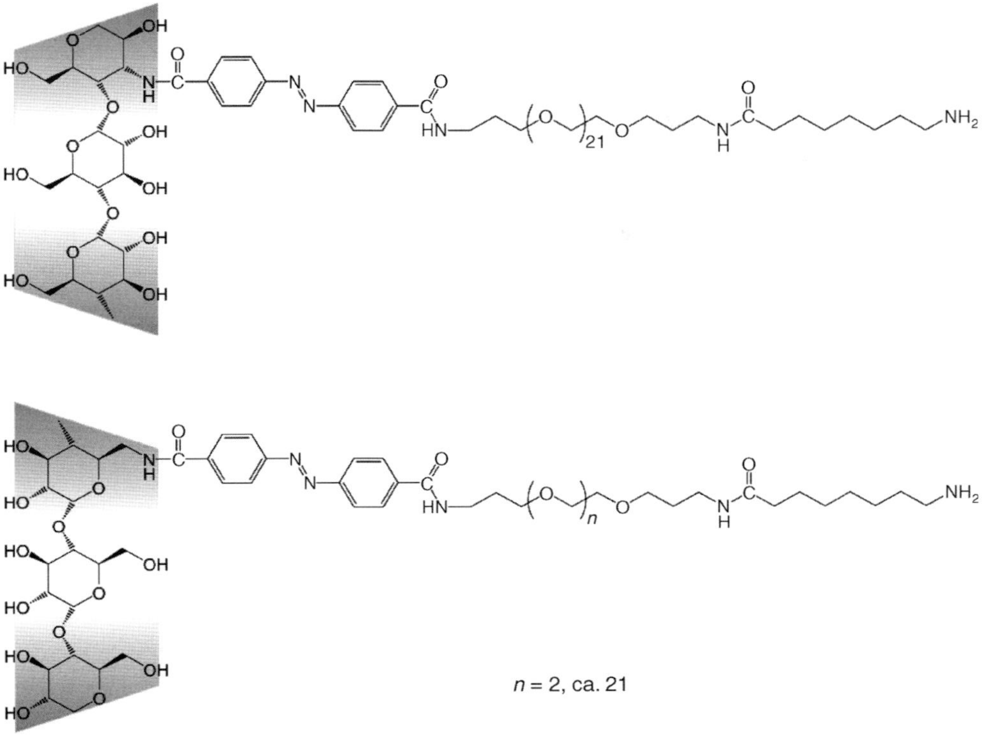

Scheme 11.26

$n = 2$, ca. 21

The pair of α-CD and azobenene is also useful for photo-responsive hydrogels or networks. Tomatsu and coworkers [84] investigated the photo-responsive viscoelastic properties of a ternary mixture of dodecyl-modified poly(acrylic acid), α-CD, and a water soluble azobenzene derivative, that is, 4,4'-azodibenzoic acid (ADA) (Scheme 11.27). Irradiating the ternary gel mixture with UV light isomerized ADA from trans to cis, and the mixture underwent a gel-to-sol transition because α-CD formed inclusion complexes more favorably with dodecyl side chains than with *cis*-ADA. Irradiating with visible light isomerized ADA from cis to trans, and the mixture underwent a sol-to-gel transition. Tomatsu and coworkers [85] expanded their study on the photo-responsive hydrogel to polymer-polymer systems. On the basis of their study [86], polymer-polymer interactions for mixtures of poly(acrylic acid) bearing azobenzene through a dodecamethylene linker and two types of poly(acrylic acid) carrying α-CD, in which α-CD was attached to the main chain through the 3- and 6-positions of the CD, were investigated by viscometry and NOESY (Scheme 11.28). These mixtures exhibited contrasting changes upon UV irradiation; the mixture thinned for the 3-position-modified α-CD polymer, but thickened for the mixture of

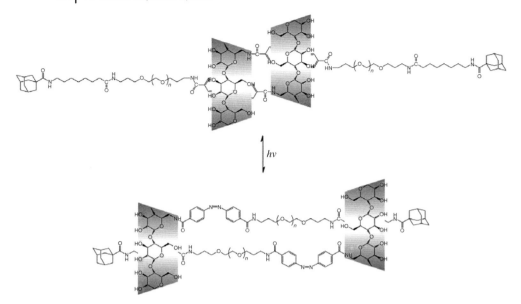

Figure 11.8 Conceptual illustration of a photo-responsive contraction and expansion of Janus [2] rotaxanes of α-CD derivatives possessing azobenzene and heptamethylene moieties linked with oligo(ethylene glycol) linkers.

the 6-position-modified α-CD polymer. After photoisomerization from trans to cis, the inclusion complexes dissociated for the mixture of the 3-position-modified α-CD polymer, whereas interlocked complexes were formed for the mixture of the 6-position-modified α-CD polymer. Peng and coworkers [87] utilized a photo-respon-

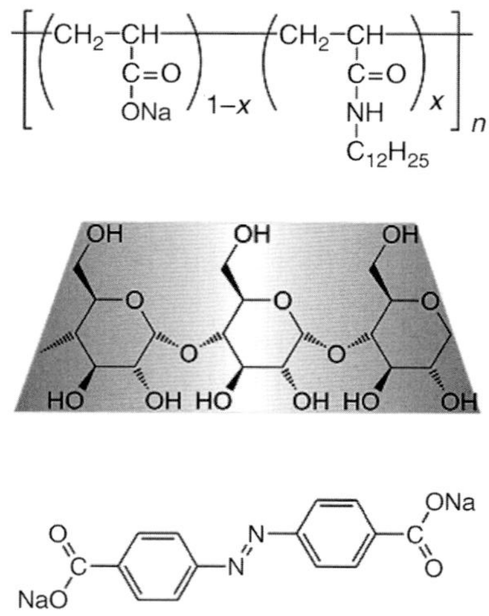

Scheme 11.27

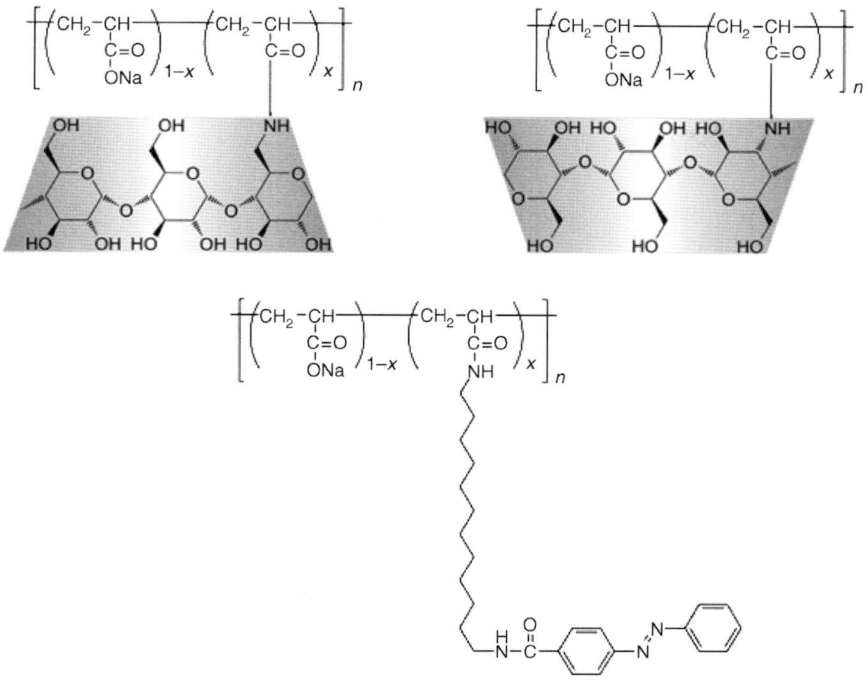

Scheme 11.28

sive hydrogel formed from dextran-based azobenzene and α-CD polymers for the controlled release of green fluorescent protein.

Ferris and coworkers [88] utilized the photo-responsive molecular recognition of β-CD and azobenzene for the valve of mesoporous silica nanoparticle containers. Rhodamine B was loaded into the pores of silica nanoparticles modified with azobenzene derivatives by soaking the particles in a solution of Rhodamine B. Pseudo-rotaxanes of β-CD, pyrene-carrying β-CD, or their mixture with the trans-azobenzene moieties on nanoparticles were subsequently formed (Scheme 11.29). UV irradiation caused trans-to-cis photoisomerization of the azobenzene moieties, which dissociated the pseudo-rotaxanes and opened the valve, releasing the Rhodamine B loaded in the pores.

11.3.6
Redox-Responsive Systems

Guo and coworkers [89] utilized a pair of p-sulfonatocalix[4]arene and viologen to construct a redox-responsive supramolecular polymer system because the binding constant of p-sulfonatocalix[4]arene is very sensitive to the redox process of viologen. A water-soluble supramolecular polymer was formed from the p-sulfonatocalix[4]-

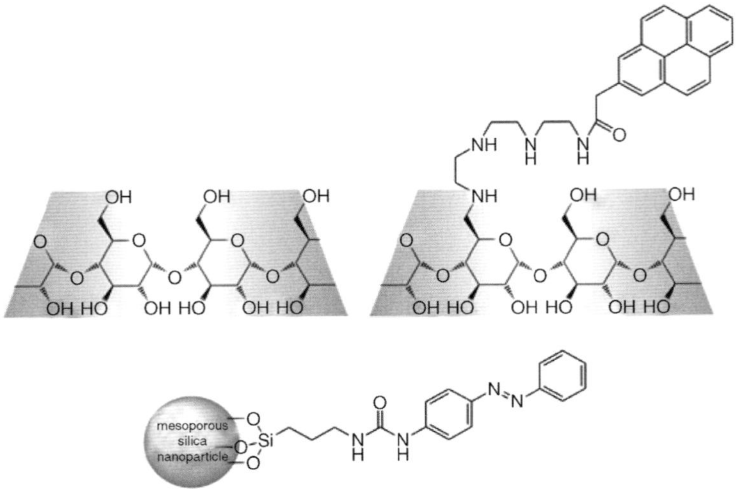

Scheme 11.29

arene dimer and viologen dimer (Scheme 11.30). The supramolecular polymer was dissociated upon the reduction of the viologen moieties, and reversible fomation was controlled electrochemically.

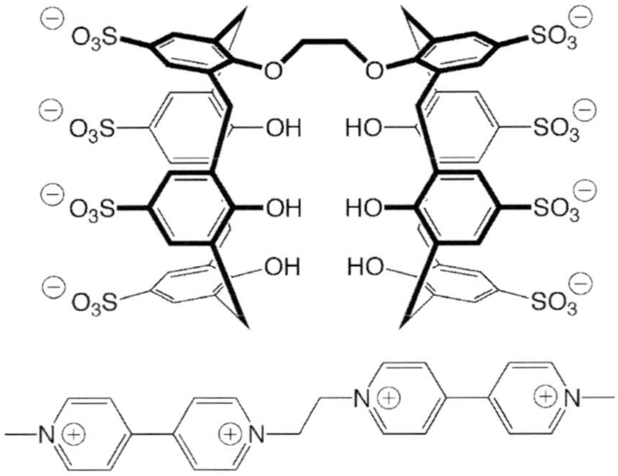

Scheme 11.30

Kawano and coworkers [40] reported a redox-responsive organogel using the different coordination modes of Cu(I) and Cu(II). An organogel was prepared utilizing the formation of a coordination complex of Cu(I) with cholesteryl- or 2-ethylhexyl-modified 4,4′-bipyridine (Scheme 11.31). Upon oxidation of Cu(I) with nitrosyl fluoroborate, the organogel mixture turned into a sol mixture containing a

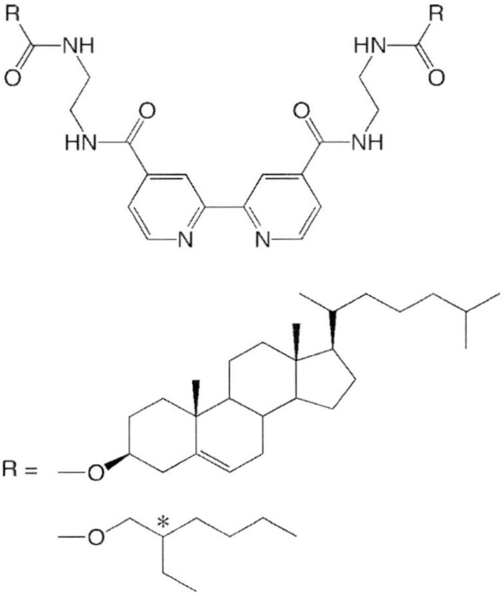

Scheme 11.31

small amount of a pale blue precipitate. However, upon reduction with ascorbic acid, the sol mixture turned back into the gel mixture.

Boweman and Nilsson [90] demonstrated the reduction-responsive gelation of oligopeptide containing a (Phe-Lys-Phe-Glu)$_2$-NH$_2$ sequence, which has a strong tendency to form water soluble amyloid like fibrillar bilayers. The designed oligopeptide, containing two Cys residues, Ac-Cys-(Phe-Lys-Phe-Glu)$_2$Cys-Gly-NH$_2$, was cyclized via the formation of disulfide bonding between the Cys residues. The cyclic oligopeptide did not assume the β-sheet conformation because of the conformational restriction. The cyclic oligopeptide was converted into the linear form by reduction of this disulfide bond with tris(2-carboxyethyl)phosphine. The linear oligopeptide took a β-sheet conformation and formed fibrous supramolecular structures, which resulted in the formation of hydrogels.

Ambrogio and coworkers [91] utilized the redox-responsive cleavage of a disulfide bond for the valve of mesoporous silica nanoparticle containers. Rhodamine B was loaded into the pores of silica nanoparticles bearing azide moieties by soaking the particles in a solution of Rhodamine B. Rotaxanes of CB[6] or α-CD were subsequently formed by click chemistry to cap with an adamantane derivative bearing an alkyne moiety (Scheme 11.32). Cleavage of the disulfide bonds by adding dithiothreitol or 2-mercaptoethanol caused the rotaxanes to dissociate, which opened up the valve and released the Rhodamine B loaded in the pores of the nanopartiles.

β-CD and ferrocene are a popular pair for redox-responsive molecular recognition; β-CD includes ferrocene but not ferrocenium. Tomatsu and coworkers [92] realized a redox-responsive hydrogel through a combination of β-CD, dodecyl-modified

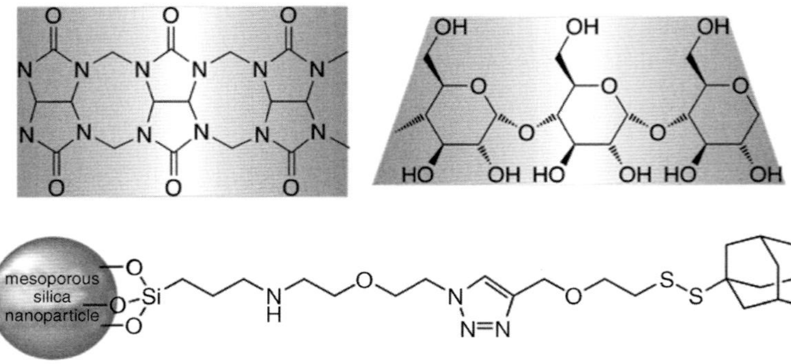

Scheme 11.32

poly-(sodium acrylate), and a water-soluble ferrocene derivative, that is, ferrocene-caroboxylic acid (FCA) (Scheme 11.33). A ternary mixture of the reduced state of FCA

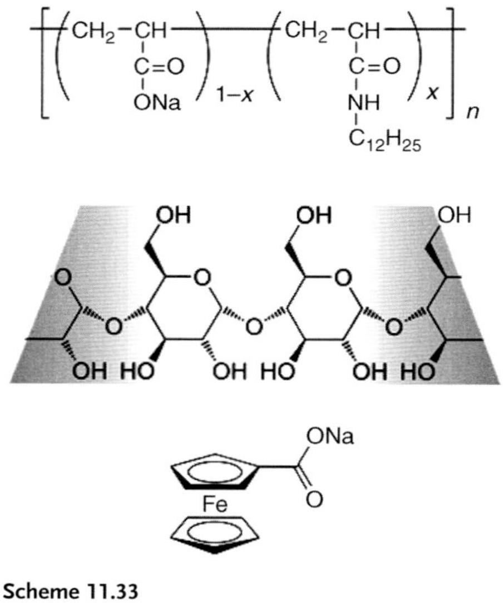

Scheme 11.33

exhibited a gel-like behavior, whereas a mixture in the oxidized state displayed a sol behavior. Yan and coworkers [93] reported a supramolecular block copolymer with an inclusion complex of β-CD and ferrocene moieties, which formed redox-responsive vesicles. A supramolecular block copolymer was formed from a β-CD-attached polystyrene of $M_n = 1.86 \times 10^4$ and a ferrocene-tagged poly(ethylene glycol) of $M_n = 1.2 \times 10^3$ and assembled into vesicles (Scheme 11.34). The vesicles disassembled upon electrochemically oxidizing the ferrocene moiety at +1.5 V because of

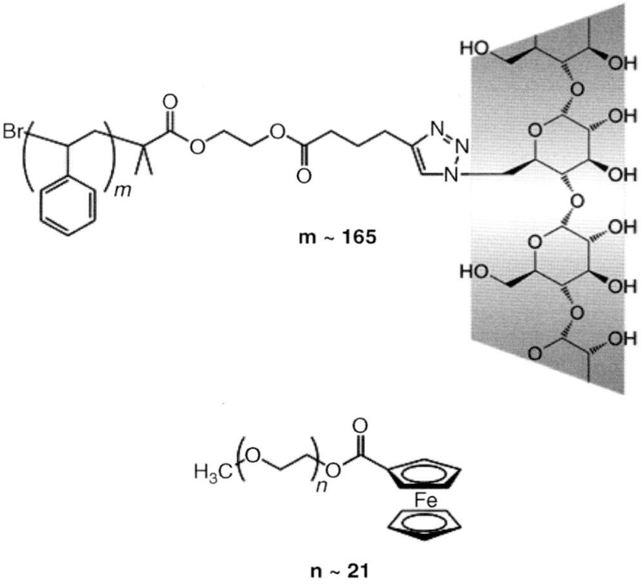

Scheme 11.34

dissociation of supramolecular polymers. These assembling and disassembling processes of the vesicles were reversible. The redox-responsive vesicles were utilized for controlled release of Rhodamine B.

11.3.7
Multi-Stimuli-Responsive Systems

In recent years, the numbers of publications on supramolecular polymer systems responsive to two or more stimuli has increased. Most of these examples deal with combinations of temperature and another stimulus: for example, temperature- and pH-responsive systems [94–97], temperature- and chemical-responsive systems [98, 99], temperature- and photo-responsive systems [100–103], and temperature- and redox-responsive systems [41].

Supramolecular coordination polymer systems are responsive to multiple stimuli, for example, temperature, stress, chemical, and redox. Rowan and coworkers [104–108] reported multi-stimuli-responsive supramolecular coordination polymer systems based on combinations of transition and lanthanoid metal ions and telechelic polymers carrying tridentate ligand moieties, that is, 2,6-bis(1′-methylbenzimidazolyl)-4-pyridyl groups, at both the ends (Scheme 11.35). Very recently, He and coworkers [109] reported a similar multi-stimuli-responsive supramolecular coordination polymer gel based on a binaphthylbisbipyridine-Cu(I) complex (Scheme 11.36).

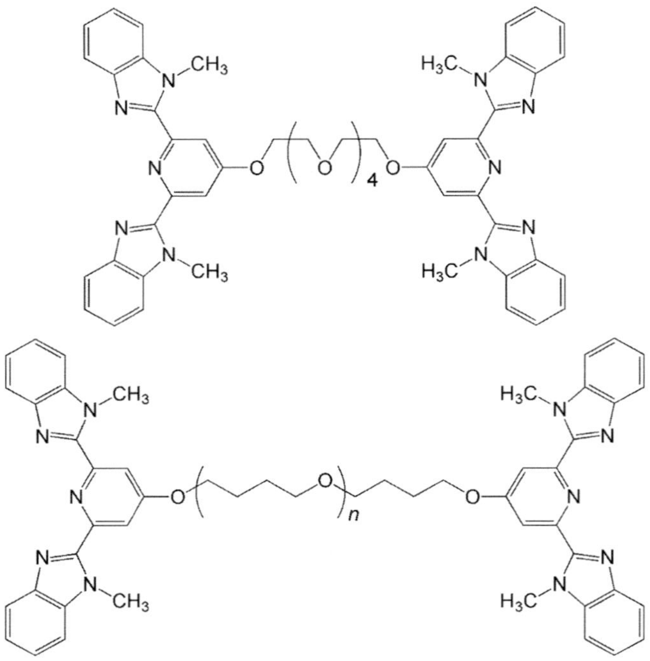

Scheme 11.35

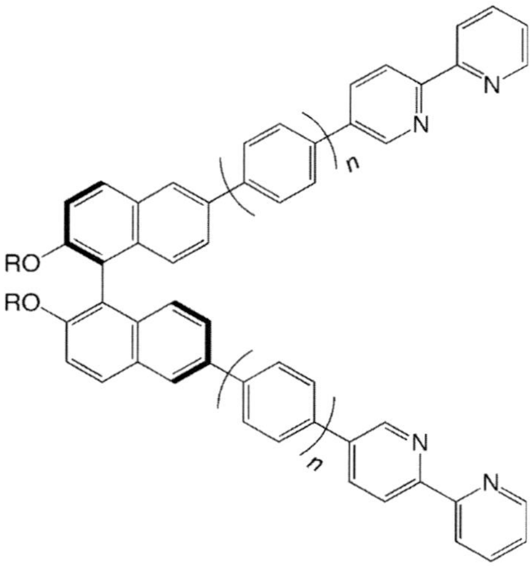

R = -CH₂OCH₃, n = 1
R = -C₂H₅, n = 0
R = -C₂H₅, n = 1
R = -C₆H₁₃-n, n = 1
R = -C₆H₁₃-n, n = 2

Scheme 11.36

11.4
Concluding Remarks

This chapter reviewed stimuli-responsive supramolecular polymer systems. Section 11.2 overviewed stimuli and responses for stimuli-responsive supramolecular polymer systems. Section 11.3 dealt with recent examples of stimuli-responsive supramolecular polymer systems. As described in the introduction, living organisms are integrated stimuli-responsive supramolecular polymer systems based on biological macromolecules that realize macroscopic responses based on a number of microscopic chemical events. Although numerous research groups have striven to construct highly functional artificial stimuli-responsive supramolecular polymer systems, a large gap remains between biological and artificial systems. To bridge this gap, scientists should accumulate and combine their knowledge for a deeper understanding of Mother Nature, because this will be a necessary step toward realizing a sustainable society.

Acknowledgment

The authors would like to express their acknowledgment to Professor Takahiro Sato, Associate Professor Hiroyasu Yamaguchi, and Assistant Professor Yoshinori Takashima, Department of Macromolecular Science, Graduate School of Science, Osaka University, and Dr. **Cyrus Anderson**, Department of Chemistry, University of Illinois at Urbana-Champaign, for fruitful discussion and suggestions.

References

1 Voet, D. and Voet, J.G. (2004) *Biochemistry*, 3rd edn, Wiley & Sons, New York.
2 Berg, J.M., Tymoczko, J.L., and Stryer, L. (2007) *Biochemistry*, 6th edn, W. H. Freeman, New York.
3 Alberts, B., Johnson, A., Lewis, J., Raff, M., Roberts, K., and Walter, P. (2008) *Molecular Biology of the Cell*, 5th edn, Garland Publishing Inc., New York.
4 Ciferri, A. (2005) *Supramolecular Polymers*, 2nd edn, CRC Press, Boca Raton, FL.
5 Ishi-i, T. and Shinkai, S. (2005) Dye-based organogels: stimuli-responsive soft materials based on one-dimensional self-assembling aromatic dyes. *Top. Curr. Chem.*, **258**, 119–160.
6 Nagasaki, T. and Shinkai, S. (2007) The concept of molecular machinery is useful for design of stimuli-responsive gene delivery systems in the mammalian cell. *J. Inclusion Phenom. Macrocycl. Chem.*, **58** (3), 205–219.
7 Paulusse, J.M.J. and Sijbesma, R.P. (2006) Molecule-based rheology switching. *Angew. Chem. Int. Ed.*, **45** (15), 2334–2337.
8 Choi, H.S. and Yui, N. (2006) Design of rapidly assembling supramolecular systems responsive to synchronized stimuli. *Prog. Polym. Sci.*, **31** (2), 121–144.
9 Serpe, M.J. and Craig, S.L. (2006) Physical organic chemistry of supramolecular polymers. *Langmuir*, **23** (4), 1626–1634.
10 Bouteiller, L. (2007) Assembly via hydrogen bonds of low molar mass compounds into supramolecular

polymers. *Adv. Polym. Sci.*, **207**, 79–112.

11 Rieth, S., Baddeley, C., and Badjić, J.D. (2007) Prospects in controlling morphology, dynamics and responsiveness of supramolecular polymers. *Soft Matter*, **3** (2), 137–154.

12 Dankers, P.Y.W. and Meijer, E.W. (2007) Supramolecular biomaterials. A modular approach towards tissue engineering. *Bull. Chem. Soc. Jpn.*, **80** (11), 2047–2073.

13 Champin, B., Mobian, P., and Sauvage, J.-P. (2007) Transition metal complexes as molecular machine prototypes. *Chem. Soc. Rev.*, **36** (2), 358–366.

14 Hoogenboom, R., Fournier, D., and Schubert, U.S. (2008) Asymmetrical supramolecular interactions as basis for complex responsive macromolecular architectures. *Chem. Commun. (Cambridge, U. K.)*, (2), 155–162.

15 Yagai, S. and Kitamura, A. (2008) Recent advances in photoresponsive supramolecular self-assemblies. *Chem. Soc. Rev.*, **37** (8), 1520–1529.

16 Cotí, K.K., Belowich, M.E., Liong, M., Ambrogio, M.W., Lau, Y.A., Khatib, H.A., Zink, J.I., Khashab, N.M., and Stoddart, J.F. (2009) Mechanised nanoparticles for drug delivery. *Nanoscale*, **1** (1), 16–39.

17 Leung, K.C.-F., Chak, C.-P., Lo, C.-M., Wong, W.-Y., Xuan, S., and Cheng, C.H.K. (2009) pH-controllable supramolecular systems. *Chem. Asian J.*, **4** (3), 364–381.

18 Schneider, H.-J. and Strongin, R.M. (2009) supramolecular interactions in chemomechanical polymers. *Acc. Chem. Res.*, **42** (10), 1489–1500.

19 Fox, J.D. and Rowan, S.J. (2009) supramolecular polymerizations and main-chain supramolecular polymers. *Macromolecules*, **42** (18), 6823–6835.

20 van de Manakker, F., Vermonden, T., van Nostrum, C.F., and Hennink, W.E. (2009) Cyclodextrin-based polymeric materials: synthesis, properties, and pharmaceutical/biomedical applications. *Biomacromolecules*, **10** (12), 3157–3175.

21 Moughton, A.O. and O'Reilly, R.K. (2010) Using metallo-supramolecular block copolymers for the synthesis of higher order nanostructured assemblies. *Macromol. Rapid Commun*, **31** (1), 37–52.

22 Tsitsilianis, C. (2010) Responsive reversible hydrogels from associative 'smart' macromolecules. *Soft Matter*, **6** (11), 2372–2388.

23 Hashidzume, A., Tomatsu, I., and Harada, A. (2006) Interaction of cyclodextrins with side chains of water soluble polymers: a simple model for biological molecular recognition and its utilization for stimuli-responsive systems. *Polymer*, **47** (17), 6011–6027.

24 Harada, A., Hashidzume, A., and Takashima, Y. (2006) Cyclodextrin-based supramolecular polymers. *Adv. Polym. Sci.*, **201**, 1–43.

25 Harada, A. (2006) Supramolecular polymers based on cyclodextrins. *J. Polym. Sci., Part A: Polym. Chem.*, **44** (17), 5113–5119.

26 Harada, A., Takashima, Y., and Yamaguchi, H. (2009) Cyclodextrin-based supramolecular polymers. *Chem. Soc. Rev.*, **38** (4), 875–882.

27 Harada, A., Hashidzume, A., Yamaguchi, H., and Takashima, Y. (2009) Polymeric rotaxanes. *Chem. Rev. (Washington, DC, U. S.)*, **109** (11), 5974–6023.

28 Harada, A. and Hashidzume, A. (2010) Supramolecular polymers based on cyclodextrins and their derivatives. *Aust. J. Chem.*, **63** (4), 599–610.

29 Miyako, E., Nagata, H., Hirano, K., and Hirotsu, T. (2008) Photodynamic thermoresponsive nanocarbon-polymer gel hybrids. *Small*, **4** (10), 1711–1715.

30 Mizuno, K. (2009) Photochemistry of aromatic compounds, in *Photochemistry* (ed. A. Albini), RSC Publishing, Cambridge, United Kingdom, pp. 175–212.

31 Jiang, Y., Wan, P., Xu, H., Wang, Z., Zhang, X., and Smet, M. (2009) Facile reversible UV-controlled and fast transition from emulsion to gel by using a photoresponsive polymer with a malachite green group. *Langmuir*, **25** (17), 10134–10138.

32 Holtz, J.H. and Asher, S.A. (1997) Polymerized colloidal crystal hydrogel films as intelligent chemical sensing

33 Swager, T.M. (1998) The molecular wire approach to sensory signal amplification. *Acc. Chem. Res.*, **31** (5), 201–207.
34 Haupt, K. and Mosbach, K. (2000) Molecularly imprinted polymers and their use in biomimetic sensors. *Chem. Rev. (Washington, DC, U. S.)*, **100** (7), 2495–2504.
35 Lavigne, J.J. and Anslyn, E.V. (2001) Sensing a paradigm shift in the field of molecular recognition: from selective to differential receptors. *Angew. Chem. Int. Ed.*, **40** (17), 3118–3130.
36 James, T.D. (2005) Boronic acid-based receptors and sensors for saccharides, in *Boronic Acids. Preparation, Applications in Organic Synthesis and Medicine* (ed. D.G. Hall), Wiley-VCH, Weinheim, pp 441–479.
37 James, T.D., Phillips, M.D., and Shinkai, S. (2006) *Boronic Acids in Saccharide Recognition*, The Royal Society of Chemistry, Cambridge, UK.
38 Kuroiwa, K., Shibata, T., Takada, A., Nemoto, N., and Kimizuka, N. (2004) Heat-set gel-like networks of lipophilic Co(II) triazole complexes in organic media and their thermochromic structural transitions. *J. Am. Chem. Soc.*, **126** (7), 2016–2021.
39 Ogoshi, T., Takashima, Y., Yamaguchi, H., and Harada, A. (2006) Cyclodextrin-grafted poly(phenylene ethynylene) with chemically-responsive properties. *Chem. Commun. (Cambridge, U. K.)* (35), 3702–3704.
40 Kawano, S.-i., Fujita, N., and Shinkai, S. (2004) A coordination gelator that shows a reversible chromatic change and sol–gel phase-transition behavior upon oxidative/reductive stimuli. *J. Am. Chem. Soc.*, **126** (28), 8592–8593.
41 Kawano, S.-i., Fujita, N., and Shinkai, S. (2005) Quater-, quinque-, and sexithiophene organogelators: unique thermochromism and heating-free sol-gel phase transition. *Chem. Eur. J.*, **11** (16), 4735–4742.
42 Woodward, P., Merino, D.H., Hamley, I.W., Slark, A.T., and Hayes, W. (2009) Thermally responsive elastomeric supramolecular polymers featuring flexible aliphatic hydrogen-bonding end-groups. *Aust. J. Chem.*, **62** (8), 790–793.
43 Sivakova, S., Bohnsack, D.A., Mackay, M.E., Suwanmala, P., and Rowan, S.J. (2005) Utilization of a combination of weak hydrogen-bonding interactions and phase segregation to yield highly thermosensitive supramolecular polymers. *J. Am. Chem. Soc.*, **127** (51), 18202–18211.
44 Kiyonaka, S., Shinkai, S., and Hamachi, I. (2003) Combinatorial library of low molecular-weight organo- and hydrogelators based on glycosylated amino acid derivatives by solid-phase synthesis. *Chem. Eur. J.*, **9** (4), 976–983.
45 Kiyonaka, S., Sugiyasu, K., Shinkai, S., and Hamachi, I. (2002) First thermally responsive supramolecular polymer based on glycosylated amino acid. *J. Am. Chem. Soc.*, **124** (37), 10954–10955.
46 Xu, D., Hawk, J.L., Loveless, D.M., Jeon, S.L., and Craig, S.L. (2010) Mechanism of shear thickening in reversibly cross-linked supramolecular polymer networks. *Macromolecules*, **43** (7), 3556–3565.
47 Xu, D. and Craig, S.L. (2010) Multiple dynamic processes contribute to the complex steady shear behavior of cross-linked supramolecular networks of semidilute entangled polymer solutions. *J. Phys. Chem. Lett.*, **1** (11), 1683–1686.
48 Zhao, Y., Beck, J.B., Rowan, S.J., and Jamieson, A.M. (2004) Rheological behavior of shear-responsive metallo-supramolecular gels. *Macromolecules*, **37** (10), 3529–3531.
49 Paulusse, J.M.J. and Sijbesma, R.P. (2004) Reversible mechanochemistry of a Pd(II) coordination polymer. *Angew. Chem. Int. Ed.*, **43** (34), 4460–4462.
50 Paulusse, J.M.J., Huijbers, J.P.J., and Sijbesma, R.P. (2006) Quantification of ultrasound-induced chain scission in Pd^{II}-phosphine coordination polymers. *Chem. Eur. J.*, **12** (18), 4928–4934.
51 Paulusse, J.M.J., van Beek, D.J.M., and Sijbesma, R.P. (2007) Reversible switching of the sol-gel transition with ultrasound in rhodium(I) and iridium(I)

coordination networks. *J. Am. Chem. Soc.*, **129** (8), 2392–2397.

52 Naota, T. and Koori, H. (2005) Molecules that assemble by sound: an application to the instant gelation of stable organic fluids. *J. Am. Chem. Soc.*, **127** (26), 9324–9325.

53 Isozaki, K., Takaya, H., and Naota, T. (2007) Ultrasound-induced gelation of organic fluids with metalated peptides. *Angew. Chem. Int. Ed.*, **46** (16), 2855–2857.

54 Jones, J.W., Bryant, W.S., Bosman, A.W., Janssen, R.A.J., Meijer, E.W., and Gibson, H.W. (2003) Crowned dendrimers: pH-responsive pseudorotaxane formation. *J. Org. Chem.*, **68** (6), 2385–2389.

55 Ghoussoub, A. and Lehn, J.-M. (2005) Dynamic sol-gel interconversion by reversible cation binding and release in G-quartet-based supramolecular polymers. *Chem. Commun. (Cambridge, U. K.)*, (46), 5763–5765.

56 Martens, A.A., Portale, G., Werten, M.W.T., de Vries, R.J., Eggink, G., Cohen Stuart, M.A., and de Wolf, F.A. (2009) Triblock protein copolymers forming supramolecular nanotapes and pH-responsive gels. *Macromolecules*, **42** (4), 1002–1009.

57 Zhou, S.-L., Matsumoto, S., Tian, H.-D., Yamane, H., Ojida, A., Kiyonaka, S., and Hamachi, I. (2005) pH-responsive shrinkage/swelling of a supramolecular hydrogel composed of two small amphiphilic molecules. *Chem. Eur. J.*, **11** (4), 1130–1136.

58 Barbé, C., Bartlett, J., Kong, L., Finnie, K., Lin, H.Q., Larkin, M., Calleja, S., Bush, A., and Calleja, G. (2004) Silica particles: a novel drug-delivery system. *Adv. Mater. (Weinheim, Ger.)*, **16** (21), 1959–1966.

59 Park, C., Oh, K., Lee, S.C., and Kim, C. (2007) Controlled release of guest molecules from mesoporous silica particles based on a pH-responsive polypseudorotaxane motif. *Angew. Chem. Int. Ed.*, **46** (9), 1455–1457.

60 Angelos, S., Yang, Y.-W., Patel, K., Stoddart, J.F., and Zink, J.I. (2008) pH-responsive supramolecular nanovalves based on cucurbit[6]uril pseudorotaxanes. *Angew. Chem. Int. Ed.*, **47** (12), 2222–2226.

61 Angelos, S., Khashab, N.M., Yang, Y.-W., Trabolsi, A., Khatib, H.A., Stoddart, J.F., and Zink, J.I. (2009) pH clock-operated mechanized nanoparticles. *J. Am. Chem. Soc.*, **131** (36), 12912–12914.

62 Choi, H.S., Ooya, T., Lee, S.C., Sasaki, S., Kurisawa, M., Uyama, H., and Yui, N. (2004) pH dependence of polypseudorotaxane formation between cationic linear polyethylenimine and cyclodextrins. *Macromolecules*, **37** (18), 6705–6710.

63 Collin, J.-P., Dietrich-Buchecker, C., Gaviña, P., Jimenez-Molero, M.C., and Sauvage, J.-P. (2001) Shuttles and muscles: Linear molecular machines based on transition metals. *Acc. Chem. Res.*, **34** (6), 477–487.

64 Dawson, R.E., Lincoln, S.F., and Easton, C.J. (2008) The foundation of a light driven molecular muscle based on stilbene and α-cyclodextrin. *Chem. Commun. (Cambridge, U. K.)* (34), 3980–3982.

65 Tsukagoshi, S., Miyawaki, A., Takashima, Y., Yamaguchi, H., and Harada, A. (2007) Contraction of supramolecular double-threaded dimer formed by α-cyclodextrin with a long alkyl chain. *Org. Lett.*, **9** (6), 1053–1055.

66 Deng, W., Yamaguchi, H., Takashima, Y., and Harada, A. (2007) A chemical-responsive supramolecular hydrogel from modified cyclodextrins. *Angew. Chem. Int. Ed.*, **46** (27), 5144–5147.

67 Deng, W., Yamaguchi, H., Takashima, Y., and Harada, A. (2008) Construction of chemical-responsive supramolecular hydrogels from guest-modified cyclodextrins. *Chem. Asian J.*, **3** (4), 687–695.

68 Ogoshi, T., Takashima, Y., Yamaguchi, H., and Harada, A. (2007) Chemically-responsive sol-gel transition of supramolecular single-walled carbon nanotubes (SWNTs) hydrogel made by hybrids of SWNTs and cyclodextrins. *J. Am. Chem. Soc.*, **129** (16), 4878–4879.

69 Gu, H., Chao, J., Xiao, S.-J., and Seeman, N.C. (2010) A proximity-based programmable DNA nanoscale assembly

line. *Nature (London, U. K.)*, **465** (7295), 202–205.
70. Lund, K., Manzo, A.J., Dabby, N., Michelotti, N., Johnson-Buck, A., Nangreave, J., Taylor, S., Pei, R., Stojanovic, M.N., Walter, N.G., Winfree, E., and Yan, H. (2010) Molecular robots guided by prescriptive landscapes. *Nature (London, U. K.)*, **465** (7295), 206–210.
71. Kuad, P., Miyawaki, A., Takashima, Y., Yamaguchi, H., and Harada, A. (2007) External stimulus-responsive supramolecular structures formed by a stilbene cyclodextrin dimer. *J. Am. Chem. Soc.*, **129** (42), 12630–12631.
72. Frkanec, L., Jokic, M., Makarevic, J., Wolsperger, K., and Zinic, M. (2002) Bis(PheOH) maleic acid amide-fumaric acid amide photoizomerization induces microsphere-to-gel fiber morphological transition: The photoinduced gelation system. *J. Am. Chem. Soc.*, **124** (33), 9716–9717.
73. Koumura, N., Kudo, M., and Tamaoki, N. (2004) Photocontrolled gel-to-sol-to-gel phase transitioning of meta-substituted azobenzene bisurethanes through the breaking and reforming of hydrogen bonds. *Langmuir*, **20** (23), 9897–9900.
74. Eastoe, J., Sanchez-Dominguez, M., Wyatt, P., and Heenan, R.K. (2004) A photo-responsive organogel. *Chem. Commun. (Cambridge, U. K.)* (22), 2608–2609.
75. Matsumoto, S., Yamaguchi, S., Ueno, S., Komatsu, H., Ikeda, M., Ishizuka, K., Iko, Y., Tabata, K.V., Aoki, H., Ito, S., Noji, H., and Hamachi, I. (2008) Photo gel-sol/sol-gel transition and its patterning of a supramolecular hydrogel as stimuli-responsive biomaterials. *Chem. Eur. J.*, **14** (13), 3977–3986.
76. Yamauchi, K., Takashima, Y., Hashidzume, A., Yamaguchi, H., and Harada, A. (2008) Switching between supramolecular dimer and supramolecular self-assembly consisting of stilbene amide α-cyclodextrin by photoirradiation. *J. Am. Chem. Soc.*, **130** (15), 5024–5025.
77. Wang, Y., Ma, N., Wang, Z., and Zhang, X. (2007) Photocontrolled reversible supramolecular assemblies of an azobenzene-containing surfactant with α-cyclodextrin. *Angew. Chem. Int. Ed.*, **46** (16), 2823–2826.
78. Nolan, D., Darcy, R., and Ravoo, B.J. (2003) Preparation of vesicles and nanoparticles of amphiphilic cyclodextrins containing labile disulfide bonds. *Langmuir*, **19** (10), 4469–4472.
79. Ravoo, B.J., Jacquier, J.-C., and Wenz, G. (2003) Molecular recognition of polymers by cyclodextrin vesicles. *Angew. Chem. Int. Ed.*, **42** (18), 2066–2070.
80. Falvey, P., Lim, C.W., Darcy, R., Revermann, T., Karst, U., Giesbers, M., Marcelis, A.T.M., Lazar, A., Coleman, A.W., Reinhoudt, D.N., and Ravoo, B.J. (2005) Bilayer vesicles of amphiphilic cyclodextrins: host membranes that recognize guest molecules. *Chem. Eur. J.*, **11** (4), 1171–1180.
81. Nalluri, S.K.M. and Ravoo, B.J. (2010) Light-responsive molecular recognition and adhesion of vesicles. *Angew. Chem. Int. Ed.*, **49** (31), 5371–5374.
82. Li, S., Taura, D., Hashidzume, A., Takashima, Y., Yamaguchi, H., and Harada, A. (2010) Photocontrolled size changes of doubly-threaded dimer based on an α-cyclodextrin derivative with two recognition sites. *Chem. Lett.*, **39** (3), 242–243.
83. Li, S., Taura, D., Hashidzume, A., and Harada, A. (2010) Light-switchable janus [2]rotaxanes based on α-cyclodextrin derivatives bearing two recognition sites linked with oligo(ethylene glycol). *Chem. Asian J.*, **5** (10), 2281–2289.
84. Tomatsu, I., Hashidzume, A., and Harada, A. (2005) Photoresponsive hydrogel system utilizing molecular recognition of α-cyclodextrin. *Macromolecules*, **38** (12), 5223–5227.
85. Tomatsu, I., Hashidzume, A., and Harada, A. (2006) Contrast viscosity changes upon photoirradiation for mixtures of poly(acrylic acid)-based α-cyclodextrin and azobenzene polymers. *J. Am. Chem. Soc.*, **128** (7), 2226–2227.
86. Tomatsu, I., Hashidzume, A., and Harada, A. (2006) Cyclodextrin-based side chain polyrotaxane with

unidirectional inclusion in aqueous media. *Angew. Chem. Int. Ed.*, **45** (28), 4605–4608.

87 Peng, K., Tomatsu, I., and Kros, A. (2010) Light controlled protein release from a supramolecular hydrogel. *Chem. Commun. (Cambridge, U. K.)*, **46** (23), 4094–4096.

88 Ferris, D.P., Zhao, Y.-L., Khashab, N.M., Khatib, H.A., Stoddart, J.F., and Zink, J.I. (2009) Light-operated mechanized nanoparticles. *J. Am. Chem. Soc.*, **131** (5), 1686–1688.

89 Guo, D.-S., Chen, S., Qian, H., Zhang, H.-Q., and Liu, Y. (2010) Electrochemical stimulus-responsive supramolecular polymer based on sulfonatocalixarene and viologen dimers. *Chem. Commun. (Cambridge, U. K.)*, **46** (15), 2620–2622.

90 Bowerman, C.J. and Nilsson, B.L. (2010) A reductive trigger for peptide self-assembly and hydrogelation. *J. Am. Chem. Soc.*, **132** (28), 9526–9527.

91 Ambrogio, M.W., Pecorelli, T.A., Patel, K., Khashab, N.M., Trabolsi, A., Khatib, H.A., Botros, Y.Y., Zink, J.I., and Stoddart, J.F. (2010) Snap-top nanocarriers. *Org. Lett.*, **12** (15), 3304–3307.

92 Tomatsu, I., Hashidzume, A., and Harada, A. (2006) Redox-responsive hydrogel system using the molecular recognition of β-cyclodextrin. *Macromol. Rapid Commun.*, **27** (4), 238–241.

93 Yan, Q., Yuan, J., Cai, Z., Xin, Y., Kang, Y., and Yin, Y. (2010) Voltage-responsive vesicles based on orthogonal assembly of two homopolymers. *J. Am. Chem. Soc.*, **132** (27), 9268–9270.

94 Zhang, Y., Gu, H., Yang, Z., and Xu, B. (2003) Supramolecular hydrogels respond to ligand-receptor interaction. *J. Am. Chem. Soc.*, **125** (45), 13680–13681.

95 Choi, H.S., Huh, K.M., Ooya, T., and Yui, N. (2003) pH- and thermosensitive supramolecular assembling system: rapidly responsive properties of β-cyclodextrin-conjugated poly(ε-lysine). *J. Am. Chem. Soc.*, **125** (21), 6350–6351.

96 Choi, H.S., Yamamoto, K., Ooya, T., and Yui, N. (2005) Synthesis of poly(ε-lysine)-grafted dextrans and their pH and thermosensitive hydrogelation with cyclodextrins. *ChemPhysChem*, **6** (6), 1081–1086.

97 Ge, Z., Hu, J., Huang, F., and Liu, S. (2009) Responsive supramolecular gels constructed by crown ether based molecular recognition. *Angew. Chem. Int. Ed.*, **48** (10), 1798–1802.

98 Shirakawa, M., Fujita, N., Tani, T., Kaneko, K., and Shinkai, S. (2005) Organogel of an 8-quinolinol platinum (II) chelate derivative and its efficient phosphorescence emission effected by inhibition of dioxygen quenching. *Chem. Commun. (Cambridge, U. K.)*, (33), 4149–4151.

99 Wang, F., Zhang, J., Ding, X., Dong, S., Liu, M., Zheng, B., Li, S., Wu, L., Yu, Y., Gibson, H.W., and Huang, F. (2010) Metal coordination mediated reversible conversion between linear and cross-linked supramolecular polymers. *Angew. Chem. Int. Ed.*, **49** (6), 1090–1094.

100 Ayabe, M., Kishida, T., Fujita, N., Sada, K., and Shinkai, S. (2003) Binary organogelators which show light and temperature responsiveness. *Org. Biomol. Chem.*, **1** (15), 2744–2747.

101 Suzuki, T., Shinkai, S., and Sada, K. (2006) Supramolecular crosslinked linear poly(trimethylene iminium trifluorosulfonimide) polymer gels sensitive to light and thermal stimuli. *Adv. Mater. (Weinheim, Ger.)*, **18** (8), 1043–1046.

102 Hwang, I., Jeon, W.S., Kim, H.-J., Kim, D., Kim, H., Selvapalam, N., Fujita, N., Shinkai, S., and Kim, K. (2007) Cucurbit[7]uril: a simple macrocyclic, ph-triggered hydrogelator exhibiting guest-induced stimuli-responsive behavior. *Angew. Chem. Int. Ed.*, **46** (1–2), 210–213.

103 Inoue, Y., Kuad, P., Okumura, Y., Takashima, Y., Yamaguchi, H., and Harada, A. (2007) Thermal and photochemical switching of conformation of poly(ethylene glycol)-substituted cyclodextrin with an azobenzene group at the chain end. *J. Am. Chem. Soc.*, **129** (20), 6396–6397.

104 Beck, J.B. and Rowan, S.J. (2003) Multistimuli, multiresponsive metallo-supramolecular polymers. *J. Am. Chem. Soc.*, **125** (46), 13922–13923.

105 Rowan, S.J. and Beck, J.B. (2005) Metal-ligand induced supramolecular polymerization: A route to responsive materials. *Faraday Discuss.*, **128**, 43–53.

106 Weng, W., Beck, J.B., Jamieson, A.M., and Rowan, S.J. (2006) Understanding the mechanism of gelation and stimuli-responsive nature of a class of metallo-supramolecular gels. *J. Am. Chem. Soc.*, **128** (35), 11663–11672.

107 Weng, W., Li, Z., Jamieson, A.M., and Rowan, S.J. (2009) Control of gel morphology and properties of a class of metallo-supramolecular polymers by good/poor solvent environments. *Macromolecules*, **42** (1), 236–246.

108 Kumpfer, J.R., Jin, J., and Rowan, S.J. (2010) Stimuli-responsive europium-containing metallo-supramolecular polymers. *J. Mater. Chem.*, **20** (1), 145–151.

109 He, Y., Bian, Z., Kang, C., Cheng, Y., and Gao, L. (2010) Chiral binaphthylbisbipyridine-based copper(I) coordination polymer gels as supramolecular catalysts. *Chem. Commun. (Cambridge, U. K.)*, **46** (20), 3532–3534.

12
Physical Organic Chemistry of Supramolecular Polymers
Stephen L. Craig and Donghua Xu

12.1
Introduction and Background

Individually and collectively, the chapters in this book illustrate the rich molecular toolkit that can be applied within the broad field of supramolecular polymers (SPs). As seen in the many examples presented, the nature of the central supramolecular motif plays a critical role in the material properties of the polymers. In many cases, the influence of the motif is tied to its chemistry, particularly its sensitivity to its environment and external stimuli, as explored, for example, by Hashidzume and Harada in Chapter 11. In the present chapter, we consider the impact of a critical feature of many supramolecular interactions – their dynamic nature – on the mechanical properties of supramolecular polymers (Figure 12.1). The contributions of reversible dynamics on the molecular level to polymer properties on the macroscopic level is a topic in which our research group has been interested for some time, and it is convenient to couch our discussion in the context of our prior work in the area.

In framing this discussion, we turn initially to a definition of supramolecular polymers that is particularly well suited for our intended purpose. It is useful to consider polymers in the context of entanglements – defined here very broadly as any intermolecular interactions that transfer mechanical force from one molecule to the next. Supramolecular polymers, in this framework, are reasonably viewed as systems in which specific intermolecular interactions create entanglements that would not exist in the absence of those interactions; the individual molecular constituents are too small, for example, to be physically entangled or to bridge between surfaces. One sees rather quickly that the mechanical properties of supramolecular polymers (defined in this way) are necessarily responsive to external stimuli to the same extent that the defining supramolecular interaction is responsive. Polymer properties can be turned 'on' and 'off' – toggling between polymeric and small-molecular behavior (and points in between) and creating materials that are more easily processed or recycled. Furthermore, the responsiveness need not be limited to on/off states; it could also be tuned through synthesis and environment to create 'smart' materials whose properties adjust along a wide continuum.

Supramolecular Polymer Chemistry, First Edition. Edited by Akira Harada.
© 2012 Wiley-VCH Verlag GmbH & Co. KGaA. Published 2012 by Wiley-VCH Verlag GmbH & Co. KGaA.

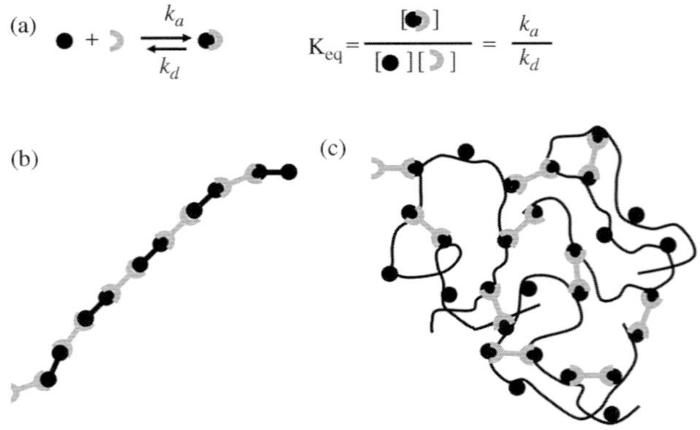

(a) ●+⟩ ⇌ ●● with k_a, k_d; $K_{eq} = \dfrac{[\text{●●}]}{[\text{●}][\text{⟩}]} = \dfrac{k_a}{k_d}$

(b) (c)

Figure 12.1 Schematic representation of (a) a reversible assembly motif and its related equilibrium and kinetic rate constants, and (b) linear and (c) networked supramolecular polymers formed from that motif.

The fields of polymer physics and the physical organic chemistry of supramolecular interactions therefore are very similar, and in this chapter we hope to demonstrate not only that similarity, but its utility. As such, the 'physical organic chemistry' of supramolecular polymers offers the opportunity to rationally and precisely control polymer properties. Not only that, but the ability to systematically vary the properties of the supramolecular interaction offers a potentially powerful tool for mechanistic studies of polymer physics, and examples of complex material behavior for which supramolecular approaches might be useful in that regard will be discussed.

12.2
Linear Supramolecular Polymers

12.2.1
N,C,N-Pincer Metal Complexes

The most fundamental difference between supramolecular polymers (SPs) and covalent polymers is the transient nature of the reversible supramolecular bond. Because of their significance as the critical feature of these systems, experimental strategies that allow one to directly probe the impact of reversible kinetics are desirable. Ideally, one would like to make a subtle structural change to the supramolecular motif and induce a dramatic change in its dissociation and formation kinetics. One can change on/off rates with temperature, of course, but with several limitations. For example, changing temperature might change the association constant of the interaction, and as a result also change the supramolecular polymer structure. Additionally, the accessible temperature range is typically limited (either by the thermal sensitivity/stability of the polymer system or by the ability to generate the necessary thermal environment), perhaps to only tens of degrees or less, meaning that in general a fairly modest range of rates can be achieved. Finally, changing the

12.2 Linear Supramolecular Polymers | 271

temperature complicates the study of complex dynamic situations in which multiple processes contribute to material behavior, because the competing processes will have different activation energies and so different temperature dependences. The idealized molecular strategy described above, however, allows the rates of dissociation and association to be varied across orders of magnitude, while still keeping the other critical features of a supramolecular polymer essentially constant: the equilibrium association constant of the supramolecular interaction, the overall structure of the supramolecular polymer, the chemical and thermal environment, and so on. ...

A particularly well suited motif is found in N, C, N-pincer metal–ligand coordination complexes such as 1•2 (Figure 12.2). Pincer compounds 1 and analogs have been synthesized and studied extensively by van Koten and coworkers [1–4], and they, with numerous other organometal coordination systems, have been used in SPs [3]. Added bulk in the N–alkyl substituents R, therefore, slows the exchange while exerting a lesser effect on the relative energy of the roughly isosteric endpoints. This concept has been reduced to practice within a series of N,C,N-pincer Pd(II) and Pt(II)

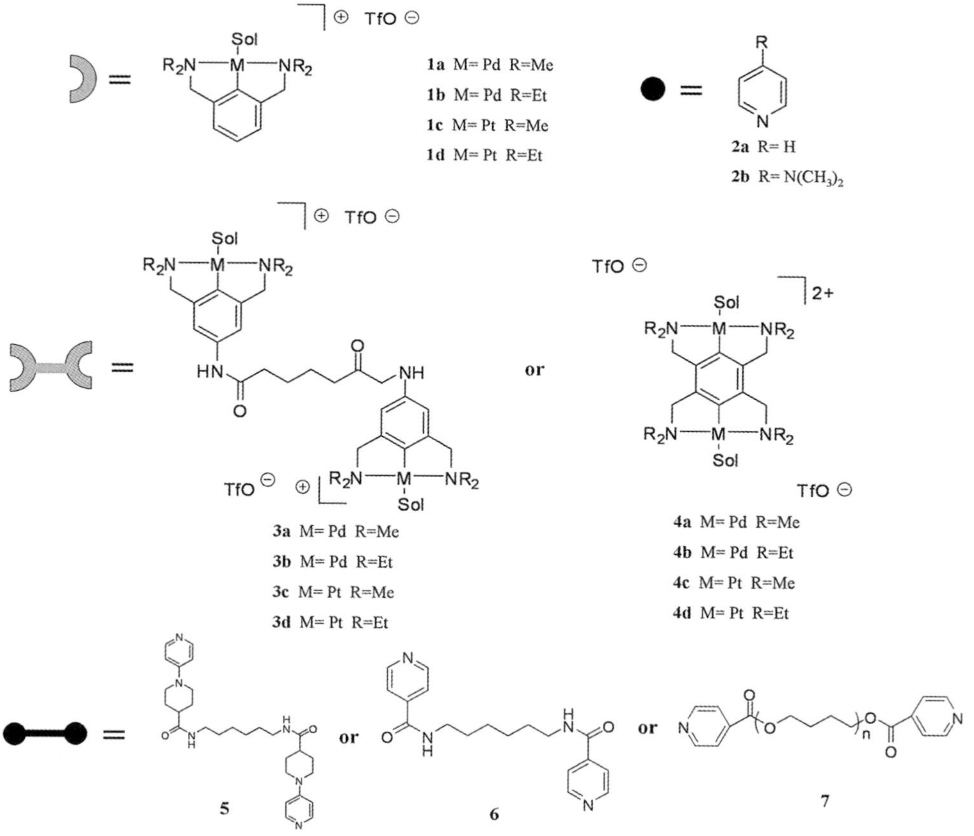

Figure 12.2 N,C,N-pincer metal complexes and ligands for supramolecular polymerizations.

complexes; the dissociation rate of a metal–pyridine bond can be changed by orders of magnitude independently of the association constant by change the N–alkyl substituents R [5, 6]. For example, the association constants for **1a•2b** and **1b•2b** in DMSO are very similar (1.6×10^3 and 1.3×10^3 M^{-1}, respectively), but the rates of ligand exchange differ by nearly two orders of magnitude (70~100 s^{-1} for **1a•2b** and 1.0 s^{-1} for **1b•2b**) [5]. This methodology has been applied in our laboratory to explore the molecular mechanisms underlying the basic rheological properties of linear supramolecular polymers in solution and the solid state [5, 7], supramolecular polymers networks [8–12], and tribological properties of cross-linked polymer brushes [13]. These SPs are best classified as metallo-supramolecular polymers, a class of SPs which, since the first reports by Rehahn [14], has grown to comprise a wide range of structural motifs and associated functions [15–25]. The metallo-supramolecular character of the pincer complexes, however, is secondary to our purposes, and in the remainder of this chapter we explore the structure–activity relationships that have been facilitated by the pincer functionality.

12.2.2
Linear SPs

The **1•2** motif can be incorporated into both linear and networked supramolecular polymers. An example of a linear SP (Figure 12.1b) is **3•5**. Because the association constant for **3a•5** and **3b•5** are effectively identical, the two combinations should form polymers of effectively identical length [26]. That this is the case is confirmed by quasi-elastic light scattering [5]. The hydrodynamic radii (R_h) of a 1:1 solution of **3a•5** and a similar solution of **3b•5** (each at 2.8 wt %) are indistinguishable, with an average effective R_h value of 26 nm. An increase in the concentration of **3a/b•5** to 4.6 wt % leads to a ~40% increase in the R_h values of both SPs. The observed change in hydrodynamic radius reflects a reasonable concentration dependence of the size of the SPs. The identical R_h values of SPs **3a•5** and **3b•5** illustrate that they have the same molecular weight at the same concentration.

The viscosities of 4.6 wt % **3a•5** and **3b•5** are essentially identical within experimental error, and the same correlation is found for related linear polymers based on Pt pincer complexes and telechelic pyridine-functionalized poly(tetrahydrofuran) (**4c:7** = 1 : 1, **4d:7** = 1 : 1), across a range of concentrations from dilute to semi-dilute and overlapped [7]. In Figure 12.3, zero-shear viscosities of supramolecular polymer solutions **4c•7** and **4d•7** with different concentrations of polymer (from 20 wt % to 1.25 wt %) are shown. The viscosities of 20 wt % **4c/d•7** solutions were ~4 Pa s, and they decreased upon dilution to ~4×10^{-3} Pa s at 1.25 wt %, a viscosity that is close to that of the maximum concentration studied for the **3a/b•5** polymer solutions [5]. The latter solutions were in the dilute concentration regime, where interactions between polymers are negligible. For SPs **4c/d•7**, scaling laws suggest that the SPs are in the semi-dilute regime above 15 wt % [7], and so overlap between polymers contributes to the viscosity of the solutions; the chains must move through and around each other under flow. Because the association constants for **3a/b•5** or **4c/d•7** are similar within their respective families, the equilibrium size of the SPs is expected to be essentially

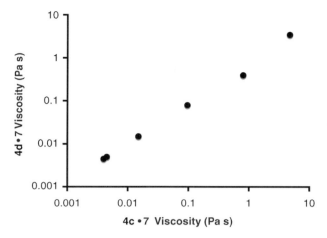

Figure 12.3 Relative viscosities of reversible polymer solutions **4c•7** vs. **4d•7** at different weight % (20%, 15%, 10%, 5%, 2.5% and 1.25%) in 1,2-dichlorobenzene with a small portion of DMSO (w/w, 95/5). Reproduced from *Supramol. Chem.* **2010**. by Jeon et al. with permission from Copyright © 2010 Taylor & Francis.

identical. The identical viscosity for same concentration of **3a/b•5** or **4c/d•7**, respectively, is expected to be attributed to the hydrodynamics of intact, equilibrium SPs structures.

It is important to make clear the concept of viscosity. Viscosity (η) is a measure of the resistance of a fluid that is being deformed by either shear stress or tensile stress, and it depends on the relaxation time (τ_0) of the materials as $\eta = G\tau_0$, where G is the shear modulus [27]. The resistance to deformation is therefore inherently kinetic, and the kinetics are determined by the process that controls the rate of material response. In the polymer solutions, the similarity in properties between **3a** vs **3b** and/or **4c** vs **4d** solutions shows that the viscosity-determining process does not involve dissociation along the SP main chain. Rather, the longest relaxation time is the so-called Rouse time of the polymer chain (τ_R), which depends on the degree of polymerization of the polymer chain. The mean length of a supramolecular polymer, in turn, is determined by the association constant between the defining units (here, the metal and ligand). In other words, the thermodynamics of the association dictate the size of supramolecular polymers, which here determines the Rouse time of the polymer chain, even though it is not accurate to state that the rheological properties of the linear supramolecular polymers are controlled by the thermodynamic. Actually, rheological properties of the samples always depend on the relaxation model of the materials. On the other hand, even if the lengths of supramolecular polymers are identical, their viscosities may be different if the concentration of the supramolecular polymer solution is above the critical concentration of entanglement. This will be discussed in the next section.

12.2.3
Theory Related to the Properties of Linear SPs

In order to explore the behavior of linear SPs, we begin by reviewing the relationship of the number-average degree of polymerization ($\langle N \rangle$) of a linear supramolecular polymers with the equilibrium constant (K_{eq}), monomer concentration (C) and the concentration of chain stoppers. We follow here the work of Knoben et al. [26], who considered supramolecular polymers formed by association of monomers bearing self-complementary end groups (bifunctional monomer 2,4-bis(2-ethylhexylureido) toluene). Although the equations will change for AB type pairwise-complementary monomers, or AA + BB (as in the example of the pincer-ligand systems), the key conclusions of the analysis are independent of the quantitative details, and we therefore use Knoben's example for convenience. When there are no chain stoppers, the number-average degree of polymerization ($\langle N \rangle$) has the following relationship with K_{eq} and C [26].

$$\langle N \rangle = \frac{1 + \sqrt{1 + 16 K_{eq} C}}{2} \approx 2\sqrt{K_{eq} C} \tag{12.1}$$

where the approximation is justified if $K_{eq} C \gg 1$.

The interplay of SP size and structure (determined by the thermodynamics of association) with dissociation dynamics (determined by the kinetics of association) depends on the concentration regime of the SP, as polymer concentration has a critical effect on the relaxation time of the polymer. As discussed above, in an unentangled polymer solution, the longest relaxation time is the Rouse time of the polymer chain (τ_R), which depends on the degree of polymerization. The specific viscosity (η_{sp}) of the polymer solution varies with the degree of polymerization as $\eta_{sp} \approx N\phi^2$ in the semidilute unentangled solution (the relationship can depend on the solvent, and we assume here a theta solvent although the critical points are quite general) [27], where N is the number of Kuhn monomers (related to polymer length) and ϕ is the volume concentration of polymer solution. Specific viscosity (η_{sp}) is generally used to characterize the polymer contribution to the viscosity of polymer solution. η_{sp} is defined as $\eta_{sp} = (\eta/\eta_s) - 1$, where η is the viscosity of polymer solution and η_s is the viscosity of pure solvent. Because the degree of polymerization of supramolecular polymer depends on the K_{eq} and concentration, η_{sp} is the same for supramolecular polymer formed by units have the same association constant.

Increasing the concentration of linear supramolecular polymers above the critical concentration of entanglements complicates the molecular mechanisms that might contribute to the rheological properties of the SPs. The dynamics of stress relaxation in a dense system of linear supramolecular polymers has been explored by Cates [28]. Two time scales are found to control the rheological properties of the concentrated supramolecular polymer. The dominant pathway for stress relaxation in conventional polymers is reptation – a snakelike movement of the polymeric chains. Stress relaxation in supramolecular polymers also occurs by reptation, and the reptation time (τ_{rep}) has a relationship with the mean chain length of supramolecular polymers (L), as $\tau_{rep} \sim L^3$, and accordingly, the equilibrium constant (K_{eq}) of the supramolecular polymers. Additionally, supramolecular chains can release strain by breaking and then recombining with other, strain-free chain-ends. So there is another time scale in Cates's work: the mean time for such a chain to break into two pieces (τ_{break}). τ_{break} has a relationship with dissociation rate of cross-linkers and mean chain length of supramolecular polymers as $\tau_{break} = 1/(k_d \bullet L)$.

According to Cates' results, the dominant stress-relaxation mechanism of linear supramolecular polymers is reptation if the reptation time (τ_{rep}) of the polymer is far less than τ_{break} ($\tau_{break}/\tau_{rep} \geq 1$). Zero-shear viscosity (η_0) has a relationship with the mean length of the living polymer chain (L) as $\eta_0 \sim L^3$. Under conditions in which $\tau_{break}/\tau_{rep} < 1$ and in which the average chain length is far larger than the entanglement length, it is further predicted that stress relaxation is characterized by a new intermediate time scale, $\tau_1 = (\tau_{break}\tau_{rep})^{1/2}$, associated with a process whereby the chain breaks at a point close enough to a given segment of tube for reptative relaxation of that segment to occur before the new chain end is lost by recombination. In this case, zero-shear viscosity (η_0) has a relationship with the mean length of the living polymer chain (L) as $\eta_0 \sim L$. If the chain length of the supramolecular polymer is less than the entanglement length, the dominant relaxation mechanism is breathing modes of the chain or local Rouse-like motion. Here, zero-shear viscosity (η_0) has a relationship with the mean length of the living polymer chain (L) as $\eta_0 \sim L$.

The above discussion can be applied to systems like **3a/b•5** and **4c/d•7** above, in which the two associations that define the different linear SPs have the same association constant (K_{eq}) but different dissociation rate (k_d). It is possible in this general case that for one supramolecular interaction, $\tau_{break}/\tau_{rep} \geq 1$, while for the other $\tau_{break}/\tau_{rep} < 1$. In this case, the zero-shear viscosities of the two supramolecular polymers are expected not only to differ, but to do so as a result of different underlying molecular mechanisms. We are aware of no experimental results that currently reflect this situation. Returning to the experimental observations discussed in the preceding section for SPs **4c•7** and **4d•7** in the semi-dilute entangled regime (concentration above 15 wt%) [7], the complexes have very low dissociation rates (0.026 s^{-1} and 6 × 10^{-4} s^{-1}) [6], and we infer that τ_{break}/τ_{rep} is greater than 1 for both SPs. Thus, $\eta_0 \sim L^3$, and accordingly the two SPs have the same viscosities even in the semidilute entangled regime. Finally, we point out that the control of reversibility in the pincer-ligand systems can be employed in this context as a mechanistic probe of relative time scales. Properties that do not depend on k_d reflect a situation in which the dominant relaxation pathway is not uniquely determined by k_d. If k_d is known, it provides a sort of internal 'clock' for the critical processes that might influence the polymer properties.

12.2.4
Linear SPs in the Solid State

Supramolecular polymers often have impressive solid-state mechanical properties [19, 20, 22, 29, 30], and probing the mechanisms (if any) by which the reversibility of the supramolecular unit contributes to those properties is of great interest. Rheological characterization [31, 32] offers some insight. In the case of the pincer-based

SPs, reasonably strong and quite tough fibers of **4c/d•7** have been hand-drawn from solution. Crude examination of mechanical properties reveal tensile strengths of ~100 MPa and strains at break of roughly 70 %. Importantly, the tensile strength of **4c•7** fibers are identical within experimental error to those of the **4d•7** fibers, providing compelling evidence that the nucleophilic displacement mechanism that governs failure under load in solution (as measured by single-molecule force spectroscopy [33]), is not operative in the solid state [7]. The absence of the typical associative ligand displacement pathway in the fibers is likely tied to their significant tensile strength, since the likely alternative failure mechanisms (chain disentanglement and/or purely dissociative scission of the coordination bond) would be comparable to those in conventional, covalent polymers [7].

12.3
Cross-Linked SPs Networks

12.3.1
Reversibility in Semidilute Unentangled SPs Networks

More recently, the kinetic control has been applied to SPs networks (Figure 12.1c), for example poly(4–vinylpyridine) (PVP), which is cross-linked by bis(M(II)–pincer) compounds **3a-d** (Figure 12.2) [8, 9]. The same simple steric effects in the pincer alkylamino ligands are used, both within the Pd complexes **3a–b** and the slower, more strongly coordinating Pt complexes **3c–d**. As with the linear SPs, the independent control of kinetics is particularly significant; cross-linkers **3a** and **3b** are structurally identical components within the network, and so their similar thermodynamics ($K_{eq} = \sim 30\,M^{-1}$ for **1a•2a** and **1b•2a**) ensure that the extent and nature of cross-linking is essentially the same in the two samples (or between Pt(II) pincer molecules **3c–d**; $K_{eq} = 8 \times 10^3\,M^{-1}$ for **1c•2a** and $4 \times 10^3\,M^{-1}$ for **1d•2a**).

The effect of cross-linking concentration on viscosity is shown in Figure 12.4 [11]. Below a critical concentration of cross-linker, C_{cr} (about 0.8%, reported as functional group equivalents of metal to pyridine), the four mixtures behave as free-flowing sols and their viscosities are very similar to that of a comparable 10 wt % solution of PVP in DMSO (approaching 0.02 Pa s); there is a negligible difference in viscosity between the mixtures formed from the four different cross-linkers. It should be emphasized that the viscosity increases with cross-linkers content in this regime; the added cross-linkers create larger (branched) polymer chains, and like the linear supramolecular polymers discussed in the previous section, the critical determinant of the properties (here, viscosity) is the Rouse time of the polymer chain (τ_R), which depends on the size of the polymer chain.

Above C_{cr}, however, the absolute and relative behaviors of the mixtures are qualitatively and quantitatively quite different. As seen in Figure 12.4, the viscosity increases – dramatically in some cases – over a relatively narrow range of cross-linking content to produce materials that behave like weak gels. This pronounced change in behavior is attributed to a sol-to-gel transition brought about by percolation

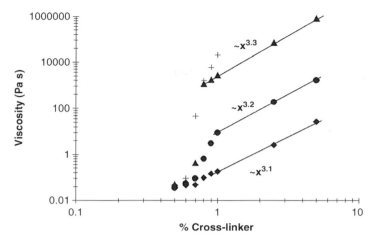

Figure 12.4 The effect of cross-linking percentage on viscosity for the different transient networks. Each network was made at a total SPs network weight percentage of 10% in DMSO. The viscosity was measured by either steady or oscillatory shear rheology. (♦) 4a•PVP, (●) 4b•PVP, (▲) 4c•PVP, and (+) 4d•PVP. Scaling laws fitted to the high cross-linking regime are shown for 4a-c•PVP. Dynamic viscosities are reported for the lowest accessible frequency (0.001 s^{-1}), at which viscosity is independent of decreasing frequency for all samples except 4d•PVP. Reproduced from *J. Mater. Chem.* **2007**, *17*, 56–61 by Loveless et al. with permission from Copyright © 2007 The Royal Society of Chemistry.

to form a pseudo-infinite transient network. In this limit, the critical molecular relaxations require that the cross-links dissociate – the chains cannot move in significant amounts without dissociation, because they are effectively 'pinned' into place by the cross-link.

Once the concentration of the cross-linkers is above this critical concentration C_{cr}, the relative viscosities of the samples are proportional to the lifetime ($\tau = 1/k_d$) of the coordinative metal–pyridine bonds that define the cross-links [8, 11] the zero shear viscosity of 100 mg mL^{-1} 4a•PVP is a factor of 80 less than that of the isostructural network 4b•PVP (Figure 12.4). The difference of a factor of ~80 in the network viscosities is within experimental error of the difference in the ligand exchange rates, and a full spectrum of dynamic mechanical behavior is similarly well correlated. The frequency-dependent storage and loss moduli, G' and G'', for networks of either 3•PVP or the related 4•PVP, are reduced to a single master plot when scaled by the corresponding ligand exchange rates, measured on model systems (data for 3•PVP are shown in Figure 12.5) [8]. That the relationships are quantitatively valid shows that the dynamics of the cross-linkers determine the dynamic mechanical response of the materials. The quantitative agreement spans the Pd (e.g., 3a/b) and Pt (3c/d) cross-linkers, even though the equilibrium association constants for the latter are roughly a factor of 10^2 higher than those of the former.

The terms storage and loss moduli, used above, are staples of polymer physics, but they are less familiar to many organic chemists involved in supramolecular polymer

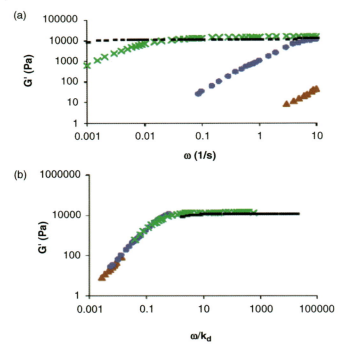

Figure 12.5 (a) Storage moduli (G') of networks of PVP and 5% **3a** (▲), **3b** (●), **3c** (×), and **3d** (−) at 100 mg mL^{-1} in DMSO at 20 °C. (b) Storage moduli of the same networks as a function of frequency scaled by k_d. Reproduced from *J. Am. Chem. Soc.* **2005**, *127*, 14488–14496 by Yount et al. with permission from Copyright © 2005 American Chemical Society.

chemistry. The definition of these terms is therefore given here for convenience. The modulus is the ratio of stress to strain, and essentially reflects how much energy (really force per area, or stress) is required to deform a material by a certain strain (percentage of its original size in direction in which force is applied). The energy that is put into material deformation can have either of two fates: it can be stored in a manner that is recovered when the force is removed (as in the case of stretching a rubber band), or it can be dissipated as heat into the environment (for example, stirring a thick batter). Materials in which deformation energy is stored are known as elastic, and those in which the energy is lost are viscous. Viscoelastic materials exhibit behavior somewhere in between that of purely viscous and purely elastic materials, and most polymers have contributions from both. Viscoelasticity is often studied using dynamic mechanical analysis where an oscillatory strain is applied to a material and the resulting stress is measured. The storage and loss modulus in viscoelastic solids measure the stored energy (reflected by the storage modulus G') and the energy dissipated as heat (given by the loss modulus, G'') [34]. A critical point is that the fraction of deformation energy that is stored is time-dependent. If the polymer flows under the stress, there is no mechanism for recovery; the structural 'memory' of where to return has been lost. By looking at the storage and loss moduli as a function of time scale (in practice, the frequency of the applied oscillatory stress), the time scales that dictate material relaxation (flow, or internal reorganization) can be deduced.

These general points are well illustrated in the supramolecular polymer networks discussed here. The cross-links critically define the network, and so deformations to the network are largely stored as long as the cross-linkers remain intact. Once the cross-linkers break, the deformation energy is lost. In principle, cross-linkers can break in one of two ways. First, they can dissociate via the usual thermally activated pathway (i.e., the applied stress has no significant impact on the rate of dissociation). In this scenario, a time scale of deformation that is shorter than the lifetime of cross-linkers ($\tau = 1/k_d$) means that the cross-linkers remain intact after deformation and the material will exhibit high storage modulus (G') and low loss modulus (G''), while if the time scale of deformation is larger than τ, the stress can be relaxed by dissociation of cross-linkers [35], and the material will exhibit low storage modulus (G') and high loss modulus (G''). This is the behavior observed in Figure 12.5 [8].

Second, one may reasonably ask how the rate of bond dissociation depends on the stress experienced by the network and the effect of this force dependence on SP properties. In the specific example of our 'case study' (the pincer-ligand complexes), the force-induced bond rupture has been examined experimentally using single-molecule force spectroscopy [33]. As in the macroscopic material measurements, this is truly a kinetic measurement: the experiment probes the probability of the bond breaking (vs remaining intact) under an applied load. Here, the load is applied not within a macroscopic network, but along a single polymer chain that is tethered at one end to the tip of an atomic force microscope (AFM) and at the other to an anchored substrate, with the two halves of the polymer bridged at the center by a bis-functional pincer-ligand complex. A stress (load) is applied, and that energy is stored in the intact, elongated polymer until the metal–ligand bond breaks, at which point the energy is dissipated. The energy is stored in the elastic deformation of the polymer coupled to the deflected AFM tip, and so the release of the tip signals the breaking of the metal–ligand bond.

As with the rheology, the probability that energy is stored over the entire course of a deformation event (here on a molecule-by-molecule, rather than ensemble, basis) depends on the time scale of the deformation, but it also depends on the load. These two factors (time scale and load) are combined in the loading rate of the experiment: the change in force as a function of time, which is determined by the rate at which the polymer is strained (the velocity at which the tip and substrate anchors are pulled away from each other). The details of the underlying physics and theoretical treatments have been discussed in detail by others [36], and we will not delve into them here. The key physical insight is that the faster that force is loaded into the bonds, the more likely it is that they survive to experience a higher force, where the probability of bond rupture now increases. Thus, the force at which a given bond is most likely to break (F^*, Figure 12.6) increases with loading rate.

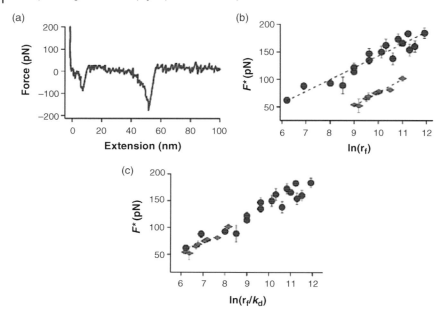

Figure 12.6 (a) A representative force–separation curve showing bond rupture in **3b•5** during retraction of the AFM tip. (b) Most probable force versus loading rate of the **3b•5** (○) and **3b•6** (♦) coordination systems. (c) Master graph of force versus loading rate scaled by thermal dissociation rates k_d measured independently by NMR. Reproduced from *J. Am. Chem. Soc.* **2006**, *128*, 3886–3887 by Kersey et al. with permission from Copyright © 2006 American Chemical Society.

The force spectroscopy results confirm that the rate of bond rupture increases with the applied force. There are two important points to be taken from the data. First, the magnitude of the force-induced acceleration is likely to play a major role only under extremely high strains. Vaccaro et al. have estimated the elastic energy of a polymer chain needed to pull off cross-linkers [37]. As long as the chain remains Gaussian, the elastic tension F in the chain grows proportionally with the end-to-end distance R according to

$$F(R) = \frac{3K_B T}{Nb^2} R \tag{12.2}$$

where b and N are the length and number of the chain Kuhn segments, respectively; K_B is the Boltzmann constant. In the linear region, R remains much smaller than the fully extended chain length Nb, and the tension F remains well under $k_B T/b$, or on the order of several pN. In Figure 12.6, the force spectroscopy work shows that more than 50 pN of force are required to accelerate the rate of cross-link dissociation by a factor of 10. These strongly suggest that forced dissociation of the cross-links is likely to be important only in concert with the stretching of polymer chain segments into the non-Gaussian range.

The second critical feature of these systems, shown in Figure 12.6, is that the probability of bond survival depends on the time scale of the deformation scaled by the stress-free lifetime of the bonds. So, in general, contributions from force-induced rupture will not be manifested as major deviations from the scaling behavior shown in prior Figures. The stress-free rates are reliable predictors of behavior in all but the most extreme, macroscopic, stress-bearing conditions.

Experiments on SP networks formed from multiple types of cross-linkers show that the response to an applied stress occurs through sequential, individual dissociation and re-association events [10]. Discrete contributions from each type of cross-linker are evident in the mechanical properties, rather than an average of the contributing species. For example, the dynamic viscosity and storage modulus G' as a function of frequency were compared for five different networks: PVP with 5% of either **4b** or **4c**, 2.5% of either **4b** or **4c**, and 2.5% each of **4b** and **4c**, all at the same concentration of 10% by total weight in DMSO. Below $\omega = 0.1\,\mathrm{s}^{-1}$, the dynamic viscosity of the mixed network closely mimics that of the network, consisting entirely of the slower component **4c**•PVP at a concentration of 2.5% (Figure 12.7). As ω increases to values greater than $0.1\,\mathrm{s}^{-1}$, the dynamic viscosity of the mixed network increases and eventually plateaus at that of a network comprising PVP and 5% of the faster cross-linking component **4b**. Similar effects are observed in the storage modulus and for different mixtures of networks (including those with three different cross-linkers). In all cases, the frequency onsets and magnitudes of the transitions are anticipated by the behavior of networks with a single cross-linker, and therefore on the intrinsic properties of the individual supramolecular interactions themselves. This illustrates an important point (and good news) for the

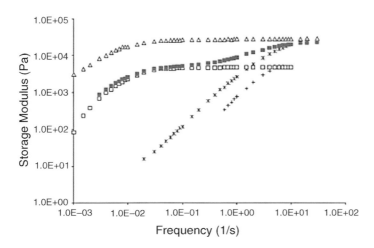

Figure 12.7 Storage modulus, G', versus frequency for (■) 2.5% + 2.5% (**4b** + **4c**)•PVP, (□) 2.5% **4c**•PVP, (△) 5% **4c**•PVP, (+) 2.5% **4b**•PVP, and (*) 5% **4b**•PVP. All networks 10% by total weight in DMSO, 20 °C. Reproduced from *Macromolecules* **2005**, *38*, 10771–10777 by Loveless *et al.* with permission from Copyright © 2005 American Chemical Society.

supramolecular chemist: an understanding of isolated molecular behavior is often directly relevant to SP environments in which there is an interplay of multiple interactions and time scales.

This behavior is essentially that of transient network models, in which the independent relaxations of stress-bearing entanglements determine the dynamic mechanical response of a network [38–40]. Without considering either the exact structure of the networks, the detailed mechanism of relaxation [41], or the extent of cooperativity in the associations, individual dissociation events clearly dominate the mechanical properties; no significant averaging or summation of different components is observed. The independence of the cross-links has significant consequences for the rational, molecular engineering of viscoelastic properties. When entanglements are defined by very specific interactions, the chemical control of properties follows. As long as the strength of the association is great enough to render associated a significant fraction of the cross-linkers, the dynamics of cross-link dissociation, rather than further details of their thermodynamics, are the key design criterion, and quite complex viscoelastic behavior can be engineered given suitable knowledge of the small molecules.

These low-strain, oscillatory rheology experiments reflect what is known as the linear dynamic mechanical properties of the network. The strains are small, and so the forces along the network chains are modest, and there is little to no reorientation or reorganization of the network under the deformation. While the linear viscoelastic properties of materials are important, however, many technological applications depend on more complex, nonlinear properties such as shear thinning (the decrease of viscosity under shear) or shear thickening (a corresponding increase in viscosity). The molecular mechanisms that underlie such nonlinear behaviors are often poorly understood, and the rational design of systems with programmable shear thinning and/or shear thickening behavior therefore remains an ongoing challenge. We now turn our discussion to the nonlinear rheology of these same supramolecular networks as an example of the structure–activity relationships that can be enabled with the supramolecular approach.

The nonlinear rheological properties of metallo-supramolecular networks formed by the reversible cross-linking of semi-dilute unentangled solutions of poly(4-vinylpyridine) (PVP) in dimethyl sulfoxide (DMSO) have been observed and are quite rich [12]. When the networks are formed from solutions of PVP in the semidilute unentangled regime (that is, at concentrations in which polymer chains are not entangled with each other), the application of steady shear above a critical shear rate leads to shear thickening (Figure 12.8). As with the linear rheology, a change in cross-link dissociation rate provides immediate insight into the phenomenon, because the critical shear rate for shear thickening is experimentally correlated with the stress-free lifetime of the metal–ligand bond. The onset and magnitude of the shear thickening depends on the amount of cross-linkers added, and further experimental work attributes the primary mechanism of shear thickening to the shear-induced transformation of intramolecular cross-linking to intermolecular cross-linking, rather than nonlinear high tension along chains that are stretched beyond the Gaussian range.

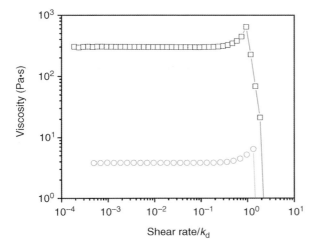

Figure 12.8 Steady shear viscosity versus scaled shear rate for ~0.1 g mL^{-1} PVP/DMSO solution with 2% of **4a** or **4b**. Circles and squares represent networks formed from cross-linker **4a** and **4b**, respectively. $T = 25\,°C$. Reproduced from *J. Phys. Chem. Lett.* **2010**, *1*, 1683–1686 by Xu et al. with permission from Copyright © 2010 American Chemical Society.

The formation of the nascent intramolecular cross-links, and of similar cyclic species, is a well-known occurrence in many supramolecular systems [42]. This kind of structure provides access to polymeric fluids with significant magnitudes of thickening under shear, a development that could have important technological consequences.

Recall that the linear oscillatory rheology of 3•PVP has been shown to scale with the values of k_d measured independently on small molecular model systems [8]. Similar scaling behavior has been observed for shear thickening under steady shear [12]; the steady shear behavior of **4a**•PVP and **4b**•PVP networks in semi-dilute unentangled concentrations of PVP (0.1 g mL^{-1}) is shown in Figure 12.8. The superposition of the rheological data upon scaling by k_d is consistent with transient network models (linear oscillatory rheology) and models for shear thickening based on either network reorganization or non-Gaussian overstretching (non-Newtonian steady shear).

12.3.2
Properties of Semidilute Entangled SPs Networks

As the concentration of the covalent polymer component increases, there are possible contributions (highlighted above) from the relaxation mechanisms of the polymers themselves. The entanglements between individually polymer chains, in other words, might provide mechanisms through which to store deformation energy and resist flow. We therefore turn our attention to SP networks formed from polymers in the semidilute entangled concentration regime, in which most polymers are

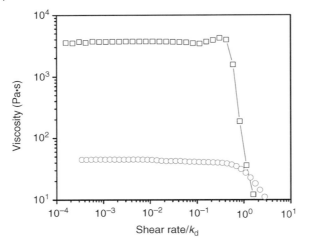

Figure 12.9 Steady shear viscosity versus scaled shear rate for ~0.26 g mL^{-1} PVP/DMSO solution with 2% of **4a** or **4b**. Circles and squares represent networks formed from cross-linker **4a** and **4b**, respectively. $T = 25\,°C$. Reproduced from *J. Phys. Chem. Lett.* **2010**, *1*, 1683–1686 by Xu et al. with permission from Copyright © 2010 American Chemical Society.

entangled. Under steady shear, however, the scaling relationships depicted previously in the semidilute unentangled regime (Figure 12.8) break down dramatically. Shear thinning is generally observed for samples with cross-linkers that have a faster dissociation rate (e.g., **4a**), while shear thickening is still observed for cross-linkers that have a slower dissociation rate (**4b**) (Figure 12.9). The breakdown requires the contributions of a mechanism that is fundamentally different from that underlying the physics discussed in the previous section [43].

The most likely explanation for the divergent behavior in the semidilute entangled regime is the presence of topological entanglements in addition to associative cross-linkers [44]. The mechanism under the divergent steady shear behavior for networks with different cross-linkers in semidilute entangled solution is further explored [45]. It is found that the divergent steady shear behavior of samples with different cross-linkers is connected to a competition between different time scales: the average time that a cross-linker remains open (τ_1) and the local relaxation time of a segment of polymer chain ($\tau_{segment}$). When τ_1 is larger than $\tau_{segment}$, the polymer chain segment disentangles and/or slips along the entanglement and orients under shear after time $\tau_{segment}$, and recombination after time τ_1 results in the transfer from intrachain to interchain cross-linkers. Because the total number of elastically active chain increases, shear thickening results [12, 45]. When τ_1 is smaller than $\tau_{segment}$, the segment of polymer chain does not have enough time to slip or disentangle altogether before the cross-linker recombines after time τ_1. In this case, the lack of reorganization does not permit cross-linkers to form interchain cross-linkers, instead remaining intrachain cross-linkers, which are not elastically active. As a result, no shear thickening is observed [45].

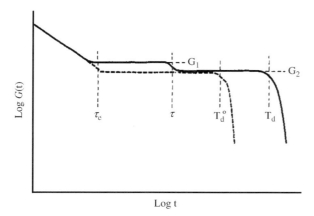

Figure 12.10 Schematic comparison of the time-dependent relaxation moduli for reversible networks comprising linear chains with stickers (solid line) and linear chains without stickers (dashed line). Reproduced from *Macromolecules* **1991**, *24*, 4701–4707 by Leibler et al. with permission from Copyright © 1991 American Chemical Society.

A competition between cross-linker recombination and polymer segment orientation is implicated as the determinant of the divergent behavior observed as a function of cross-linker dynamics [45]. Two time scales—the average time that a cross-linker remains open (τ_1) and the local relaxation time of a polymer chain segment ($\tau_{segment}$)—are used to compare the time scale of cross-linker recombination and polymer segment orientation, respectively. When τ_1 is larger than $\tau_{segment}$ ($\tau_1 > \tau_{segment}$), shear thickening is observed. When τ_1 is smaller than $\tau_{segment}$ ($\tau 1 < \tau_{segment}$), no shear thickening is observed.

12.3.3
The Sticky Reptation Model

The complexity of the dynamic interplay, discussed above in largely qualitative terms, can be treated more quantitatively. To capture the dynamics of entangled networks made up of linear chains with many temporary cross-links, Leibler et al. have presented a 'Sticky Reptation' model [41]. The overall schematic representation of stress relaxation moduli [$G(t)$] for a reversible gel is compared to that of an identical linear polymer without stickers in Figure 12.10. The four regimes of stress relaxation are clearly delineated in Figure 12.10, separated by the three important time scales in the reversible gel: τ_e, τ, and T_d. τ_e is the Rouse time of an entanglement strand (previous term τ_R is the Rouse time of the whole polymer chain, not just an entanglement strand), τ is the lifetime of a cross-linker, and T_d is the terminal relaxation time of the reversible gel. The terminal relaxation time of ordinary linear polymers (T_d^0) is also shown in Figure 12.10 for comparison.

There are four time regimes that are important for the stress relaxation modulus of reversible gels. At times shorter than the Rouse time of an entanglement strand τ_e, relaxation is indistinguishable from that in the polymer without stickers. Since the average number of monomers along the chain between stickers (N_s) is assumed to be larger than the number of monomers in an entanglement strand (N_e), Rouse motions of the entanglement strands are unaffected by the presence of the stickers. Under conditions where $\tau_e < t < \tau$, there will be a rubbery plateau analogous to that observed in permanently cross-linked networks, with a modulus that has contributions from both cross-links and entanglements. If $\tau < t < T_d$, the stress held by the stickers relaxes, and the modulus drops to the level of the identical linear chain system without stickers. This second plateau persists until the terminal relaxation time of the reversible gel, T_d, which depends on the chain length (N), the number of cross-linking groups per chain (S), and the lifetime (τ) and probability (p) of formation of cross-links [41].

$$T_d = \left(\frac{N}{Ne}\right)^{1.5} \frac{2S^2\tau}{1 - 9/p + 12/p^2} \tag{12.3}$$

So at times shorter than the lifetime (τ) of a cross-link, the networks behave as elastic rubbers (gels). On longer time scales the successive breaking of only a few cross-links allows the chain to diffuse along its confining tube. The motion of a chain in this hindered reptation model is controlled by the concentration and lifetime of tie points. These features are evident in the examples discussed previously in the chapter, and reflect an important link between the 'physical organic chemistry' of supramolecular polymers and theoretical polymer physics [46].

12.4
Hybrid Polymer Gels

Autonomously self-repairing polymers [47] are promising 'next-generation' materials with a myriad potential applications, and rapidly reversible, noncovalent interactions [48] such as hydrogen bonds or metal–ligand coordination have been suggested to be possible tools for endowing materials with self-healing character that does not require an external stimulus. While conceptually attractive, the SP strategy to self-healing materials is challenged in practice by the simultaneous requirements of reversibility and memory of an initial structural state; water is self-repairing, but fairly useless as a structural material. In the language used previously in this chapter, self-repair requires the breaking and reformation of the supramolecular bond, but the loss of the supramolecular bond also leads to energy dissipation rather than recovery. To preserve a desired structure in a self-healing context, reversible interactions might be combined in a composite material with components that preserve a memory of the desired state: encasement within an exoskeleton, tethering to an endoskeleton, or with an additional 'fixed' system in an interpenetrating network. Among the additional interesting avenues for future research, therefore, are hybrid systems comprising both permanent scaffolds and supramolecular polymer networks.

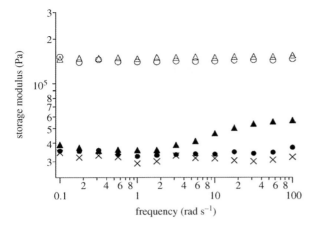

Figure 12.11 Storage modulus as measured by the oscillatory rheology of covalent and hybrid DMSO gels: 'pure' covalent gel, **8** (×); **8•4a** (●); **8•4b** (▲); **8•4c** (△); **8•4d** (○). Reproduced from *J. R. Soc. Interface* **2007**, *4*, 373–380 by Kersey *et al.* with permission from Copyright © 2006 The Royal Society.

These same pincer motifs that recur throughout this chapter have been incorporated into a family of hybrid polymer gels in which covalent cross-links create a permanent, stiff scaffold onto which the reversible metal–ligand coordinative cross-links are added [49]. The hybrid gels exhibit frequency-dependent mechanical properties that are different from those of the parent, covalent-only gel (Figure 12.11). At low oscillatory frequencies, the storage modulus is that of the unmodified covalent gels, but at higher oscillatory frequencies, the storage modulus increases measurably with increasing frequency. The underlying relaxation is directly attributed to the dissociation and reassociation of the supramolecular group by comparing the hybrid gel containing **4b** to that formed with **4a**. The structural similarity of **4a** and **4b**, along with the similarities in their complexation thermodynamics to pyridine, ensures that the equilibrium structures of the two gels are effectively identical. The oscillatory rheology of the **4a** hybrid gel, however, shows a storage modulus that is nearly identical to that of the covalent network alone up to 100 rad s^{-1}, where a slight positive deviation occurs. Inclusion of the slower Pt complexes **4c** and **4d** yields higher storage moduli than each of the Pd cross-linkers, showing frequency-dependent behavior that is consistent with the observations made on networks with mixed cross-linkers discussed above. Even in the hybrid covalent systems, the individual supramolecular cross-links bear stress and act as largely independent contributors to the dynamic mechanical properties.

These and related hybrid networks offer the opportunity for future mechanistic studies of important, but complex, processes whose mechanistic origins are still poorly understood, including: gel fracture [50], self-repair [51], and energy dissipation [52]. Such investigations would be part of a natural evolution of the structure–activity studies we have dubbed the 'physical organic chemistry' of supramolecular polymers. Previous work has been (and should continue to be) devoted to mapping molecular processes onto what are generally well-established areas of polymer physics (e.g., Newtonian behavior and transient network models), but the success of these studies provides the foundation for using the molecular control that comes with a supramolecular polymer approach to probe the mechanisms and behavior of increasingly complex material processes. The non-Newtonian shear thinning and shear thickening described in earlier sections is an additional example of one such opportunity. In that case, as well as the cases of gel fracture and rational energy dissipation discussed here, there is the chance to not only shed new fundamental insight, but to then use that insight in conjunction with the impressive SP toolkit highlighted throughout this book to create a new generation of polymer-based materials with improved properties that are critical to a variety of technological applications. Although not highlighted in this chapter, we profess an ongoing fascination for the properties of SPs at surfaces and interfaces as another example of an opportunity for important contributions of both a fundamental and applied nature [53–55].

12.5
Conclusion

Our perspective is that supramolecular polymers provide a unique opportunity to blend very specific aspects of molecular interactions with polymer physics, and in so doing to provide clear insight into material properties and their eventual manipulation. We note that, to first order, the critical molecular elements in these relationships are often not the 'organo-centric' structural components, but instead are the 'physico-centric' k_d's and K_{eq}'s. The benefit of the suitability of this reductionist view is that it serves the purpose of providing insights that can be applied across multiple SP platforms. The 'cost' is mainly esthetic, as the need for exotic (and often beautiful) supramolecular structures is seemingly reduced. As we have pointed out previously [46], however, the diversity in structure illustrated in the examples found in this text is as important as diversity in kinetics and thermodynamics, to the extent that structure impacts the environmental responsiveness of the recognition event. The chemistry of these and other recognition motifs is then directly translated into the environmental responsiveness of SPs. We further conclude that the accomplishments in the field of supramolecular polymers and SP polymer physics, which extend far beyond the limited examples from our own lab that have been employed in this chapter, justify the relevance of a small-molecule view of SPs, and they speak to the potential of reversible and specific intermolecular interactions as the critical interactions that create the entanglements necessary to transmit force in materials.

References

1 Rietveld, M.H.P., Grove, D.M., and van Koten, G. (1997) Recent advances in the organometallic chemistry of aryldiamine anions that can function as n,c,n'- and c,n,n'-chelating terdentate 'pincer' ligands: An overview. *New J. Chem.*, **21**, 751–771.

2 Albrecht, M. and van Koten, G. (2001) Platinum group organometallics based on 'pincer' complexes: Sensors, switches, and catalysts. *Angew. Chem. Int. Ed.*, **40**, 3750–3781.

3 Rodriguez, G., Albrecht, M., Schoenmaker, J., Ford, A., Lutz, M., Spek, A.L., and van Koten, G. (2002) Bifunctional pincer-type organometallics as substrates for organic transformations and as novel building blocks for polymetallic materials. *J. Am. Chem. Soc.*, **124**, 5127–5138.

4 Slagt, M.Q., van Zwieten, D.A.P., Moerkerk, A., Gebbink, R., and van Koten, G. (2004) Ncn-pincer palladium complexes with multiple anchoring points for functional groups. *Coord. Chem. Rev.*, **248**, 2275–2282.

5 Yount, W.C., Juwarker, H., and Craig, S.L. (2003) Orthogonal control of dissociation dynamics relative to thermodynamics in a main-chain reversible polymer. *J. Am. Chem. Soc.*, **125**, 15302–15303.

6 Jeon, S.L., Loveless, D.M., Yount, W.C., and Craig, S.L. (2006) Thermodynamics of pyridine coordination in 1,4-phenylene bridged bimetallic (Pd, Pt) complexes containing two N,C,N′ Motifs, 1,4-M_2-$[C_6(CH_2NR_2)_4$-2,3,5,6]. *Inorg. Chem*, **45**, 11060–11068.

7 Jeon, S.L., Loveless, D.M., Yount, W.C., and Craig, S.L. (2010) Main-chain dynamics in metallo-supramolecular polymers: From solution to elastomeric fibers. *Supramol. Chem.*, **22**, 697–703.

8 Yount, W.C., Loveless, D.M., and Craig, S.L. (2005) Small-molecule dynamics and mechanisms underlying the macroscopic mechanical properties of coordinatively cross-linked polymer networks. *J. Am. Chem. Soc.*, **127**, 14488–14496.

9 Yount, W.C., Loveless, D.M., and Craig, S.L. (2005) Strong means slow: Dynamic contributions to the mechanical properties of supramolecular networks. *Angew. Chem. Int. Ed.*, **44**, 2746–2748.

10 Loveless, D.M., Jeon, S.L., and Craig, S.L. (2005) Rational control of viscoelastic properties in multicomponent associative polymer networks. *Macromolecules*, **38**, 10171–10177.

11 Loveless, D.M., Jeon, S.L., and Craig, S.L. (2007) Chemoresponsive viscosity switching of a metallo-supramolecualr polymer network near the percolation threshold. *J. Mater. Chem.*, **17**, 56–61.

12 Xu, D., Hawk, J., Loveless, D.M., Jeon, S.L., and Craig, S.L. (2010) Mechanism of shear thickening in reversibly cross-linked supramolecular polymer networks. *Macromolecules*, **43**, 3556–3565.

13 Loveless, D.M., Abu-Lail, N.I., Kaholek, M., Zauscher, S., and Craig, S.L. (2006) Molecular contributions to the mechanical properties of reversibly cross-linked surface grafted polymer brushes. *Angew. Chem. Int. Ed.*, **45**, 7812–7814.

14 Knapp, R., Schott, A., and Rehahn, M. (1996) A novel synthetic strategy toward soluble, well-defined ruthenium(II) coordination polymers. *Macromolecules*, **29**, 478–480.

15 Corbin, P.S., Webb, M.P., McAlvin, J.E., and Fraser, C.L. (2001) Biocompatible polyester macroligands: New subunits for the assembly of star-shaped polymers with luminescent and cleavable metal cores. *Biomacromolecules*, **2**, 223–232.

16 Lohmeijer, B.G.G. and Schubert, U.S. (2002) Supramolecular engineering with macromolecules: An alternative concept for block copolymers, *Angew. Chem., Int. Ed.*, **41**, 3825–3829.

17 Vermonden, T., van Steenbergen, M.J., Besseling, N.A.M., Marcelis, A.T.M., Hennink, W.E., Sudhölter, E.J.R., and Cohen Stuart, M.A. (2004) Linear rheology of water-soluble reversible neodymium (III) coordination polymers. *J. Am. Chem. Soc.*, **126**, 15802–15808.

18 Zhao, Y., Beck, J.B., Rowan, S.J., and Jamieson, A.M. (2004) Rheological behavior of shear-responsive metallo-supramolecular gels. *Macromolecules*, **37**, 3529–3531.

19 Beck, J.B., Ineman, J.M., and Rowan, S.J. (2005) Metal/Ligand-Induced Formation of Metallo-Supramolecular Polymers. *Macromolecules*, **38**, 5060–5068.

20 Weng, W., Beck, J.B., Jamieson, A.M., Stuart, J., and Rowan, S.J. (2006) Understanding the mechanism of

gelation and stimuli-responsive nature of a class of metallo-supramolecular gels. *J. Am. Chem. Soc.*, **128**, 11663–11672.

21 Paulusse, J.M.J. and Sijbesma, R.P. (2004) Reversible mechanochemistry of a Pd^{II} coordination polymer. *Angew. Chem. Int. Ed.*, **43**, 4460–4462.

22 Paulusse, J.M.J., Huijbers, J.P.J., and Sijbesma, R.P. (2005) Reversible, high molecular weight palladium and platinum coordination polymers based on phosphorus ligands. *Macromolecules*, **38**, 6290–6298.

23 Paulusse, J.M.J., van Beek, D.J.M., and Sijbesma, R.P. (2007) Reversible switching of the sol-gel transition with ultrasound in rhodium(I) and iridium(I) coordination networks. *J. Am. Chem. Soc.*, **129**, 2392–2397.

24 Williams, K.A., Boydston, A.J., and Bielawski, C.W. (2007) Main-chain organometallic polymers: Synthetic strategies, applications, and Perspectives. *Chem. Soc. Rev.*, **36**, 729–744.

25 Friese, V.A. and Kurth, D.G. (2008) Soluble dynamic coordination polymers as a paradigm for materials science. *Coord. Chem. Rev.*, **252**, 199–211.

26 Knoben, W., Besseling, N.A.M., and Stuart, M.A.C. (2006) Chain stoppers in reversible supramolecular polymer solutions studied by static and dynamic light scattering and osmometry. *Macromolecules*, **39**, 2643–2653.

27 Rubinstein, M. and Colby, R.H. (2003) *Polymer Physics*, Oxford University Press.

28 Cates, M.E. (1987) Reptation of living polymers – dynamics of entangled polymers in the presence of reversible chain-scission reactions. *Macromolecules*, **20**, 2289–2296.

29 Sijbesma, R.P., Beijer, F.H., Brunsveld, L., Folmer, B.J.B., Ky Hirschberg, J.H.K., Lange, R.F.M., Lowe, J.K.L., and Meijer, E.W. (1997) Reversible polymers formed from self-complementary monomers using quadruple hydrogen bonding. *Science*, **278**, 1601–1604.

30 Castellano, R.K., Clark, R., Craig, S.L., Nuckolls, C., and Rebek, J. Jr. (2000) Emergent mechanical properties of self-assembled polymeric capsules. *PNAS*, **97**, 12418–12421.

31 Sivakova, S., Bohnsack, D.A., Mackay, M.E., Suwanmala, P., and Rowan, S.J. (2005) Utilization of a combination of weak hydrogen-bonding interactions and phase segregation to yield highly thermosensitive supramolecular polymers. *J. Am. Chem. Soc.*, **127**, 18202–18211.

32 Feldman, K.E., Kade, M.J., Meijer, E.W., Hawker, C.J., and Kramer, E.J. (2009) Model transient networks from strongly hydrogen-bonded polymers. *Macromolecules*, **42**, 9072–9081.

33 Kersey, F.R., Yount, F.C., and Craig, S.L. (2006) Single-molecule force spectroscopy of bimolecular reactions: System homology in the mechanical activation of ligand substitution reactions. *J. Am. Chem. Soc.*, **128**, 3886–3887.

34 Ferry, J.D. (1980) *Viscoelastic Properties of Polymers*, 3rd edn, John Wiley & Sons, Inc.

35 Pellens, L., Corrales, R.G., and Mewis, J. (2004) General nonlinear rheological behavior of associative polymers. *J. Rheol.*, **48**, 379–393.

36 Evans, E. and Ritchie, K. (1997) Dynamic strength of molecular adhesion bonds. *Biophys. J.*, **72**, 1541–1555.

37 Vaccaro, A. and Marrucci, G. (2000) A model for the nonlinear rheology of associating polymers. *J. Non-Newtonian Fluid Mech.*, **92**, 261–273.

38 Lodge, A.S. (1956) A network theory of flow birefringence and stress in concentrated polymer solutions. *Trans. Faraday Soc.*, **52**, 120–130.

39 Tanaka, F. and Edwards, S.F. (1992) Viscoelastic properties of physically cross-linked networks – transient network theory. *Macromolecules*, **25**, 1516–1523.

40 Jongschaap, R.J.J., Wientjes, R.H.W., Duits, M.H.G., and Mellema, J. (2001) A generalized transient network model for associative polymer networks. *Macromolecules*, **34**, 1031–1038.

41 Leibler, L., Rubinstein, M., and Colby, R.H. (1991) Dynamics of reversible networks. *Macromolecules*, **24**, 4701–4707.

42 Chen, C.C. and Dormidontova, E.E. (2004) Supramolecular polymer formation by

metal-ligand complexation: Monte Carlo simulations and analytical modeling. *J. Am. Chem. Soc.*, **126**, 14972–14978.

43 Xu, D. and Craig, S.L. (2010) Multiple dynamic processes contribute to the complex steady shear behavior of cross-linked supramolecular networks of semidilute entangled polymer solutions. *J. Phys. Chem. Lett.*, **1**, 1683–1686.

44 English, R.J., Gulati, H.S., Jenkins, R.D., and Khan, S.A. (1997) Solution rheology of a hydrophobically modified alkali soluble associative polymer. *J. Rheol.*, **41**, 427–444.

45 Xu, D., Liu, C., and Craig, S.L. (2011) Divergent shear thinning and shear thickening of supramolecular polymer networks in semidilute entangled polymer solutions. *Macromolecules*, **44**, 2343–2353.

46 Serpe, M.J. and Craig, S.L. (2007) Physical organic chemistry of supramolecular polymers. *Langmuir*, **23**, 1626–1634.

47 White, S.R., Sottos, N.R., Geubelle, P.H., Moore, J.S., Kessler, M.R., Sriram, S.R., Brown, E.N., and Viswanathan, S. (2001) Autonomic healing of polymer composites. *Nature*, **409**, 794–797.

48 Moore, J.S. (1999) Supramolecular polymers. *Curr. Opin. Colloid Interface Sci.*, **4**, 108–116.

49 Kersey, F.R., Loveless, D.M., and Craig, S.L. (2007) A hybrid polymer gel with controlled rates of cross-link rupture and self-repair. *J. R. Soc. Interface*, **4**, 373–380.

50 Furukawa, A. and Tanaka, H. (2009) Inhomogeneous flow and fracture of glassy materials. *Nature Materials*, **8**, 601–609.

51 Cordier, P., Tournilhac, F., Soulié-Ziakovic, C., and Leibler, L. (2008) Self-healing and thermoreversible rubber from supramolecular assembly. *Nature*, **451**, 977–980.

52 Kong, H.J., Wong, E., and Mooney, D.J. (2003) Independent control of rigidity and toughness of polymeric hydrogels. *Macromolecules*, **36**, 4582–4588.

53 van der Gucht, J., Besseling, N.A.M., and Stuart, M.A.C. (2002) Surface forces, supramolecular polymers, and inversion symmetry. *J. Am. Chem. Soc.*, **124**, 6202–6205.

54 Chen, C.C. and Dormidontova, E.E. (2006) Monte Carlo simulations of end-adsorption of head-to-tail reversibly associated polymers. *Macromolecules*, **39**, 9528–9538.

55 Serpe, M.J., Rivera, M., Kersey, F.R., Clark, R.L., and Craig, S.L. (2008) Time and distance dependence of reversible polymer bridging followed by single-molecule force spectroscopy. *Langmuir*, **24**, 4738–4742.

13
Topological Polymer Chemistry: A Quest for Strange Polymer Rings
Yasuyuki Tezuka

13.1
Introduction

There are frequent examples in the macroscopic world of the fact that the form of an object governs its functions and properties. On the other hand, the fabrication of extremely small objects having precisely defined structures has only recently become an attractive possibility, and this is now opening the door to nanoscience and nanotechnology. In the field of synthetic polymer chemistry, the form of macromolecules has long been restricted to a linear or a mostly uncontrolled branched topology. In recent decades, however, a variety of nonlinear topologies of precisely controlled geometry have been realized owing to the remarkable progress in synthetic techniques, including a self-assembly strategy [1–6]. Nevertheless, the synthesis of a variety of cyclic and multicyclic polymers (Scheme 13.1) has been a formidable synthetic challenge, and even the practical synthesis of simple ring polymers has only recently become a subject of intensive efforts [2].

13.2
Systematic Classification of Nonlinear Polymer Topologies

A rational classification of nonlinear polymer topologies would provide an insightful means to show their structural relationships, and eventually lead to their convenient synthetic pathways. There have been, however, few attempts to devise a systematic classification of nonlinear, and in particular cyclic and multicyclic, polymer architectures composed of sufficiently long and thus flexible segment components.

We have formulated a systematic classification of a series of well-defined cyclic and branched polymer architectures by reference to the graph presentation (molecular graph) of constitutional isomers in alkanes (C_nH_{2n+2}) and mono- and polycycloalkanes (C_nH_{2n}, C_nH_{2n-2}, etc.) [7]. Each alkane and (poly)cycloalkane molecules is taken as a source to generate a unique topological construction. The total number of termini (chain ends) and of junctions (branch points) are taken as invariant (constant)

Supramolecular Polymer Chemistry, First Edition. Edited by Akira Harada.
© 2012 Wiley-VCH Verlag GmbH & Co. KGaA. Published 2012 by Wiley-VCH Verlag GmbH & Co. KGaA.

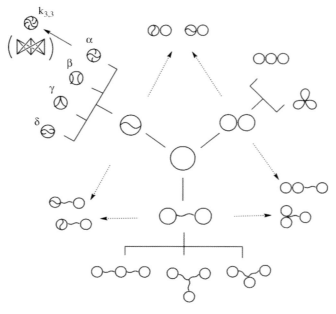

Scheme 13.1 Ring families.

geometric parameters, and the total number of branches at each junction and the connectivity of each junction are likewise maintained as invariant parameters. On the other hand, such Euclidian geometric properties as the distance between two adjacent junctions and that between the junction and terminus are taken as variant parameters. This conforms to the flexible nature of the randomly coiled polymer segments. Furthermore, topological constructions having five or more branches at one junction are allowed, while the relevant isomers having the corresponding molecular formula are absent.

In Table 13.1-A, the molecular graphs of alkanes of generic molecular formula C_nH_{2n+2} with n = 3–7 and their relevant topological constructions are listed. A hypothetical point presentation from methane (CH_4) and a line construction from ethane (C_2H_6) are not included, since the former is not significant with respect to polymer topology and the latter and propane (C_3H_8) produce an equivalent graph presentation. Two butane isomers produce a linear and a three-armed star construction, respectively. Likewise, pentane isomers produce a four-armed star construction in addition to the two produced already from the butane isomers. From five hexane isomers, two new constructions of an H-shaped (or Cayley tree) and a five-armed star architecture are produced. Heptane isomers produce the two new constructions of a super H-shaped and a six-armed star architecture. Accordingly, a series of branched polymer topologies are hierarchically ranked as shown in Table 13.1-A.

Topological constructions obtained from the graph presentation of monocycloalkane molecules of C_nH_{2n} up to n = 6 are collected in Table 13.1-I. Cyclopropane (C_3H_6) produces a simple ring topology. From the two isomers of C_4H_8, a tadpole and a

13.2 Systematic Classification of Nonlinear Polymer Topologies

Table 13.1 Classification of nonlinear polymer topologies.

A

Polymer Topology	C_nH_{2n+2}				
	n =3	4	5	6	7
$A_3(2,0)$					
$A_4(3,1)$					
$A_5(4,1)$					
$A_6(4,2)$					
$A_6(5,1)$				(*)	(*)
$A_7(5,2)$					
$A_7(6,1)$					(*)

II

Polymer Topology	C_nH_{2n-2}			Polymer Topology	C_nH_{2n-2}
	n =4	5	6		n =6
$II_4(0,2)$				$II_6(0,2)$	
				$II_6(1,1)$	(⋈)
$II_5(0,1)$				$II_6(1,2)$	
$II_5(1,2)$				$II_6(2,2)$	(⋈)
				$II_6(2,2)$	
$II_5(1,3)$				$II_6(2,3)$	
				$II_6(2,3)$	
				$II_6(2,4)$	

I

Polymer Topology	C_nH_{2n}			Polymer Topology	C_nH_{2n}
	n = 3	4	5	6	n = 6
$I_3(0,0)$					$I_6(2,2)$
$I_4(1,1)$					$I_6(3,1)$ (*)
$I_5(2,1)$					$I_6(3,2)$
$I_5(2,2)$					$I_6(3,3)$

III

Polymer Topology	C_nH_{2n-4}			Polymer Topology	C_nH_{2n-4}	Polymer Topology	C_nH_{2n-4}
	n = 4	5	6		n = 6		N = 7
$III_4(0,4)$				$III_6(0,4)$		$III_7(0,1)$	(*)
$III_5(0,2)$				$III_6(0,2)$	(⋈)	$III_7(0,2)$	
$III_5(0,3)$				$III_6(0,3)$			

IV

Polymer Topology	C_6H_6
$IV_6(0,3)$	
$IV_6(0,6)$	
$IV_6(0,6)$	$K_{3,3}$

simple ring construction are produced. Two types of new topologies are produced from C_5H_{10} isomers, and are distinguished from each other by their junction and branch structures, that is, one has two outward branches at one common junction in the ring unit (twin-tail tadpole), while the other has two outward branches located at two separate junctions in the ring unit (two-tail tadpole). Further, the four new topological constructions are produced from C_6H_{12} isomers; one having five branches at one junction is hypothetical and is therefore shown in parentheses in

Table 13.1-I. A series of 'ring with branches' constructions have thus been classified as summarized in Table 13.1-I.

Another series of topological constructions is produced from the graph presentation of bicycloalkanes of C_nH_{2n-2} up to n = 6, as listed in Table 13.1-II, where the basic three bicyclic constructions free of outward branches are included. These are designated as θ-ring (*fused* type), 8-ring (*spiro* type) and manacle-ring (bridged type), respectively. First, a θ construction is produced from bicyclo[1,1,0]butane, C_4H_6. From the five bicycloalkane isomers of C_5H_8, three new constructions are produced. One of these is an 8-shaped construction from spiro[2,2]pentane. By reference to bicyclohexane (C_6H_{10}) isomers, moreover, eight new constructions are produced. All constructions but one from bi(cyclopropane) possess outward branches emanating from either the θ-ring or the 8-ring unit shown above. In this manner, a series of *fused*, *spiro* and *bridged*-type double cyclic topological constructions are classified into the corresponding categories by reference to the bicycloalkane isomers.

Moreover, selected topological constructions obtainable from the graph presentation of tricycloalkanes of C_nH_{2n-4} and tetracycloalkanes of C_nH_{2n-6} are shown in Table 13.1-III and Table 13.1-IV, respectively. A large variety of 'rings with branches' constructions are additionally produced in these groups. In the tricyclic topologies, eight constructions are derived from θ- or 8-forms. Six of these are produced directly from the corresponding molecular graph of tricycloalkanes, while the additional two are from a hypothetical molecular formula possessing five or six branches at a junction, as given in Table 13.1-III. Further, the seven topological constructions of externally linked (or 'bridged') rings with a simple cyclic θ- or 8-form are obtainable with reference to the relevant tricycloalkane isomers. (Scheme 13.1) Finally, in Table 13.1-IV, three particularly notable constructions produced by reference to the relevant tetracycloalkane isomers of C_nH_{2n-6} are listed. These include triply-*fused* tetracyclic topology of prismane and $K_{3,3}$ type constructions, in which their internal graph connection patterns are distinct from each other.

13.3
Topological Isomerism

The concept of isomerism [8], from the Greek '*isos*' (equal) and '*meros*' (part), has been a fundamental concept in chemistry since its introduction by Berzelius in 1830 [9]. Isomers are compounds having the same chemical constitution (thus, molar mass) but different properties. The evolution of constitutional (structural) isomerism by Kekulé [10] has remarkably extended rational understanding of both the static and dynamic structures of organic substances, in which noninterconvertible three-dimensional structures of the isomers bring about distinct properties. The term 'constitutional (structural) isomers' refers to those isomers with distinctive *connectivity* of atoms or atomic groups. The term 'stereoisomers' refers to isomers with indistinguishable *connectivity* but which are distinct from each other in terms of the Euclidean geometric rigidity of the molecules, such as restriction of bond angle bending and bond rotation.

Moreover, in flexible polymer molecules, a pair of isomeric products can be formed from an identical set of telechelic (end-reactive) polymer precursors and end-linking reagents. As discussed in the next section, three units of a bifunctional polymer precursor and two units of a trifunctional end-linking reagent, or two units of a trifunctional star polymer precursor and three units of a bifunctional end-linking reagent could produce *topological isomers* possessing identical termini and junction numbers as well as branch numbers at each junction, referring to manacle- and θ-shaped polymers [1, 5]. It is noteworthy that these isomers are formed from the least common combination of the functionalities of the two components.

It is remarkable, therefore, that polymer topologies are not only distinguished by their terminus/junction numbers, that is, *the first order topological parameters*, but also by their *connectivity* or by their *combination* of flexible polymer segment components. The former, such as a pair of polymers, one manacle-shaped and the other θ-shaped, corresponds to constitutional isomers having topologically distinct molecular graphs, while the latter, such as a pair of star polymers having the same arm numbers but having different sets of arm lengths, corresponds to constitutional isomers having topologically equivalent molecular graphs. Topological isomers based on such *second-order topological parameters* uniquely occur for randomly coiled, flexible polymer molecules.

Moreover, topological isomerism involving double-cyclic polymers having θ-shaped and manacle-shaped constructions and an additional intertwined pretzel-shaped construction is worthy of discussion. (Scheme 13.2) A pair of manacle- and θ-shaped polymers and also a pair of θ- and pretzel-shaped polymers are identified as constitutional isomers with distinctive connectivity of segment components because these pairs are interconvertible with each other only by applying chain breaking at the appropriate TWO positions to rearrange the chain connectivity. Interestingly, this isomerism corresponds to a pair of a large simple ring and a catenane (more precisely [2]-catenane, Scheme 13.2). In contrast, a pair of manacle- and pretzel-shaped polymers are interconvertible with each other by applying chain breaking at the appropriate SINGLE position to rearrange the chain conformation but not the chain connectivity. This isomerism corresponds exactly to a pair of a ring and a knot [11] (Scheme 13.2).

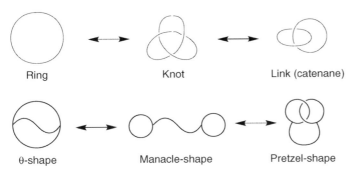

Scheme 13.2 Topological isomerism in cyclic polymers.

In general, the interconversion process in the molecular graph presentation corresponds to the isomerization in actual molecules and distinguishes between constitutional isomers and stereoisomers. Since constitutional isomers are distinguishable in terms of their *connectivity*, chain breaking with at least TWO appropriate positions is required to rearrange the segment connectivity to complete the interconversion process. In contrast, for a pair of stereoisomers, hypothetical deformation of bond angles and freedom of bond rotation are sufficient to complete the interconversion process, and the chain breaking process is not required. Thus, a single ring compound is a stereoisomer of a triangle, a square, or a randomly folded ring compound, since a ring and a triangle (and a square, etc.) are topologically equivalent to each other, interconvertible through the conceptual continuous chain deformation while retaining the total chain length, but without the chain breaking process (Scheme 13.2).

In contrast, a ring and a knot can be interconvertible with each other by only applying the chain breaking process. Significantly, chain breaking at a SINGLE position and subsequent conformational rearrangement of the chain are required to complete interconversion of this isomer pair (Scheme 13.2). This is in contrast to the classical constitutional isomers, where chain breaking with at least TWO (appropriately chosen) positions is required to complete the process. By this distinction, a ring and a knot are traditionally assigned as *topological stereoisomers* or, more specifically, since they are not mirror images of each other, as *topologically distinct diastereomers*, in contrast to a ring and a triangle (or a square, etc.) as *topologically equivalent diastereomers*. And according to the theorem of topological geometry, a pair consisting of a ring and a knot are interconvertible to each other in *4-dimensional* space, while this not intuitively imaginable in 3-dimensional space.

13.4
Designing Unusual Polymer Rings by Electrostatic Self-Assembly and Covalent Fixation

We have so far developed an 'electrostatic self-assembly and covalent fixation' technique for designing nonlinear polymer architectures [12]. Linear or star telechelic precursors having moderately strained, 5-membered cyclic ammonium salt groups, carrying appropriately reactive counteranions like pluricarboxylates, have been employed as key polymer precursors. It is notable that not only small (low-molecular-weight and water-soluble) but also large (polymeric and water-insoluble) carboxylates can be employed in this 'electrostatic self-assembly and covalent fixation' process [13]. By simple precipitation of cationic telechelic precursor into aqueous solution containing small carboxylates, or by coprecipitation of an equimolar mixture of cationic and anionic telechelic precursors, the corresponding electrostatic polymer self-assemblies are obtainable in high yields. The cations and anions always balance the charges even under dilution, and the selective nucleophilic ring-opening reaction occurs at an elevated temperature, to convert the ionic interaction into the permanent covalent linkage (Scheme 13.3). An alternative ring-emitting covalent conversion

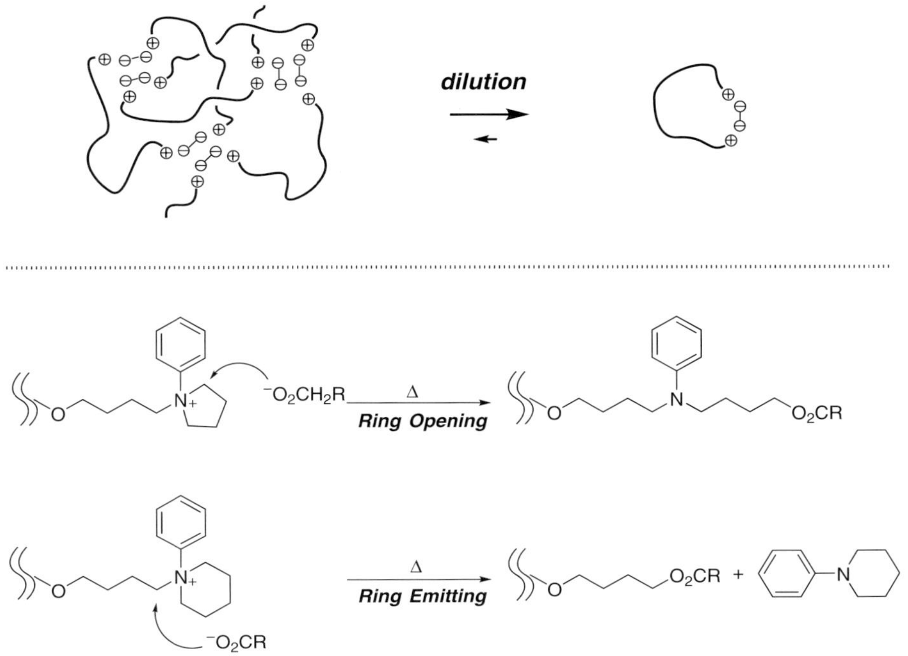

13.4 Designing Unusual Polymer Rings by Electrostatic Self-Assembly and Covalent Fixation

Scheme 13.3 Electrostatic self-assembly and covalent fixation process.

process has recently been developed by making use of telechelics having unstrained, 6-membered cyclic ammonium salt groups to give the products having a simple ester linkage [14] (Scheme 13.3).

By employing the 'electrostatic self-assembly and covalent fixation' protocol, we have achieved the efficient synthesis of various ring polymers, optionally having specific functional groups at the designated positions of the ring polymer structures (*kyklo*-telechelics and cyclic macromonomers). A series of ring with branches, that is, a simple tadpole as well as twin-tail and two-tail tadpole, polymer topologies have also been constructed by this methodology. Moreover, all of the three types of dicyclic polymers having either 8-, θ- or manacle-topologies have been successfully synthesized. (Scheme 13.4) Such a tricyclic polymer having a trefoil topology has also been produced by using a linear telechelic precursor carrying a hexafunctional carboxylate counteranion. (Scheme 13.4)

As for polymers of unique topological isomerism, a pair of θ- and manacle-shaped polymers were obtained simultaneously from an assembly comprised of bifunctional linear precursors carrying trifunctional carboxylates, or from trifunctional star-shaped precursors carrying bifunctional carboxylates [2, 12] (Scheme 13.5). The relevant polymeric constitutional isomers were also obtainable through the metathesis condensation of an H-shaped telechelic precursor having four allyl groups [15]. They are regarded as a first example of rationally synthesized '*polymeric topological*

Electrostatic polymer self-assembly

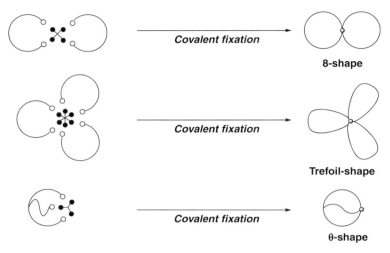

Scheme 13.4 Multicyclic polymer topologies by electrostatic polymer self-assembly.

isomers,' in which not only their molecular weights but also their chemical compositions are identical to each other, but their topological connectivities are distinct from each other. We have subsequently shown that the two isomers and other topologically unique polymers could be resolved by means of the reversed-phase chromatography (RPC) technique based on theoretical analysis [16], and a pair of topological isomers could be unequivocally assigned by employing isomers containing a cleavable olefinic group in a backbone segment obtainable through electrostatic polymer–polymer self-assembly as an intermediate [17].

As an extension of the 'electrostatic self-assembly and covalent fixation' process, we have prepared a variety of cyclic as well as multicyclic polymers having functional groups at designated positions in their unique constructions by either employing

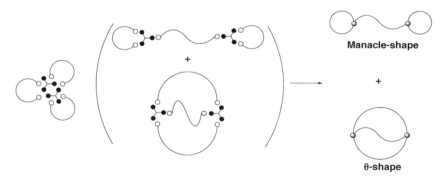

Scheme 13.5 Polymeric topological isomers through electrostatic polymer self-assembly.

functional carboxylate counteranions or functionalized polymeric precursors. In particular, we have introduced a variety of functional groups, including allyl and other olefinic groups as well as hydroxyl, bromophenyl, alkyne, and azide groups for further topological transformation, by taking advantage of metathesis condensation in the presence of a Grubbs catalyst (Scheme 13.6), a palladium-mediated coupling reaction [18] as well as the 'click' process with prepolymers having relevant functional groups. We have so far confirmed that metathesis condensation is an efficient means of causing polymer cyclization even under dilution with notable functional group tolerance. Thus, a telechelic poly(oxytetramethylene) having allyl groups was subjected to metathesis polymer cyclization (MPC) to produce a cyclic polymer having a linking oxybutene group, which was subsequently hydrogenated to form an oxytetramethylene unit. In consequence, a unique *defect-free* ring poly(THF) consisting exclusively of monomer units could be produced for rigorous inspection of the topology effect of polymers [19]. MPC has also been combined with the atom transfer radical polymerization (ATRP) process to produce a variety of telechelic polyacrylates for the synthesis of a simple cyclic [20] but also an 8-shaped polymer [21], as well as cyclic block copolymers [22]. Moreover, we have prepared an 8-shaped *kyklo*-telechelics having two allyl groups at opposite positions by 'electrostatic self-assembly and covalent fixation' with two cationic linear precursors having an allyl group and tetracarboxylate counteranion, and subjected this to the metathesis condensation to construct a polymeric δ-graph having a doubly-*fused* tricyclic topology [23].

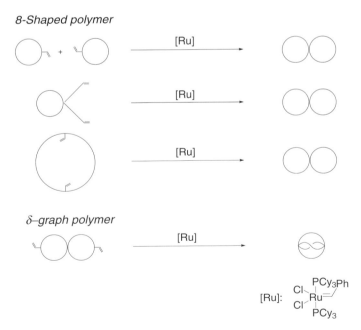

Scheme 13.6 Metathesis polymer condensation for topologically unique polymers.

An 8-shaped polymer having a metathesis-cleavable olefinic group at the focal position has also been prepared for further topological conversion to form simple ring polymers and for elucidating polymer catenane formation through the entanglement of the two polymer precursor segments geometrically placed close by each other [24]. An alternative synthetic means to effectively produce polymer catenanes has been developed with the 'electrostatic self-assembly and covalent fixation' process by using polymer precursors having a hydrogen-bonding unit [25].

Furthermore, the 'click' process using cyclic prepolymers having alkyne or azide groups as well as an 8-shaped bicyclic precursor having two alkyne groups at the opposite positions of the two ring units has been performed to produce a series of *bridged-* and *spiro*-multicyclic polymers as well as topological block copolymers consisting of linear/ring and star/ring polymer components [26].

13.5
Conclusion and Future Perspectives

Numerous future opportunities are anticipated along with ongoing progress in *topological polymer chemistry*. A new concept of topological isomerism is now established, and a pair of topological isomers uniquely occurring in flexible, nonlinear polymer architectures has been synthesized by the *electrostatic self-assembly and covalent fixation* process with newly designed telechelic polymer precursors having moderately strained cyclic ammonium salt groups. Further synthetic challenges include such significant polymer topologies as tricyclic α, β, γ, and δ constructions as well as a tetracyclic $K_{3,3}$ construction by employing specifically designed telechelic precursors of Cayley-tree constructions.

In addition, *topological polymer chemistry* is now offering unique research opportunities by systematically providing a variety of topologically defined polymers. Together with the progress in theory and simulations, we expect to reveal unique topological effects in static and dynamic polymer properties [27, 28] that rely, in particular, on *the second-order topological parameters*.

Acknowledgments

The author is grateful to the Yamada Science Foundation for financial support for the participation in the Yamada Conference 2008, 'Topological Molecules.' The author thanks to Professor A. Harada and the organizing committee for their kind invitation to this outstanding symposium. Financial support for this work from the Mitsubishi Foundation, the Sekisui Chemical Grant Program for Research on Manufacturing Based on Learning from Nature, and in the form of a Tokyo Tech Innovative Research Engineering Award are gratefully acknowledged. This work was also supported in part by a grant from the Ministry of Education, Culture, Sports, Science and Technology, Japan, through the Japanese Society for the Promotion of Science (17350054).

References

1. Tezuka, Y. (2009) *Topology Designing*, NTS, Tokyo.
2. Adachi, K. and Tezuka, Y. (2009) *J. Synth. Org. Chem. Jpn,*, **67**, 1136.
3. Tezuka, Y. (2005) *Chem. Rec.*, **5**, 17.
4. Tezuka, Y. (2003) *J. Polym. Sci., Part-A: Polym. Chem.*, **41**, 2905.
5. Tezuka, Y. and Oike, H. (2002) *Prog. Polym. Sci.*, **27**, 1069.
6. Tezuka, Y. and Oike, H. (2001) *Macromol. Rapid Commun.*, **22**, 1017.
7. Tezuka, Y. and Oike, H. (2001) *J. Am. Chem. Soc.*, **123**, 11570.
8. Rouvray, D.H. (1974) *Chem. Soc. Rev.*, **3**, 355 and references cited therein.
9. Berzelius, J.J. (1830) *Pogg. Ann.*, **19**, 305.
10. Kekulé, F.A. (1858) *Ann. Chem.*, **106**, 129.
11. Sauvage, J.-P. and Dietrich-Buchecker, C. (eds) (1999) *Molecular Catenanes, Rotaxanes and Knots*, Wiley-VCH, Weinheim.
12. Oike, H., Imaizumi, H., Mouri, T., Yoshioka, Y., Uchibori, A., and Tezuka, Y. (2000) *J. Am. Chem. Soc.*, **122**, 9592.
13. Adachi, K., Irie, H., Sato, T., Uchibori, A., Shiozawa, M., and Tezuka, Y. (2005) *Macromolecules*, **38**, 10210.
14. Adachi, K., Takasugi, H., and Tezuka, Y. (2006) *Macromolecules*, **39**, 5585.
15. Tezuka, Y. and Ohashi, F. (2005) *Macromol. Rapid Commun.*, **26**, 608.
16. Vakhrushev, A., Gorbunov, A., Tezuka, Y., Tsuchitani, A. and Oike, H. (2008) *Anal. Chem.*, **80**, 8153.
17. Tezuka, Y., Takahashi, N., Sato, T., and Adachi, K. (2007) *Macromolecules*, **40**, 7910.
18. Tezuka, Y., Ohtsuka, T., Adachi, K., Komiya, R., Ohno, N., and Okui, N. (2008) *Macromol. Rapid Commun.*, **29**, 1237.
19. Tezuka, Y., Komiya, R., Ido, Y., and Adachi, K. (2007) *React. Funct. Polym.*, **67**, 1233.
20. Hayashi, S., Adachi, K., and Tezuka, Y. (2007) *Chem. Lett.*, **36**, 982.
21. Hayashi, S., Adachi, K., and Tezuka, Y. (2008) *Polym. J.*, **40**, 572.
22. Adachi, K., Honda, S., Hayashi, S., and Tezuka, Y. (2008) *Macromolecules*, **41**, 7898.
23. Tezuka, Y. and Fujiyama, K. (2005) *J. Am. Chem. Soc.*, **127**, 6266.
24. Ishikawa, K., Yamamoto, T., and Tezuka, Y., (2010) *Macromolecules*, **43**, 7062.
25. Ishikawa, K., Yamamoto, T., Asakawa, M., and Tezuka, Y. (2010) *Macromolecules*, **43**, 168.
26. Sugai, N., Heguri, H., Ohta, K., Meng, Q., Yamamoto, T., and Tezuka, Y., (2010) *J. Am. Chem. Soc.*, **132**, 14790.
27. Habuchi, S., Satoh, N., Yamamoto, T., Tezuka, Y., and Vacha, M. (2010) *Angew. Chem., Int. Ed.*, **49**, 1418.
28. Honda, S., Yamamoto, T., and Tezuka, Y. (2010) *J. Am. Chem. Soc.*, **132**, 10251.

14
Structure and Dynamic Behavior of Organometallic Rotaxanes

Yuji Suzaki, Tomoko Abe, Eriko Chihara, Shintaro Murata,
Masaki Horie, and Kohtaro Osakada

14.1
Introduction

Rotaxanes and pseudorotaxanes have attracted significant attention not only for their unique interlocked structures but also for their properties and functions [1–4]. Their potential utility as materials for molecular mechanisms has been investigated by a number of research groups since the first discovery of the operation of a rotaxane as a switchable molecular shuttle by Stoddart et al. in 1994 [5]. Recent important topics of this research area involve synthesis of multiple-component (higher-order) rotaxanes as well as controlled aggregation of the rotaxanes in designated directions in order to effectively amplify physical motion of the one-dimensional supramolecules. Oriented films of rotaxanes spread over a solid surface were reported to perform as molecular muscles and molecular devices, and to cause transposition of a liquid droplet upon photo-irradiation [6–9]. The supramolecules, which include various kinds of rotaxanes as well as the more complicated interlocked molecules including molecular Borromean rings and Solomon's knot, are obtained in the form of single crystals [10, 11]. The crystalline rotaxanes may be regarded as assemblies of these supramolecules with the positions and orientations regulated by crystallographic symmetry. Recent reports on unique crystalline rotaxanes are summarized below.

Paraquat-type cyclic molecules having two viologen units and the axle molecules bearing three aromatic stations form [2]rotaxane or [4]rotaxane selectively [12, 13]. Two different rotaxanes are obtained as crystals, depending on the ratios of the axle and cyclic compounds in the mixture as well as the conditions of crystallization [13]. An MOF (Metal-Organic Framework) was reported in the crystalline materials formed from metal centers and bifunctional ligands. Loeb et al. prepared an MOF using rotaxanes, namely an MORF (Metal-Organic Rotaxane Framework), and discovered the assembly of rotaxane ligands of dibenzo[24]crown-8-ether and bis(N-alkyl-4,4′-bipyridinium) with transition metals such as Cu(II), Cd(II), and Ni(II) (Scheme 14.1) [14]. Recently Yaghi, Stoddart and their coworkers have reported

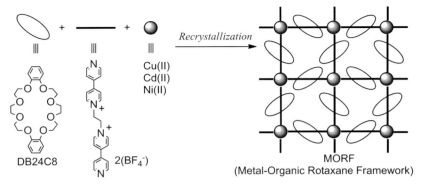

Scheme 14.1 MORF (Metal-Organic Rotaxane Framework).

MOFs having macrocyclic polyethers as recognition modules for relatively large guest molecules such as paraquat dication [15].

The rotaxanes with dialkylammonium salt and crown ethers, mostly dibenzo[24]crown-8-ether (DB24C8), have become one of the most common supramolecules (Figure 14.1) [16]. A solution of DB24C8 and dialkylammonium contains a pseudorotaxane which is in equilibrium with the component molecules. Pseudorotaxane formation is preferred in solvents with relatively lower polarity such as dichloromethane and chloroform and is not observed in highly polar solvents such as dmso. A number of research groups have reported crystal structures of these pseudorotaxanes composed of DB24C8 and dialkylammonium, and the number of crystals registered in the Cambridge database reached almost to 60 by the end of 2010. Interactions between the axle and macrocyclic molecules in these pseudorotaxanes were investigated in detail both in the solution and in the solid state [17]. N−H···O hydrogen bonds between the axle and cyclic components bind these molecules strongly. The interaction is evidenced by crystallography and by IR and NMR spectroscopy. Attractive interaction between C−H bonds of the axle component and oxygen atoms of the macrocycle may be possible because X-ray results of many rotaxanes point to such interaction. Additional interaction is observed between aromatic groups of the axle and cyclic components. Two aromatic planes in Figure 14.1 are parallel, and the distance suggests π···π interaction between them.

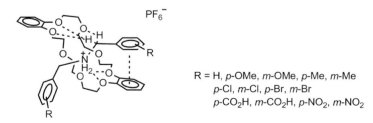

R = H, p-OMe, m-OMe, p-Me, m-Me
p-Cl, m-Cl, p-Br, m-Br
p-CO$_2$H, m-CO$_2$H, p-NO$_2$, m-NO$_2$

Figure 14.1 [2]Pseudorotaxane composed of DB24C8 and dialkylammonium.

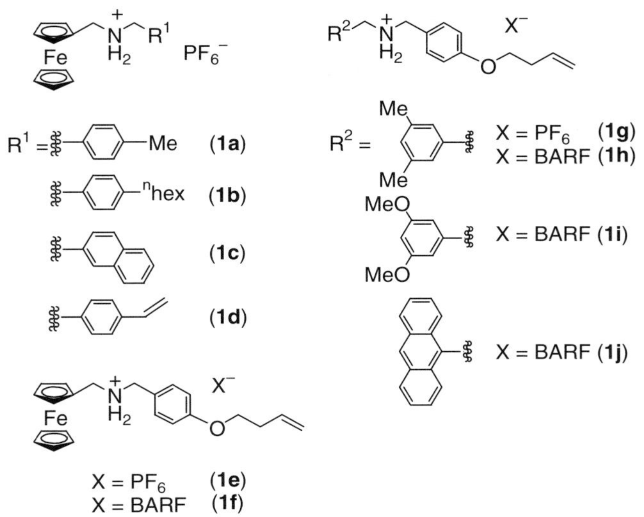

Figure 14.2 Structure of the dialkylammonium used for the axle component of the pseudorotaxanes.

X-ray crystallography is an efficient tool for characterization of these crystalline supramolecules, and it often provides a *frozen* structure in the cage of the crystal lattice. In this chapter we describe our results on the crystalline pseudorotaxane composed of DB24C8 and ferrocenylmethylammonium, which indicate change of structure in the crystalline states. Synthesis and properties of ferrocene-containing lower- and higher- order rotaxanes including [2]-, [3]- and polyrotaxane are also described [18].

14.1.1
Crystals of Pseudorotaxanes [20–22]

Figure 14.2 lists the dialkylammoniums used as the axle components of the pseudorotaxanes, [R^{1}CH$_2$NH$_2$CH$_2$R^2]X (**1a-1j**, R^1, R^2 = Fc (Fc = Fe(C$_5$H$_4$)(C$_5$H$_5$)), C$_6$H$_4$-4-Me, C$_6$H$_4$-4-nHex, C$_{10}$H$_7$, C$_6$H$_4$-4-CHCH$_2$, C$_6$H$_4$-4-OCH$_2$CH$_2$CHCH$_2$, C$_6$H$_3$-3,5-Me$_2$, C$_6$H$_3$-3,5-(OMe)$_2$, 9-anthryl; X = PF$_6$, BARF (BARF = B{C$_6$H$_3$-3,5-(CF$_3$)$_2$}$_4$). [FcCH$_2$NH$_2$CH$_2$C$_6$H$_4$-4-Me]PF$_6$ (**1a**) was prepared from ferrocene carboxaldehyde and H$_2$NCH$_2$C$_6$H$_4$-4-Me. These compounds underwent a condensation reaction to form FcCH=NCH$_2$C$_6$H$_4$-4-Me, which was followed by reduction of the N=C double bond by NaBH$_4$. The FcCH$_2$NHCH$_2$C$_6$H$_4$-4-Me thus obtained underwent protonation of the amine by HCl and exchange of the counter anion with NH$_4$PF$_6$. Other amines have been prepared by similar procedures. All these compounds have methylene and aryl groups which may stabilize the pseudorotaxane structure formed with benzo-crown ethers by C−H ··· π as well as the π–π attractive interaction.

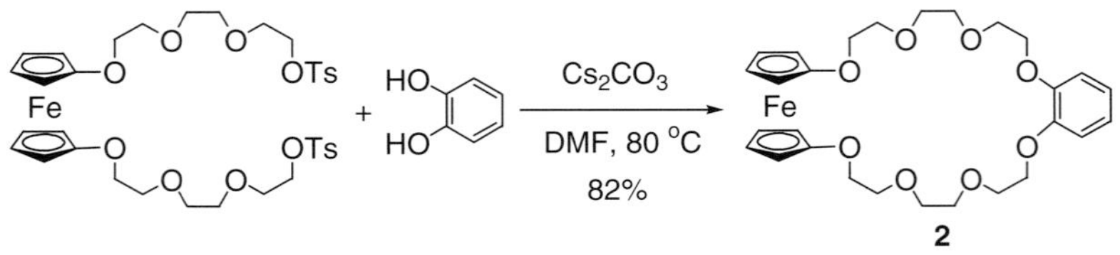

Scheme 14.2 Synthesis of ferrocene-containing crown ether **2**.

Scheme 14.2 shows the synthesis of the 1,1′-ferrocenylene-containing crown ether **2**. The cyclizative Williamson's ether synthesis of Fe{C_5H_4(OCH_2CH_2)$_3$OTs}$_2$ (Ts = MeC_6H_4-4-SO_3-) and ortho-catechol yields 1,1′-ferrocenylene-containing crown ether **2** in 82% yield. The [22]ferrocenophane **2** has the cyclopentadienyl ligands connected with an oligo(ethyleneglycol) chain, and has a similar structure to DB24C8. The circumferential bond lengths of **2** (37.02(8) Å), determined by X-ray crystallography, are greater than that of DB24C8 (34.16(6) Å).

Slow addition of Et_2O to a CH_2Cl_2 solution of **1a** and DB24C8 causes growth of orange-colored single crystals of the corresponding pseudorotaxane **1a·DB24C8** (Scheme 14.3). Crystals of **1b·DB24C8** and **1c·DB24C8** were also obtained by slow evaporation of an MeOH/$CHCl_3$ solution containing **1b** and DB24C8 and an acetone/CH_2Cl_2 solution containing **1c** and DB24C8, respectively. Table 14.1 summarizes IR data of the compounds. Peaks due to asymmetric and symmetric N−H vibration, ν(asym. N−H) and ν(sym. N−H), of **1a**, **1b**, and **1c** were observed in the range of 3258–3266 and 3231–3235 cm^{-1}, respectively. The corresponding peaks of **1a·DB24C8**, **1b·DB24C8** and **1c·DB24C8** were shifted to lower wavenumbers (ν(asym. N−H) = 3166–3193 cm^{-1} and ν(sym. N−H) = 3067–3076 cm^{-1}), which indicates the existence of N−H···O hydrogen bonds between the ammonium hydrogens and oxygens of DB24C8.

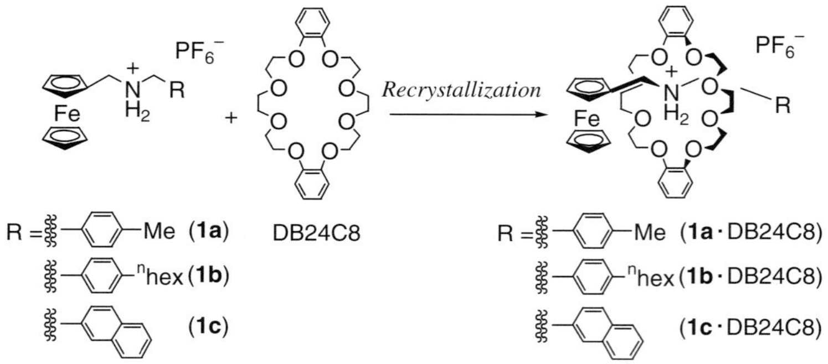

Scheme 14.3 Formation of crystalline [2]pseudorotaxanes.

Figure 14.3 depicts the molecular structures of **1a·DB24C8** and **1b·DB24C8** determined by single-crystal X-ray crystallography. **1b·DB24C8** crystallizes in a space group of $P2/a$ (No. 13) which differs from that of **1a·DB24C8** (P-1 (No. 2)). **1b·DB24C8** has a similar co-conformation to that of **1a·DB24C8** in the solid state

Table 14.1 IR data of the axles, **1a–1c**, and pseudorotaxanes, **1a·DB24C8**, **1b·DB24C8**, and **1c·DB24C8** (KBr disk, r. t.).

Compound	ν(asym. N-H)/cm^{-1}	ν(sym. N-H)/cm^{-1}
1a	3266	3233
1b	3260	3235
1c	3258	3231
1a·DB24C8	3166	3067
1b·DB24C8	3193	3076
1c·DB24C8	3187	3067

as described below. Cyclopentadienyl ligands of the ferrocenyl group of **1b·DB24C8** are in the eclipsed conformation with a dihedral angle of 4.4°. The conformation of the C(aryl)-CH$_2$-NH$_2$-CH$_2$-C(cyclopentadienyl) adopts a near-planar all-anti geometry. Torsion angles calculated from C10-C11-N1-C12 and C13-C12-N1-C11 are 175° and 178°, respectively. DB24C8 in the pseudorotaxane adopts a chair-like conformation. Two aromatic rings of DB24C8 (A and B rings in Figure 14.3) are almost parallel. The tilt angle and interplanar distance between A and B rings are 11.4(2)° and 11.094(2) Å, respectively. The NH$_2$ group of the axle component shows short contacts with oxygen atoms of DB24C8 (N1-H1···O5, 2.562 Å and N1-H2···O3,

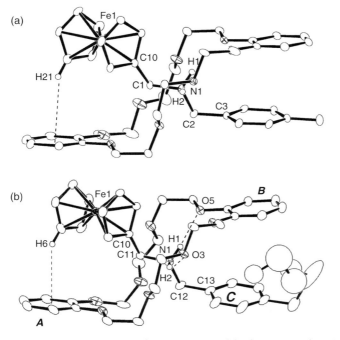

Figure 14.3 ORTEP drawing of (a) **1a·DB24C8** and (b) **1b·DB24C8** with 30% probability. PF$_6$ anions were omitted. Positions of the hydrogen atoms are obtained by calculation.

Table 14.2 TGA data of the compounds (scan rate = 5 °C min^{-1}).

Compound	$T_{d5}/°C$[a]	$\Delta T_{d5}/°C$
1a	217	—
1b	208	—
1c	205	—
1a + DB24C8[b]	231	—
1b + DB24C8[b]	216	—
1c + DB24C8[b]	227	—
1a·DB24C8	237	20[c]
1b·DB24C8	222	14[d]
1c·DB24C8	242	37[e]
DB24C8	295	—

[a] 5% weight loss temperature.
[b] physical mixture of the compounds (axle and DB24C8).
[c] $T_{d5}(1a·DB24C8) - T_{d5}(1a)$.
[d] $T_{d5}(1b·DB24C8) - T_{d5}(1b)$.
[e] $T_{d5}(1c·DB24C8) - T_{d5}(1c)$.

2.116 Å; N1-H1-O5, 151° and N1-H2-O3, 143°), suggesting the presence of N−H···O hydrogen bonds. The distance between N1 and the centroid of ring A (5.052(4) Å) is shorter than that between N1 and the centroid of ring B (6.195(3) Å), probably due to the hydrogen bonds between N1-H1···O5 and N1-H2···O3. A cyclopentadienyl hydrogen H6 is close to the ring A of DB24C8. The distance between H6 and the centroid of ring A (2.99 Å) indicates C−H···π attractive interaction. The interplanar distance between the B and C rings (4.507(3) Å) is longer than the corresponding distance between p-methylphenyl and the C_6H_4 planes in 1a·DB24C8 (3.725(5) Å).

Thermal properties of the pseudorotaxanes in the solid state were investigated by thermal gravimetric analysis (TGA) and differential scanning calorimetry (DSC). TGA of the pseudorotaxanes 1a·DB24C8, 1b·DB24C8, and 1c·DB24C8 shows 5% weight loss in the range of T_{d5} = 222–242 °C, which is higher than that of 1a, 1b and 1c (205–217 °C) and than that of the physical mixture of the axle (1a, 1b and 1c) and DB24C8 (216–231 °C) (Table 14.2). These results indicate that thermal degradation of the ferrocenylmethylammonium was retarded by the formation of the pseudorotaxane.

DSC of 1a·DB24C8 showed reversible endothermic and exothermic peaks at 125 °C and 116 °C, respectively. These peaks, with the thermodynamic parameters $\Delta H = 7.6$ kJ mol^{-1} and $\Delta S = 19.3$ J mol^{-1} K^{-1}, are assigned to reversible crystal-to-crystal phase transitions as shown in Figure 14.4 and Scheme 14.4, which is characterized by variable-temperature X-ray crystallography [22].

Figure 14.4 depicts the structures of 1a·DB24C8 at 30 and 128 °C respectively. At both temperatures, two 1a·DB24C8 molecules in the unit cell are located at C/2 symmetric positions (P-1 (No. 2), Z = 2). The structure at 30 °C shown in Figure 14.4, is stabilized by N−H···O and C−H···O hydrogen bonds between the dialkylammonium group and the oxygen atoms in DB24C8. The aromatic ring of the p-xylyl

14.1 Introduction | 311

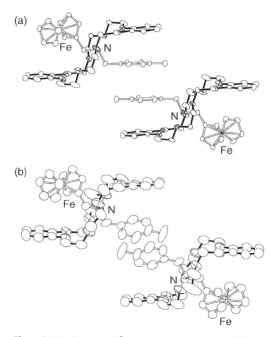

Figure 14.4 Structure of **1a**·DB24C8 (a) at 30 °C and (b) at 128 °C obtained by X-ray crystallography.

group of the axle component of **1a**·DB24C8 also has intrarotaxane π–π stacking interactions with the aromatic ring of the catechol group of DB24C8 and an interrotaxane π–π stacking interaction with the aromatic ring of the axle component, resulting in layer stacking of these aromatic rings in the crystal. **1a**·DB24C8, at 128 °C shown in Figure 14.4, also has the pseudorotaxanes structure stabilized by intrarotaxane N−H···O and C−H···O hydrogen bonds. The aromatic ring of the axle molecule rotates by 37°, and the π–π stacking interaction in pseudorotaxane switches to C−H···π interaction between aromatic hydrogen and the C_6H_4 group in DB24C8. Interrotaxane π–π stacking interactions between p-xylyl rings were also kept in the structure shown in Figure 14.4. Heating the crystal of pseudorotaxane at a

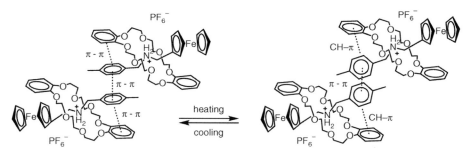

Scheme 14.4 Solid-state phase transition reaction of **1a**·DB24C8.

rate of 2 °C min^{-1} led to a change of the color of the crystal on the (001) face upon irradiation with polarized white light from green to orange at 128 °C and a change to green on cooling the crystal. The color changes of the crystal of **1a·DB24C** were reversible and were complete within 68 ms.

DSC of **1b·DB24C8** and **1c·DB24C8** did not show any reversible peaks below 130 °C, but showed irreversible endothermic peaks at 147 °C and 181 °C, respectively. DSC of **1a·DB24C8**, **1b·DB24C8** and **1c·DB24C8** did not show peaks assigned to the phase transition of free DB24C8 ($T_{endo} = 104\,°C$ and $T_{exo} = 79\,°C$), indicating that the thermal behaviors observed here are due to the pseudorotaxanes rather than the free component molecules.

14.1.2
Synthesis of Ferrocene-Containing [2]Rotaxanes by the Threading-Followed-by-End-Capping Strategy [20, 23, 24]

Suitable end-functionalization reactions of the axle of pseudorotaxane are regarded as a most convenient strategy for the synthesis of rotaxanes (Scheme 14.5) and are described as the threading-followed-by-end-capping (or end-capping) method [3]. This strategy requires high association constants in the pseudorotaxane formation (Scheme 14.5a) as well as a selective bond formation reaction (Scheme 14.5b) under the same reaction conditions. The obtained [2]rotaxane maintains the interlocked structure in solution because the bulky end groups prevent the macrocyclic molecule from undergoing the dethreading reaction. Several 'rotaxane-like complexes' were reported to be synthesized by a similar strategy, which causes a slow dethreading

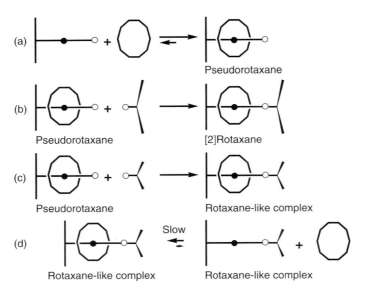

Scheme 14.5 Threading-followed-by-end-capping strategy for [2]rotaxane and rotaxane-like complex and dethreading reaction of the rotaxane-like complex.

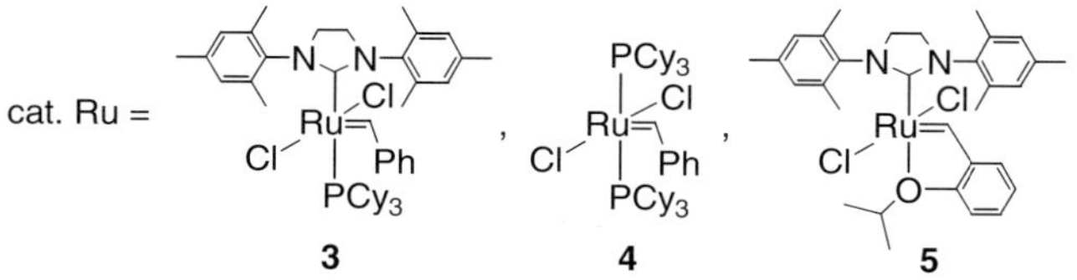

Figure 14.5 Ru-carbene complexes, **3**, **4** and **5**, for olefin metathesis reaction.

reaction if the size of the end groups is similar to that of the inner cavity of the macrocyclic molecule (Scheme 14.5c). These pseudorotaxanes, also known as 'rotaxane-like complexes', cause the dethreading reaction under harsh conditions, for example, in polar solvents and at higher temperatures, but are stable enough to be isolated under normal conditions (Scheme 14.5d) [25]. The pseudorotaxanes of dialkylammonium and DB24C8 have been used for this purpose because of their high association constant in the solvent with lower polarity. Many bond-forming reactions, such as esterification [26], urethane formation [27], Wittig reaction [27], hydrosilylation [29], propargylic substitution [30], Sonogashira coupling [31], cycloaddition [32], and Diels–Alder reaction [33], have been used for the threading-followed-by-end-capping method. In most of these reports, reactions that take place under neutral or acidic conditions were chosen to prevent the ammonium group from being neutralized.

The Ru-carbene complex-catalyzed olefin metathesis reaction has advantages for rotaxane synthesis via the end-capping strategy because this catalyst has a high tolerance to various functional groups and because it promotes the reaction in high efficiency under neutral condition, in non-polar solvents, and at relatively low temperature [34]. Previous studies revealed that the Ru-carbene complexes, **3**, **4**, and **5** (Figure 14.5) catalyze the homo-metathesis reaction of olefins as well as cross-metathesis reactions of two olefins with different reactivities.

Pseudorotaxane formation of dialkylammoniums, **1a**, **1d**, **1e**, and DB24C8 is a prerequisite for the [2]rotaxane synthesis by the end-capping method, and this was confirmed by ^{1}H NMR and mass spectroscopy of the mixture (Scheme 14.6). The ^{1}H NMR spectrum of the CD$_3$CN solution of **1a** and DB24C8 (10 mM for each

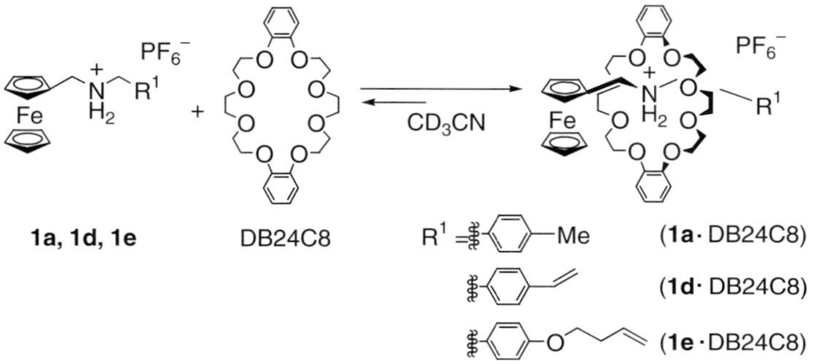

Scheme 14.6 Pseudorotaxane formation of DB24C8 and dialkylammonium in solution.

compound) shows signals assigned to pseudorotaxane **1a·DB24C8** as well as **1a** and DB24C8, the pseudorotaxane formation and the disaggregation reactions being slower than the ^{1}H NMR time scale. The molar ratio between **1a·DB24C8** and **1a** is 51:49 at room temperature. The solution containing **1d** (or **1e**) instead of **1a** yields corresponding pseudorotaxanes **1d·DB24C8** ([**1d·DB24C8**]/[**1d**] = 64/36 at 20 °C) and **1e·DB24C8** ([**1e·DB24C8**]/[**1e**] = 43/57 at 20 °C).

Scheme 14.7 summarizes the synthesis of rotaxanes, **6–15**, by the end-capping strategy and subsequent N-acylation of the axle component affording neutral rotaxanes **16–18**. The cross-metathesis reaction, catalyzed by (H$_2$IMes)(PCy$_3$) Cl$_2$Ru=CHPh (H$_2$IMes = N,N-bis(mesityl)-4,5-dihydroimidazol-2-ylidene) (**3**), of **1e** and 3,5-dimethylphenyl acrylate in the presence of DB24C8 yields [2]rotaxane **6** in 72% isolated yield. Similar reaction using ferrocenyl acrylate gives rotaxane **7** having two ferrocenyl moieties at each end of the axle component. The vinylbenzene group in **1d** undergoes neither reaction with 3,5-dimethylphenyl acrylate nor homometathesis reaction under similar conditions, which is ascribed to steric hindrance around the vinyl group by DB24C8. Figure 14.6 summarizes the structure and yields of the [2]rotaxanes synthesized similarly to the method for **6**. The conformations of the internal olefins in these rotaxanes, **6–9**, formed by cross-metathesis reaction, was controlled to trans when acrylates and styrene derivatives were employed for the end-capping group. **9** is obtained in higher yield (41%) from the reaction catalyzed by **3** than that catalyzed by (PCy$_3$)$_2$Cl$_2$Ru=CHPh (**4**) (23%). The vinylborane and vinylsilane derivatives are also useful as end-capping groups of the heteroatom-functionalized [2]rotanxaes, **10** and **11**, which are obtained as a mixture of the cis and trans isomers (cis/trans = 1/9). [2]Rotaxane synthesis using ferrocene-containing crown ether **2** instead of DB24C8 yields **8** and **12–15**, having a ferrocenylene group in the macrocyclic component with various axle molecules. Rotaxane **15** with a long chain, –(CH$_2$)$_{11}$O(CH$_2$)$_{11}$O– unit, in the axle molecule is synthesized by

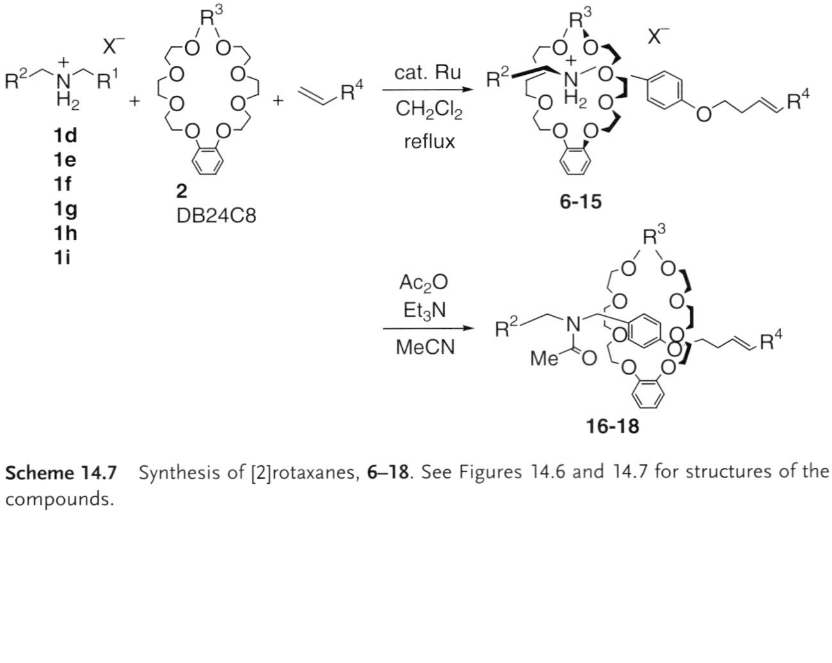

Scheme 14.7 Synthesis of [2]rotaxanes, **6–18**. See Figures 14.6 and 14.7 for structures of the compounds.

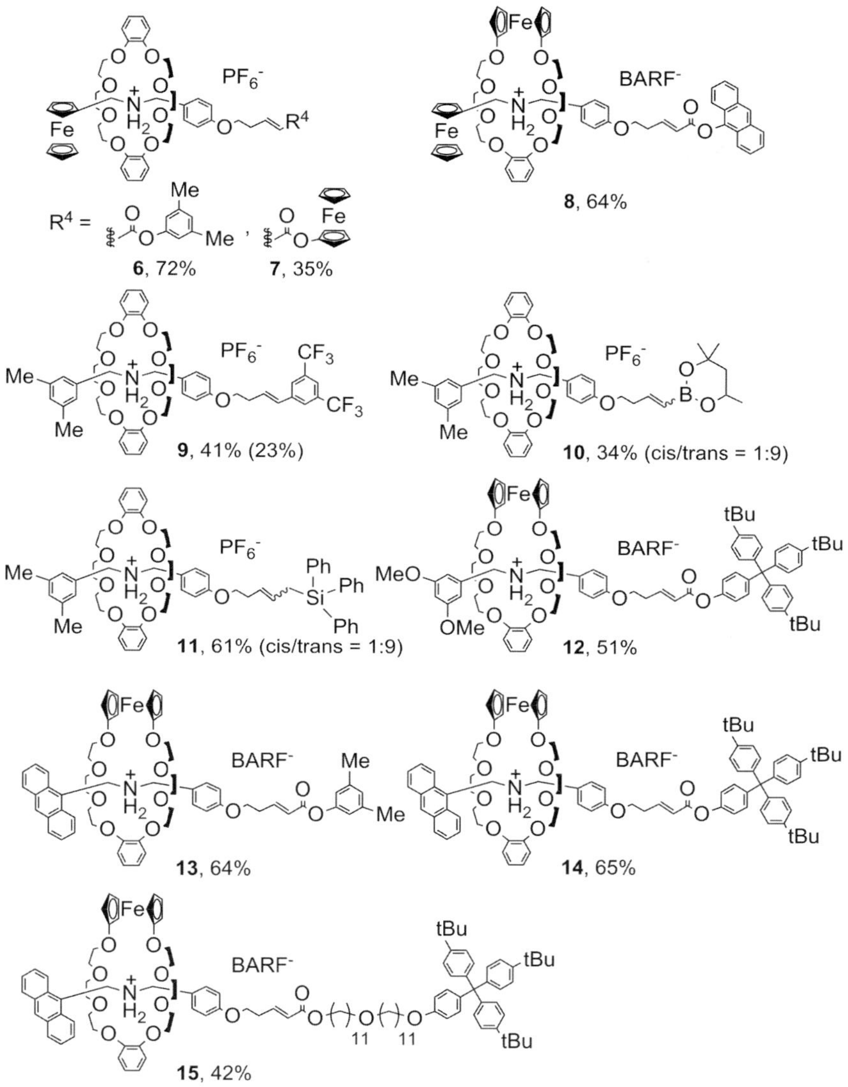

Figure 14.6 Structure and yield of [2]rotaxanes, **6–15**.

catalyst **5**. ^{1}H NMR and IR spectroscopic studies of these compounds indicate that the crown ethers of these rotaxanes are located around the ammonium unit by hydrogen bonding between hydrogens of the $CH_2NH_2CH_2$ unit and oxygens of the crown ethers.

N-acylation of rotaxanes **6**, **14** and **15** is achieved by the reaction reported by Takata et al. using Ac_2O under basic condition, affording neutral rotaxanes **16**, **17**, and **18** (Scheme 14.7, Figure 14.7). All of these rotaxanes keep their interlocked structure in

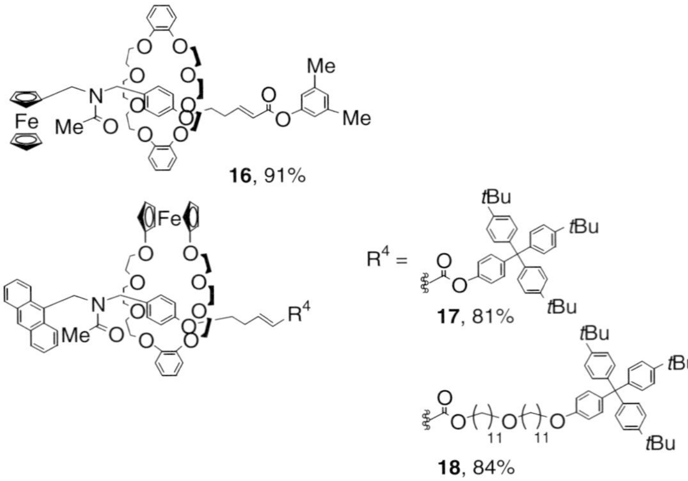

Figure 14.7 Structure and yield of neutral [2]rotaxanes, **16–18**.

the solvents with relatively lower polarity, such as $CHCl_3$ and CH_2Cl_2. Several rotaxanes, **8**, **10**, **12**, **13**, and **14**, cause dethreading of the axle molecule from the macrocyclic component. This is described below.

14.1.3
Dethreacting Reaction of Rotaxane-Like Complex [20, 24]

Pseudorotaxanes having linear-shaped axle molecules are generally in rapid equilibrium with the mixture of the uncomplexed component molecules. Rotaxane-like complex as mentioned above has the end groups of the axle molecule, whose size is similar to the cavity of the macrocycle, and causes slow dethreading reaction only under severe conditions. The rate of the dethreading reactions was found to depend on the temperature as well as the solvents used and pH. The controlled dethreading reactions of rotaxane-like complex are receiving increasing attention because of the controlled release of the component molecules upon application of stimulus and their potential application to the drug delivery system (DDS) [35]. Synthesis of such a rotaxane-like complex was still difficult because the simple reaction of the component molecules in solution yield the desired rotaxane-like complex very slowly due to the steric bulkiness of the end groups. Nevertheless, the thermal stability of the [2] pseudorotaxane structure is higher than that of the mixture of the component molecules [36].

Several rotaxanes, **8**, **10**, **12**, **13** and **14**, synthesized by the end-capping strategy (Scheme 14.7) were found to show the properties of a rotaxane-like complex. They were stable enough to be isolated but underwent a dethreading reaction in polar solvents such as CD_3CN and dmso-d_6. Heating the CD_3CN solution containing **8** at 50 °C yields a mixture of the component molecules, **2** and

[FcCH$_2$NH$_2$CH$_2$C$_6$H$_4$-4-OCH$_2$CH$_2$CH=CHCOOAn]BARF (Scheme 14.8). The first-order kinetic rate, k_{obs}, of the dethreading reaction of **8** is estimated to be 3.9×10^{-6} s^{-1}. The dethreading reaction of **8** is much faster in dmso-d_6 ($k_{obs} = 2.7 \times 10^{-5}$ s^{-1}). The similar dethreading reactions of **10**, **12**, **13** and **14** are also monitored and the obtained k_{obs} is summarized in Table 14.3. In all cases, the dethreading reaction in dmso-d_6 is faster than the reaction in CD$_3$CN. Table 14.4 summarizes thermodynamic parameters, $\Delta G^{\ddagger}$ kJ mol^{-1}, $\Delta H^{\ddagger}$ kJ mol^{-1} and $\Delta S^{\ddagger}$ J K^{-1} mol^{-1}, of the dethreading reaction of **8**, **12** and **14**, obtained from temperature dependence of k_{obs}. $\Delta G^{\ddagger}$ values of these dethreading reactions are similar to each other, and fall in the range of 100–110 kJ mol^{-1}. The large negative values of $\Delta S^{\ddagger}$ of the dethreading reactions indicate that reaction of two or more molecules involves

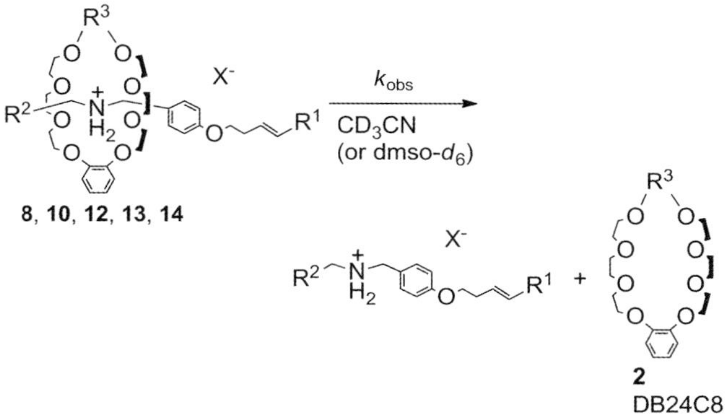

8: R^1 = C(=O)OAn, R^2 = Fc, R^3 = Fe(C$_5$H$_4$)$_2$, X = BARF
10: R^1 = -BOCMe$_2$CH$_2$CHMeO, R^2 = C$_6$H$_3$-3,5-Me$_2$, R^3 = C$_6$H$_4$-1,2-, X = PF$_6$
12: R^1 = C(=O)OC$_6$H$_4$-4-C(C$_6$H$_4$-4-tBu)$_3$, R^2 = C$_6$H$_3$-3,5-(OMe)$_2$, R^3 = Fe(C$_5$H$_4$)$_2$, X = BARF
13: R^1 = C(=O)OC$_6$H$_3$-3,5-Me$_2$, R^2 = An, Fe(C$_5$H$_4$)$_2$, X = BARF
14: R^1 = C(=O)OC$_6$H$_4$-4-C(C$_6$H$_4$-4-tBu)$_3$, R^2 = An, R^3 = Fe(C$_5$H$_4$)$_2$, X = BARF

Scheme 14.8 Dethreading reaction of rotaxane-like complexes.

Table 14.3 First-order kinetic rate constants, k_{obs}, of the dethreading reaction of rotaxane-like complex **8**, **10**, **12**, **13** and **14**.

Run	Compound	k_{obs}/s^{-1} (temperature)	
		in CD$_3$CN	in dmso-d_6
1	8	3.9×10^{-6} (45 °C)	2.7×10^{-5} (25 °C)
2	10	2.4×10^{-7} (50 °C)	2.4×10^{-6} (60 °C)
3	12	1.1×10^{-5} (50 °C)	1.1×10^{-4} (50 °C)
4	13	1.0×10^{-5} (45 °C)	4.9×10^{-3} (20 °C)
5	14	1.3×10^{-7} (25 °C)	4.3×10^{-6} (25 °C)

Table 14.4 Thermodynamic parameters, $\Delta G^{\ddagger}$, $\Delta H^{\ddagger}$ and $\Delta S^{\ddagger}$, for the dethreading reactions of rotaxane-like complex **8**, **12**, and **14**.

Compound	Solvent	$\Delta G^{\ddagger}$/kJ mol^{-1}	$\Delta H^{\ddagger}$/kJ mol^{-1}	$\Delta S^{\ddagger}$/J K^{-1} mol^{-1}
8	dmso-d_6	100	72	−90
12	CD$_3$CN	110	72	−120
12	dmso-d_6	102	32	−220
14	dmso-d_6	110	79	−91

the dethreading reaction. Figure 14.8 shows a plausible structure of the dethreading reaction of the rotaxane-like complexes where the solvent molecules and counter anion coordinate to the ammonium [36].

14.1.4
Photochemical Properties of Ferrocene-Containing Rotaxanes [20]

Rotaxanes, **13**, **14**, **15**, **17** and **18**, show photochemical properties owing to their 9-anthryl group and the ferrocene unit. These rotaxanes show characteristic absorption bands due to the 9-anthryl group at $\lambda_{max} = 368$–372 nm with a molar absorption coefficient of $\varepsilon = 6900$–9800 M^{-1} cm^{-1} (Table 14.5). The positions of the maximum fluorescence are in the range of 414–420 nm. Quantum yield of [AnCH$_2$NH$_2$CH$_2$C$_6$H$_4$-4-OCH$_2$CH$_2$CH=CH$_2$]BARF (**1j**) ($\phi = 0.56$) is higher than those of the rotaxanes ($\phi = 0.0010$–0.097). Rotaxanes **14**, **17**, and **18** having the ferrocenylene group in the cyclic unit show effective quenching of the fluorescence of the 9-anthryl stopper of the axle molecule, as described below. Figure 14.9 compares UV/Vis spectra of CHCl$_3$ solutions of **14**, **15**, **18**, and AnCH$_2$NAcCH$_2$C$_6$H$_4$-4-OCH$_2$CH$_2$CH=CHCOOC$_6$H$_4$-4-C(C$_6$H$_4$-4-tBu)$_3$ (**19**) which show the similar characteristic structural bands of the anthracene chromophoric unit. Emission spectra of the compounds upon excitation at the corresponding wavelength of maximum absorption (Figure 14.9) differ among these compounds. The intensity of the emission of **19** ($\phi = 0.344$) is significantly larger than that of rotaxane **17** ($\phi = 0.030$), although a 1:1 mixture of **2** and **19** gave an almost identical emission spectrum to that of **19** ($\phi = 0.33$). The quantum yield of fluorescence of the cationic rotaxane **14** ($\phi = 0.012$) is smaller than that of the neutral rotaxane **17** ($\phi = 0.030$).

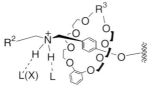

L = solvent (dmso-d_6, CD$_3$CN)
X = counter anion (PF$_6$)

Figure 14.8 Plausible intermediate of the dethreading reaction.

Table 14.5 Photochemical data of the compounds.

Compound	Absorption[a]		Emission[b]	
	λ_{max}/nm	ε/M^{-1}cm^{-1}	λ_{max}/nm	ϕ[c]
1j	371	7600	423	0.56
13	372	7000	420	0.043
14	372	6900	419	0.012
15	373	7100	414	0.0043
17	368	9800	418	0.030
18	369	9600	418	0.097
19	368	10 700	418	0.344
2 + 19	368	11 400	418	0.33

[a] [Compound] = 1.0 × 10^{-2} mM, CHCl$_3$, 25 °C.
[b] [Compound] = 2.0 × 10^{-3} mM, CHCl$_3$, 25 °C, $\lambda_{ex} = \lambda_{max}$(absorption).
[c] Quantum yield.

Degree of quenching is smaller in rotaxane **18** ($\phi = 0.097$), having a longer axle component.

The above results indicate that quenching of emission due to an excited anthryl group is caused by the ferrocenylene unit in the rotaxanes. The interlocked structure of **14** fixes the distance between the ferrocenyl and anthryl groups within the range of effective quenching. Previous studies revealed the quenching pathway of the excited anthracene by the ferrocene group via dominant energy transfer mechanisms. The quenching of rotaxane **14** is more efficient than that of **17** and **18** due to the difference in the relative positions of the anthryl and ferrocenylene groups. Since the efficiency of quenching is lower than the value expected from ideal FRET (Fluorescence Resonance Energy Transfer) procedure, the mechanism for the quenching may not be simple.

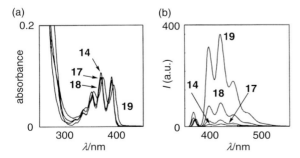

Figure 14.9 (a) UV-vis spectra ([compound] = 1.0 × 10^{-2} mM) and (b) emission spectra ([compound] = 2.0 × 10^{-3} mM, $\lambda_{ex} = \lambda_{max}$(absorption)) of CHCl$_3$ solution of **14**, **17**, **18**, and **19**.

14.1.5
Ferrocene-Containing [3]Rotaxane and Side-Chain Polyrotaxane [18, 24]

14.1.5.1 Strategies and Synthesis of [3]Rotaxanes

The higher-order rotaxanes such as [n]rotaxane (n ≥ 3) and polyrotaxane are composed of multiple component molecules, and are expected to have an integrated rotaxane function. Studies on the synthesis of [3]rotaxane could be regarded as a protocol for the stepwise synthesis of the higher-order rotaxanes with definite structures. Scheme 14.9 depicts two plausible strategies for the synthesis of [3] rotaxanes via metathesis reactions: reaction of vinyl-functionalized [2]rotaxane and [2]pseudorotaxane having a terminal olefin unit at an end of the axle molecule (Scheme 14.9a), the other plausible pathway being homometathesis reaction of the axle component of [2]pseudorotaxanes to afford symmetrical [3]rotaxane (Scheme 14.9b).

[2]Rotaxane **20** bearing an olefin group is synthesized by acylation of the nitrogen of cationic rotaxane **6** with acryloyl chloride in THF under basic condition in 48% yield Scheme 14.10. Similar reaction using methacryloyl chloride gave **21**. **20** and **21**, existing as a mixture of s-cis and s-trans isomers of the amide group in solution. These neutral rotaxanes keep an interlocked structure in toluene-d_8 and in dmso-d_6 at 100 °C for 24 h.

The Ru-carbene complex **3**-catalyzed cross-metathesis reaction of **20** and [FcCH$_2$NH$_2$CH$_2$C$_6$H$_4$-4-OCH$_2$CH$_2$CH=CH$_2$]PF$_6$ (**1e**) in the presence of DB24C8 yields [3]rotaxane **22** in 33% yield (Scheme 14.11). The mass spectrum of **22** shows a peak $m/z = 1823$ which is assigned to the cationic rotaxane, [{FcCH$_2$NH$_2$CH$_2$C$_6$H$_4$-4-OCH$_2$CH$_2$CH=CHCON(CH$_2$Fc)CH$_2$C$_6$H$_4$-4-OCH$_2$CH$_2$CH=CHCOO(C$_6$H$_3$-3,5-Me$_2$)}(DB24C8)$_2$]$^+$. The ^{1}H NMR spectrum of **22** shows signals at δ 5.67, 7.72 (in CDCl$_3$) which are assigned to the olefinic hydrogens formed by the cross-metathesis reaction. The obtained [3]rotaxane **22** has an ammonium group.

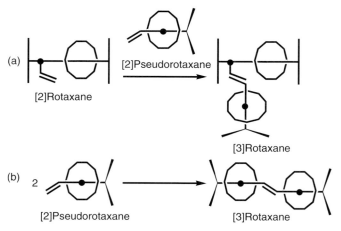

Scheme 14.9 Strategies of [3]rotaxane synthesis.

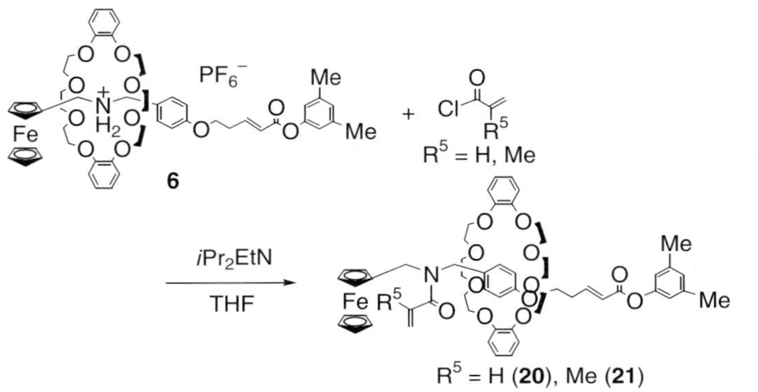

Scheme 14.10 Synthesis of [2]rotaxane with N-acryloyl group.

Further N-acylation followed by cross-metathesis reaction with [2]pseudorotaxane of **1e** and DB24C8 could become a scheme for the [n]rotaxane synthesis (n > 4).

Homo-metathesis reaction of $[(C_6H_3\text{-}3,5\text{-}Me_2)CH_2NH_2CH_2C_6H_4\text{-}4\text{-}OCH_2CH_2CH=CH_2]PF_6$ (**1g**) in the presence of DB24C8 yields symmetric [3]rotaxane **23** composed of two DB24C8s and diammonium in 29% yield (Scheme 14.12). The internal olefin of the [3]rotaxane is reduced by reaction with H_2 catalyzed by PtO_2 to **24**, while the [3]rotaxane **23** was obtained as a mixture of the cis and trans isomers. [3] Rotaxane synthesis by similar strategy as well as by the 'magic rod' strategy involving the reversible olefin metathesis reaction is also reported by Leigh et al. [38].

14.1.5.2 Strategies and Synthesis of Side-Chain Type Polyrotaxane [20, 24, 39]

Polyrotaxane is mainly divided into two classes, namely main-chain type and side-chain type polyrotaxanes. Main-chain polyrotaxane has a rotaxane structure on the polymer main chain. Side-chain type polyrotaxane has the rotaxane structures on the side chain of the main polymer chain. Scheme 14.13 shows examples of strategies for

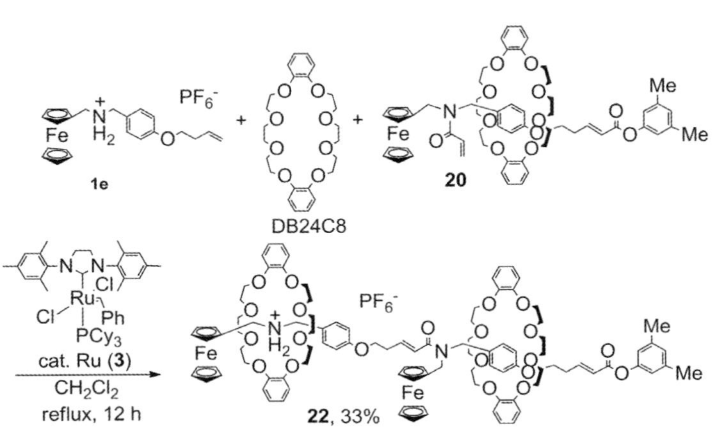

Scheme 14.11 Synthesis of [3]rotaxane **22** via cross-metathesis reaction.

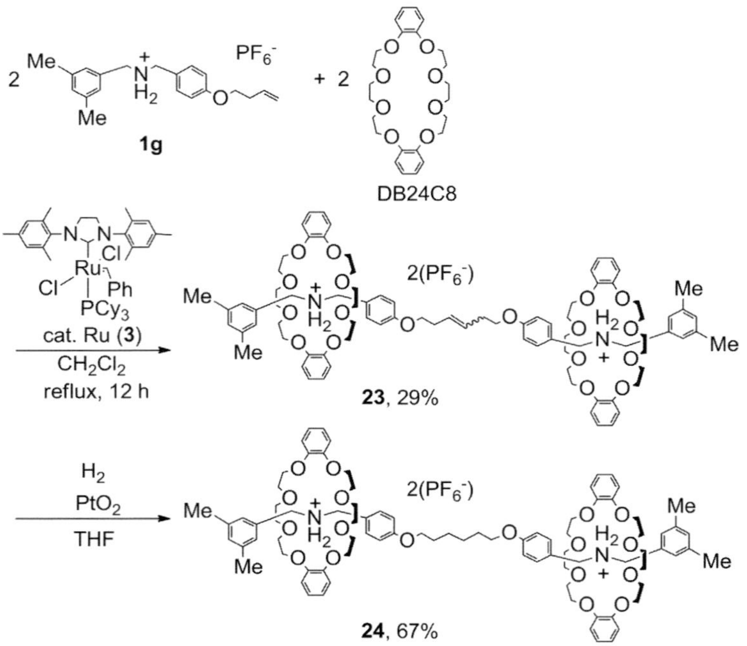

Scheme 14.12 Synthesis of [3]rotaxanes, **23** and **24**, via homo-metathesis reaction.

side-chain polyrotaxane. Polymerization of [2]pseudorotaxane, which has a polymerizable monomer unit as an axle molecule, gives polyrotaxane. The obtained polyrotaxane can have a side chain without a rotaxane structure because the [2] pseudorotaxane monomer is in equilibrium with the free macrocyclic and axle molecule which is also incorporated in the polymer chain (Scheme 14.13a) [40, 41].

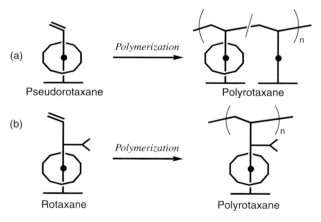

Scheme 14.13 Strategies for side-chain polyrotaxanes.

[2]Rotaxane is also an attractive monomer for the side-chain polyrotaxane because of the interlocked structure. Polymerization of [2]rotaxane monomer yields the side-chain polyrotaxane without defect of the rotaxane structure (Scheme 14.13b).

Free-radical copolymerization of **20** with styrene induced by AIBN (azobisiso-butyronitrile) yields polyrotaxane **25** composed of two randomly polymerized monomers (Scheme 14.14). The ratio of the rotaxane unit to the styrene unit in polyrotaxane **25**, x, varied in the range 0.18–0.48 as well as in molecular weight ($M_n = 2000$–6700), depending on the initial concentration of the monomers, **20** and styrene. Rotaxane **20** undergoes copolymerization with various styrene derivatives, as shown in Scheme 14.14, as well as the acrylamide derivative $(PhCH_2)_2NC(=O)OCH=CH_2$, to yield the corresponding side-chain polyrotaxane in 50–75% yield, the incorporation ratio of the rotaxane unit being 0.18–0.30. Rotaxane **21**, having a methacryloyl group, undergoes neither homopolymerization nor copolymerization with styrene under similar conditions to those used for **20**.

Thermal gravimetric analysis (TGA) of polyrotaxane show 5% weight loss at the temperature $T_{d5} = 249$–315 °C, which is higher than that of **20** ($T_{d5} = 244$ °C), which is ascribed to the fact that that the thermal stability of the rotaxane unit is improved by the incorporation into the thermally more stable polystyrene. The cyclic voltammogram of polyrotaxane **25a** exhibits a redox peak at $E_{1/2} = 0.08$ V (vs Fe(III)$(C_5H_5)_2$/Fe(II)$(C_5H_5)_2$) which is similar to that of the rotaxane monomer **20**.

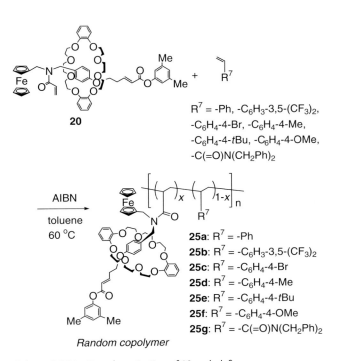

Scheme 14.14 Co-polymerization of **20** and olefins.

14.2
Conclusion

In this article, we describe the synthesis of the crystalline pseudorotaxanes composed of DB24C8 and ferrocenylmethylammonium as well as the synthesis of [2]-, [3]- and polyrotaxanes. The supramolecular structure was stabilized by N−H · · · O hydrogen bonds between the ammonium group of the axle component and oxygens of DB24C8. Thermal stability of the ferrocenylmethylammonium was improved by pseudorotaxane formation with DB24C8. The single-crystal-to-single-crystal phase transition reaction of the pseudorotaxane depends on the substituent group of the axle component of the pseudorotaxane. Ferrocene-containing crown ether in rotaxanes shows behavior as a fluorescence quencher of the emission from the anthryl group. Another integration of the rotaxane was achieved by repeated cross-metathesis reaction and copolymerization of dibenzylacryloylamide and the [2]rotaxane-bearing acryloyl group.

14.3
Appendix: Experimental Section (Data of Compounds Reported in Reference [24])

[{(C_6H_3-3,5-Me_2)$CH_2NH_2CH_2C_6H_4$-4-OCH_2CH_2CH=CHC_6H_3-3,5-$(CF_3)_2$}(DB24C8)] (PF_6) (9):^{1}H NMR (300 MHz, $CDCl_3$, r.t.): δ 2.14 (s, 6H, Me), 2.74 (dt, 2H, OCH_2CH_2, J = 12.0, 6.6 Hz), 3.45 (s, 8H, CH_2-DB24C8), 3.76 (t, 8H, CH_2-DB24C8, J = 3.6 Hz), 4.07–4.13 (10H, OCH_2CH_2, CH_2-DB24C8), 4.42 (m, 2H, NCH_2), 4.53 (m, 2H, NCH_2), 6.51 (m, 1H, CH_2C=CH), 6.61 (d, 1H, CH_2CH=CH, J = 15.9 Hz), 6.80–6.92 (13H, C_6H_4-DB24C8, C_6H_4-axle, $C_6H_3Me_2$), 7.25 (d, 2H, C_6H_4-axle, J = 9.0 Hz), 7.50 (br, 2H, NH_2), 7.69 (s, 1H, C_6H_3-para), 7.78 (s, 2H, C_6H_3-ortho). $^{13}C\{^1H\}$ NMR (75.5 MHz, $CDCl_3$, r.t.): δ 21.2 (Me), 32.8 (OCH_2CH_2), 52.1 (NCH_2), 52.4 (NCH_2), 66.8 (OCH_2-axle), 68.2 (CH_2-DB24C8), 70.1 (CH_2-DB24C8), 70.6 (CH_2-DB24C8), 112.6 (C_6H_4-DB24C8), 114.5 (C_6H_4-axle), 120.4 (C_6H_3-3,5-$(CF_3)_2$), 121.6 (C_6H_4-DB24C8), 121.9 (C_6H_3-3,5-$(CF_3)_2$), 123.7 (C_6H_3), 124.6 (C_6H_3-3,5-$(CF_3)_2$), 125.9 (C_6H_3-3,5-$(CF_3)_2$), 126.5 (C_6H_3), 129.5 (C_6H_3-3,5-$(CF_3)_2$), 130.4 (C_6H_3), 130.9 (C_6H_4-axle), 131.4 (C_6H_3), 131.8 (CH=CH), 138.1 (C_6H_4-axle), 139.4 (CH=CH), 147.4 (C_6H_4-DB24C8), 159.2 (C_6H_4-axle). FABMS: Calcd. for $C_{52}H_{60}F_6NO_9$: 956 Found: m/z = 956 ([M-PF_6]$^+$).

[{(C_6H_3-3,5-Me_2)$CH_2NH_2CH_2C_6H_4$-4-OCH_2CH_2CH=CH $BOCMe_2CH_2CHMeO$} (DB24C8)](PF_6) (10) (mixture of cis and trans (1/9)): ^{1}H NMR (300 MHz, CD_3CN, r.t.): δ 1.25 (s, 3H, OC(CH_3)$_2$), 1.27 (s, 3H, OC(CH_3)$_2$), 1.30 (s, 3H, OCHCH_3), 1.50 (dd, 1H, OC(CH_3)$_2CH_2$, J = 13.8, 11.4 Hz), 1.79 (dd, 1H, OC(CH_3)$_2CH_2$, J = 13.8, 3.0 Hz), 2.14 (s, 6H, C_6H_3-3,5-Me_2), 2.58 (dt, 2H, OCH_2CH_2, J = 12.0, 6.6 Hz), 3.46 (s, 8H, CH_2-DB24C8), 3.76 (t, 8H, CH_2-DB24C8, J = 3.6 Hz), 3.96 (t, 2H, OCH_2CH_2, J = 6.6 Hz), 4.10 (t, 8H, CH_2-DB24C8, J = 3.6 Hz), 4.21 (tdd, 1H, OCHCH_3, J = 13.2, 6.6 Hz, J = 3.3 Hz), 4.42 (s, 2H, NCH_2), 4.53 (s, 2H, NCH_2), 5.49 (d, 1H, CH_2CH = CH, J = 18 Hz), 6.57 (dt, 1H, CH_2CH = CH, J = 17.7, 6.6 Hz), 6.71 (d, 2H, C_6H_4-axle, J = 9.0 Hz), 6.80–6.92 (m, 11H, C_6H_4-DB24C8, C_6H_3-para, C_6H_3-ortho), 7.25 (d, 2H, C_6H_4-axle, J = 9.0 Hz), 7.45 (br, 2H, NH_2). $^{13}C\{^1H\}$ NMR (75.5 MHz, $CDCl_3$, r.t.): δ 21.2 (C_6H_3-3,5-Me_2), 23.2 (OCHCH_3), 28.1 (OC

(CH_3)$_2$), 31.3 (OC(CH_3)$_2$), 35.0 (OCH_2CH_2), 45.9 (OC(CH_3)$_2CH_2$), 52.2 (NCH_2), 52.4 (NCH_2), 64.7 (OCHCH$_2$), 66.7 (OCH_2CH$_2$), 68.2 (CH_2-DB24C8), 70.1 (CH_2-DB24C8), 70.7 (CH_2-DB24C8), 70.8 (OC(CH_3)$_2$), 112.6 (C_6H_4-DB24C8), 114.4 (C_6H_4-axle), 121.6 (C_6H_4-DB24C8), 123.5 (C_6H_3 or C_6H_4-axle), 126.5 (C_6H_3), 130.5 (C_6H_3), 130.8 (C_6H_4-axle), 131.5 (C_6H_3 or C_6H_4-axle), 138.2 (C_6H_3 or C_6H_4-axle), 145.5 (CH=CH), 147.4 (C_6H_4-DB24C8), 159.4 (C_6H_4-axle). IR (KBr disk, r.t.): ν 3231 (N-H), 3163 (N-H), 843 (P-F), 558 (P-F) cm^{-1}. Anal. Calc. for $C_{50}H_{69}BF_6NO_{11}P$: C, 59.12; H, 6.85; N, 1.38. Found: C, 59.15; H, 6.92; N, 1.44. FABMS: Calcd. for $C_{50}H_{69}BNO_{11}$: 871 Found: m/z = 871 ([M-PF$_6$]$^+$).

[{(C_6H_3-3,5-Me$_2$)CH$_2$NH$_2$CH$_2$C$_6$H$_4$-4-OCH$_2$CH$_2$CH=CHCH$_2$SiPh$_3$}(DB24C8)] (PF$_6$) (11) (mixture of cis and trans (1/9)): ^{1}H NMR (300 MHz, CDCl$_3$, r.t.): δ 2.14 (s, 6H, Me), 2.28–2.44 (4H, SiCH$_2$, OCH$_2$CH_2), 3.47 (s, 8H, CH$_2$-DB24C8), 3.59 (t, 2H, OCH$_2$-cis isomer, J = 7 Hz), 3.74 (t, 2H, OCH$_2$-trans isomer, J = 6 Hz), 3.76 (s, 8H, CH$_2$-DB24C8), 4.10 (s, 8H, CH$_2$-DB24C8), 4.29 (m, 2H, NCH$_2$-cis isomer), 4.41–4.45 (m, 2H, NCH$_2$), 4.55 (m, 2H, NCH$_2$-trans isomer), 5.36 (m, 2H, SiCH$_2$CH-trans isomer), 5.55–5.74 (m, 1H, OCH$_2$CH$_2$CH), 5.92 (m, 1H, SiCH$_2$CH-cis isomer), 6.66 (m, 2H, C$_6$H$_4$), 6.79–6.83 (7H, C$_6$H$_3$, C$_6$H$_4$-DB24C8), 6.86–6.91 (4H, C$_6$H$_4$-DB24C8), 7.22–7.26 (2H, C$_6$H$_4$), 7.30–7.42 (9H, SiPh$_3$), 7.45 (br, 2H, NH$_2$), 7.53 (m, 6H, SiPh$_3$). ^{13}C{^{1}H} NMR (100 MHz, CDCl$_3$, r.t.): δ 15.4 (SiCH$_2$-cis isomer), 19.5 (SiCH$_2$-trans isomer), 21.2 (Me), 27.2 (OCH$_2$CH$_2$-cis), 32.6 (OCH$_2$CH$_2$-trans), 52.1 (NCH$_2$), 52.4 (NCH$_2$), 67.2 (OCH$_2$-cis), 67.7 (OCH$_2$-trans), 68.2 (CH$_2$-DB24C8), 70.1 (CH$_2$-DB24C8), 70.6 (CH$_2$-DB24C8), 112.6 (C$_6$H$_4$-DB24C8), 114.4, 114.5, 121.6, 121.6 (C$_6$H$_4$-DB24C8), 123.5, 124.2, 126.0, 126.5, 127.7 (SiPh), 127.7, 127.8, 129.4, 129.5, 130.4, 130.7, 131.4, 134.5, 135.6 (SiPh), 138.1 (SiPh-ipso), 147.4 (C$_6$H$_4$-DB24C8-ipso), 159.2 (ipso-C$_6$H$_4$-cis), 159.4 (ipso-C$_6$H$_4$-trans). FABMS: Calcd. for $C_{63}H_{74}NO_9Si$: 1017 Found: m/z = 1017 ([M-PF$_6$]$^+$).

[{(C_6H_3-3,5-Me$_2$)CH$_2$NH$_2$CH$_2$C$_6$H$_4$-4-OCH$_2$CH$_2$CH=CHCH$_2$CH$_2$OC$_6$H$_4$-4-CH$_2$NH$_2$CH$_2$(C$_6$H$_3$-3,5-Me$_2$)}(DB24C8)$_2$](PF$_6$)$_2$ (23): ^{1}H NMR (300 MHz, CDCl$_3$, r.t.): δ 2.13 (s, 6H, Me), 2.51 (s, 2H, OCH$_2$CH_2), 3.43 (s, 8H, CH$_2$-DB24C8), 3.76 (d, 8H, CH$_2$-DB24C8, J = 3.6 Hz), 3.96 (t, 2H, OCH$_2$, J = 6.6 Hz), 4.10 (d, 8H, CH$_2$-DB24C8, J = 3.6 Hz), 4.38 (s, 2H, NCH$_2$), 4.50 (s, 2H, NCH$_2$), 5.67 (s, 1H, CH$_2$CH=), 6.77–6.90 (13H, C$_6$H$_4$-DB24C8, C$_6$H$_4$-axle, C$_6$H$_3$), 7.28 (d, 2H, C$_6$H$_4$-axle, J = 9.0 Hz), 7.45 (br, 2H, NH$_2$). ^{13}C{^{1}H} NMR Data of trans (100 MHz, CDCl$_3$, r.t.): δ 21.2 (Me), 32.4 (OCH$_2$CH$_2$), 52.2 (NCH$_2$), 52.4 (NCH$_2$), 67.5 (OCH$_2$CH$_2$), 68.2 (CH$_2$-DB24C8), 70.1 (CH$_2$-DB24C8), 70.6 (CH$_2$-DB24C8), 112.6 (C$_6$H$_4$-DB24C8), 114.5 (C$_6$H$_4$-DB24C8), 121.5 (C$_6$H$_4$-DB24C8), 123.5 (C$_6$H$_3$ or C$_6$H$_4$-axle), 126.5 (C$_6$H$_3$), 128.1 (CH$_2$CH=), 130.4 (C$_6$H$_3$), 130.9 (C$_6$H$_4$-axle), 131.5 (C$_6$H$_3$ or C$_6$H$_4$-axle), 138.1 (C$_6$H$_3$ or C$_6$H$_4$-axle), 147.4 (C$_6$H$_4$-DB24C8), 159.4 (C$_6$H$_4$-axle). ^{13}C{^{1}H} NMR Data of cis (100 MHz, CDCl$_3$, r.t.): δ 66.5 (OCH$_2$CH$_2$), 114.8 (C$_6$H$_4$-axle). IR (KBr disk, r.t.): ν 3163 (N-H), 3069 (N-H), 841 (P-F), 558 (P-F) cm^{-1}. FABMS: Calcd. for $C_{86}H_{112}N_2O_{18}$: 1462 Found: m/z = 1462 ([M-2(PF$_6$)]$^+$).

[{(C_6H_3-3,5-Me$_2$)CH$_2$NH$_2$CH$_2$C$_6$H$_4$-4-OCH$_2$CH$_2$CH$_2$CH$_2$CH$_2$CH$_2$OC$_6$H$_4$-4-CH$_2$NH$_2$CH$_2$(C$_6$H$_3$-3,5-Me$_2$)}(DB24C8)$_2$](PF$_6$)$_2$(24): ^{1}H NMR (300 MHz, CDCl$_3$, r.t.): δ 2.13 (s, 6H, Me), 2.21 (s, 2H, OCH$_2$CH$_2$CH_2), 2.51 (s, 2H, OCH$_2$CH_2), 3.44 (s, 8H, CH$_2$-DB24C8), 3.76 (d, 8H, CH$_2$-DB24C8, J = 3.6 Hz), 3.92 (t, 2H,

OCH$_2$, $J = 6.6$ Hz), 4.10 (d, 8H, CH$_2$-DB24C8, $J = 3.6$ Hz), 4.40 (s, 2H, NCH$_2$), 4.52 (s, 2H, NCH$_2$), 6.77–6.90 (13H, C$_6$H$_4$-DB24C8, C$_6$H$_4$-axle, C$_6$H$_3$), 7.28 (d, 2H, C$_6$H$_4$-axle, $J = 9.0$ Hz), 7.45 (br, 2H, NH$_2$).

Acknowledgments

We thank Dr. Takafumi Sassa, Dr. Daisuke Hashizume, and Dr. Tatsuo Wada for useful discussion. This work was supported by Grant-in-Aid for Scientific Research for Young Scientists from the Ministry of Education, Culture, Sports, Science and Technology, Japan (No. 19750044), a Global COE Program 'Education and Research Center for Emergence of New Molecular Chemistry,' and The Mazda Foundation.

References

1 Schill, G. (1971) *Catenanes, Rotaxanes, and Knots, Organic Chemistry*, vol. 22, Academic Press, New York.
2 Sauvage, J.-P. and Dietrich-Buchecker, C. (eds) (1999) *Molecular Catenanes, Rotaxanes and Knots*, Wiley-VCH, Weinheim.
3 (a) Amabilino, D.B. and Stoddart, J.F. (1995) Interlocked and intertwined structures and superstructures. *Chem. Rev.*, **95**, 2725–2828; (b) Harada, A. (1996) Preparation and structures of supramolecules between cyclodextrins and polymers. *Coord. Chem. Rev.*, **148**, 115–133; (c) Harada, A. (1997) Construction of supramolecular structures from cyclodextrins and polymers. *Carbohydr. Polym.*, **34**, 183–188; (d) Jäger, R. and Vögtle, F. (1997) A new synthetic strategy towards molecules with mechanical bonds: nonionic template synthesis of amide-linked catenanes and rotaxanes. *Angew Chem. Int. Ed. Engl.*, **36**, 930–944; (e) Nepogodiev, S.A. and Stoddart, J.F. (1998) Cyclodextrin-based catenanes and rotaxanes. *Chem. Rev.*, **98**, 1959–1976; (f) Chambron, J.-C. and Sauvage, J.-P. (1998) Functional rotaxanes: From controlled molecular motions to electron transfer between chemically nonconnected chromophores. **4**, 1362–1366; (g) Hubin, T.J. and Busch, D.H. (2000) Template routes to interlocked molecular structures and orderly molecular entanglements. *Coord. Chem. Rev.*, 200–202, 5–52; (h) Schalley, C.A., Beizai, K. and Vögtle, F. (2001) On the way to rotaxane-based molecular motors: studies in molecular mobility and topological chirality. *Acc. Chem. Res.*, **34**, 465–476; (i) Kim, K. (2002) Mechanically interlocked molecules incorporating cucurbituril and their supramolecular assemblies. *Chem. Soc. Rev.*, **31**, 96–107.
4 (a) Osakada, K., Sakano, T., Horie, M. and Suzaki, Y. (2006) Functionalized ferrocenes: Unique properties based on electronic communication between amino group of the ligand and Fe center. *Coord. Chem. Rev.*, **250**, 1012–1022; (b) Suzaki, Y., Taira, T., Osakada, K. and Horie, M. (2008) Rotaxanes and pseudorotaxanes with Fe-, Pd- and Pt-containing axles. Molecular motion in the solid state and aggregation in solution. *Dalton Trans.*, 4823–4833.
5 (a) Anelli, P.L., Spencer, N. and Stoddart, J.F. (1991) A molecular shuttle. *J. Am. Chem. Soc.*, **113**, 5131–5133; (b) Bissell, R.A., Córdova, E., Kaifer, A.E. and Stoddart, J.F. (1994) A chemically and electrochemically switchable molecular shuttle. *Nature*, **369**, 133–137.
6 Liu, Y., Flood, A.H., Bonvallet, P.A., Vignon, S.A., Northrop, B.H., Tseng, H.-R., Jeppesen, J.O., Huang, T.J., Brough, B., Baller, M., Magonov, S.,

Solares, S.D., Goddard, W.A., Ho, C.-M. and Stoddart, J.F. (2005) Linear artificial molecular muscles. *J. Am. Chem. Soc.*, **127**, 9745–9759.

7 Collier, C.P., Wong, E.W., Belohradský, M., Raymo, F.M., Stoddart, J.F., Kuekes, P.J., Williams, R.S. and Heath, J.R. (1999) Electronically configurable molecular-based logic gates. *Science*, **285**, 391–394.

8 Berná, J., Leigh, D.A., Lubomska, M., Mendoza, S.M., Pérez, E.M., Rudolf, P., Teobaldi, G. and Zerbetto, F. (2005) Macroscopic transport by synthetic molecular machines. *Nature Mater.*, **4**, 704–710.

9 Marois, J.-S. and Morin, J.-F. (2008) Synthesis and surface self-assembly of [3] rotaxane-porphyrin conjugates: toward the development of a supramolecular surface tweezer for C_{60}. *Langmuir*, **24**, 10865–10873.

10 Christinat, T., Scopelliti, R. and Severin, K. (2008) Boron-based rotaxanes by multicomponent self-assembly. *Chem. Commun.*, 3660–3662.

11 (a) Chichak, K.S., Cantrill, S.J., Pease, A.R., Chiu, S.-H., Cave, G.W.V., Atwood, J.L. and Stoddart, J.F. (2004) Molecular borromean rings. *Science*, **304**, 1308–1312; (b) Pentecost, C.D., Chichak, K.S., Peters, A.J., Cave, G.W.V., Cantrill, S.J. and Stoddart, J.F. (2007) A molecular solomon link. *Angew. Chem. Int. Ed.*, **46**, 218–222.

12 Northrop, B.H., Khan, S.J. and Stoddart, J.F. (2006) Kinetically controlled self-assembly of pseudorotaxanes on crystallization. *Org. Lett.*, **8**, 2159–2162.

13 Yoon, I., Miljanić, O.Š., Benítez, D., Khan, S.I. and Stoddart, J.F. (2008) An interdigitated functionally rigid [2] rotaxane. *Chem. Commun.*, 4561–4563.

14 (a) Loeb, S.J. and Wisner, J.A. (1998) 1,2-Bis(4,4'-dipyridinium)ethane: a versatile dication for the formation of [2]rotaxanes with dibenzo-24-crown-8 ether. *Chem. Commun.*, 2757–2758; (b) Davidson, G.J.E., Loeb, S.J., Parekh, N.A. and Wisner, J.A. (2001) Zwitterionic [2] rotaxanes utilising anionic transition metal stoppers. *J. Chem. Soc., Dalton Trans.*, 3135–3136; (c) Davidson, G.J.E. and Loeb, S.J. (2003) Iron(II) complexes utilising terpyridine containing [2] rotaxanes as ligands. *Dalton Trans.*, 4319–4323; (d) Loeb, S.J. (2007) Rotaxanes as ligands: from molecules to materials. *Chem. Soc. Rev.*, **36**, 226.

15 Li, Q., Zhang, W., Miljanić, O.Š., Sue, C.-H., Zhao, Y.-L., Liu, L., Knobler, C.B., Stoddart, J.F. and Yaghi, O.M. (2009) Docking in metal-organic frameworks. *Science*, **325**, 855–859.

16 (a) Ashton, P.R., Campbell, P.J., Chrystal, E.J.T., Glink, P.T., Menzer, S., Philp, D., Spencer, N., Stoddart, J.F., Tasker, P.A. and Williams, D.J. (1995) Dialkylammonium ion/crown ether complexes: The forerunners of a new family of interlocked molecules. *Angew Chem. Int. Ed. Engl.*, **34**, 1865–1869; (b) Ashton, P.R., Chrystal, E.J.T., Glink, P.T., Menzer, S., Schiavo, C., Spencer, N., Stoddart, J.F., Tasker, P.A., White, A.J.P. and Williams, D.J. (1996) Pseudorotaxanes formed between secondary dialkylammonium salts and crown ethers. *Chem. Eur. J.*, **2**, 709–728.

17 Ashton, P.R., Fyfe, M.C.T., Hickingbottom, S.K., Stoddart, J.F., White, A.J.P. and Williams, D.J. (1998) Hammett correlations 'beyond the molecule'. *J. Chem. Soc., Perkin Trans. 2*, 2117–2128.

18 Osakada, K., Abe, T., Chihara, E., Murata, S., Suzaki, Y. and Horie, M. (2008) Rotaxanes and pseudorotaxanes with ferrocenyl groups. Proceedings of the Yamada Conference 2008 'Topological Molecules', September 1–4, 2008, Hyogo, Japan, Yamada Science Foundation (ed. A. Harada).

19 (a) Horie, M., Suzaki, Y. and Osakada, K. (2004) Formation of pseudorotaxane induced by electrochemical oxidation of ferrocene-containing axis molecule in the presence of crown ether. *J. Am. Chem. Soc.*, **126**, 3684–3685; (b) Horie, M., Suzaki, Y. and Osakada, K. (2005) Chemical and electrochemical formation of pseudorotaxanes composed of alkyl (ferrocenylmethyl)ammonium and dibenzo[24]crown-8. *Inorg. Chem.*, **44**, 5844–5853.

20 (a) Suzaki, Y., Chihara, E., Takagi, A. and Osakada, K. (2009) Rotaxanes of a macrocyclic ferrocenophane with dialkylammonium axle components. *Dalton Trans.*, 9881–9891; (b) Suzaki, Y., Takagi, A., Chihara, E. and Osakada, K. (2010) Synthesis and dethreading reaction of a rotaxane-like complex of an octaoxa[22]ferrocenophane with dialkylammonium. *Supramol. Chem.* **23**, 2–8.

21 Suzaki, Y., Takagi, A. and Osakada, K. (2010) Synthesis, structure and properties of the macrocyclic ferrocenophanes with cyclopentadienyl ligands tethered by oligo(ethylene glycol) chain. *J. Organomet. Chem.*, **695**, 2515–2518.

22 Horie, M., Sassa, T., Hashizume, D., Suzaki, Y., Osakada, K. and Wada, T. (2007) A crystalline cupramolecular switch: controlling the optical anisotropy via the collective dynamic motion of molecules. *Angew. Chem. Int. Ed.*, **46**, 4983–4986.

23 (a) Suzaki, Y. and Osakada, K. (2006) End-capping of pseudo[2]rotaxane composed of alkyl(ferrocenylmethyl)ammonium and dibenzo[24]crown-8 via cross metathesis reactions. *Chem. Lett.*, **35**, 374–375; (b) Suzaki, Y. and Osakada, K. (2007) Ferrocene-containing [2]- and [3]rotaxanes. Preparation via end-capping cross-metathesis reaction and their electrochmical properties. *Dalton Trans.*, 2376–2383.

24 Murata, S. (2009) Synthesis of [2]-, [3]- and polyrotaxanes composed of dibenzo[24]crown-8-ether and acrylamide derivatives, Master's thesis, Tokyo Institute of Technology.

25 Ashton, P.R., Baxter, I., Fyfe, M.C.T., Raymo, F.M., Spencer, N., Stoddart, J.F., White, A.J.P. and Williams, D.J. (1998) Rotaxane or Pseudorotaxane? *J. Am. Chem. Soc.*, **120**, 2297–2307.

26 Kawasaki, H., Kihara, N. and Takata, T. (1999) High yielding and practical synthesis of rotaxanes by acetylation end-capping catalyzed by tributylphosphine. *Chem. Lett.*, **28**, 1015–1016.

27 Furusho, Y., Sasabe, H., Natsui, D., Murakawa, K., Takata, T. and Harada, T. (2004) Synthesis of [2]- and [3]rotaxanes by an end-capping approach utilizing urethane formation. *Bull. Chem. Soc. Jpn.*, **77**, 179–185.

28 Rowan, S.J., Cantrill, S.J. and Stoddart, J.F. (1999) Triphenylphosphonium-stoppered [2]rotaxanes. *Org. Lett.*, **1**, 129–132.

29 Sasabe, H., Kihara, N., Mizuno, K., Ogawa, A. and Takata, T. (2005) Efficient synthesis of [2]- and higher order rotaxanes via the transition metal-catalyzed hydrosilylation of alkyne. *Tetrahedron Lett.*, **46**, 3851–3853.

30 Tokunaga, Y., Kawai, N. and Shimomura, Y., Using ruthenium-catalysed propargylic substitutions for the efficient syntheses of rotaxanes. *Tetrahedron Lett.*, **48**, 4995–4998.

31 Giguère, J.-B., Thibeault, D., Cronier, F., Marois, J.-S., Auger, M. and Morin, J.-F. (2009) Synthesis of [2]- and [3]rotaxanes through Sonogashira Coupling. *Tetrahedron Lett.*, **50**, 5497–5500.

32 Cao, J., Fyfe, M.C.T., Stoddart, J.F., Cousins, G.R.L. and Glink, P.T. (2000) Molecular shuttles by the protecting group approach. *J. Org. Chem.*, **65**, 1937–1946.

33 Sasabe, H., Kihara, N., Furusho, Y., Mizuno, K., Ogawa, A. and Takata, T. (2004) End-capping of a pseudorotaxane via Diels-Alder reaction for the construction of C_{60}-terminated [2]rotaxanes. *Org. Lett.*, **6**, 3957–3960.

34 Trnka, T.M. and Grubbs, R.H. (2001) The development of $L_2X_2Ru=CHR$ olefin metathesis catalysts: An organometallic success story. *Acc. Chem. Res.*, **34**, 18–29.

35 Yui, N. (2002) *Supramolecular Design for Biological Applications*, CRC Press, Boca Raton, FL.

36 Chiu, S.-H., Rowan, S.J., Cantrill, S.J., Glink, P.T., Garrell, R.L. and Stoddart, J.F. (2000) A rotaxane-like complex with controlled-release characteristics. *Org. Lett.*, **2**, 3631–3634.

37 Suzaki, Y., Takagi, A. and Osakada, K. (2010) Disaggregation reaction of [2]Pseudorotaxanes Composed of Dibenzo[24]crown-8 and dialkylammonium having isopropyl end groups. *Chem. Lett.*, **39**, 510–512.

38 Hannam, J.S., Kidd, T.J., Leigh, D.A. and Wilson, A.J. (2003) 'Magic rod' rotaxanes:

The hydrogen bond-directed synthesis of molecular shuttles under thermodynamic control. *Org. Lett.*, **5**, 1907–1910.

39 Suzaki, Y., Murata, S. and Osakada, K. (2009) Ferrocene-containing side chain polyrotaxanes obtained by radical copolymerization of styrenes with acrylamide with a [2]rotaxane structure. *Chem. Lett.*, **38**, 356–357.

40 Takata, T., Kawasaki, H., Asai, S., Kihara, N. and Furusho, Y. (1999) Radically polymerizable pseudorotaxane monomers: Versatile building units for side chain polyrotaxane synthesis. *Chem. Lett.*, **28**, 111–112.

41 (a) Gibson, H.W., Liu, S., Lecavalier, P., Wu, C. and Shen, Y.X. (1995) Synthesis and preliminary characterization of some polyester rotaxanes. *J. Am. Chem. Soc.*, **117**, 852–974; (b) Yamaguchi, I., Osakada, K. and Yamamoto, T. (1996) Polyrotaxane containing a blocking group in every structural unit of the polymer chain. Direct synthesis of poly (alkylenebenzimidazole) rotaxane from Ru complex-catalyzed reaction of 1,12-dodecanediol and 3,3′-diaminobenzidine in the presence of cyclodextrin. *J. Am. Chem. Soc.*, **118**, 1811–1812; (c) Yamaguchi, I., Osakada, K. and Yamamoto, T. (1997) Introduction of a long alkyl side chain to poly(benzimidazole)s. *N*-alkylation of the imidazole ring and synthesis of novel side chain polyrotaxanes. *Macromolecules*, **30**, 4288; (d) Tamura, M. and Ueno, A. (2000) Energy transfer and guest responsive fluorescence spectra of polyrotaxane consisting of α-cyclodextrins bearing naphthyl moieties. *Bull. Chem. Soc. Jpn.*, **73**, 147–154.

15
Polyrotaxane Network as a Topologically Cross-Linked Polymer: Synthesis and Properties

Toshikazu Takata, Takayuki Arai, Yasuhiro Kohsaka, Masahiro Shioya, and Yasuhito Koyama

15.1
Introduction

Supramolecules characterized by their molecular topology, such as rotaxane, catenane, and their polymers, have attracted great interest experimentally and theoretically [1]. A typical example is a polyrotaxane network bearing rotaxane structures on the cross-linking points. These structures are characterized in terms of both their uniqueness and their specific properties based on the movable cross-links [2]. The movable cross-link points in network polymer enable of tensions due to external stimuli or stress to be equalized, affording high swelling ability with several solvents, high elasticity, and high stress-releaving ability. The strategic synthetic methods of such polyrotaxane networks are divided into two approaches [3, 4]. The first approach is through cross-linking by the direct linking of polyrotaxane at the wheels (Figure 15.1, (I)), and the second involves the precise rotaxanative cross-linking of macrocycle-containing polymer by the threading of two or more macrocycles (Figure 15.1, (II)) [2e–f]. While the former has merit of certainty of cross-linking, the latter requires no pre-synthesis of polyrotaxane. Recently we have realized the synthesis and properties of several polyrotaxane networks based on these two approaches, and these are discussed below.

15.2
Linking of Wheels of Main-Chain-Type Polyrotaxane – Structurally Defined Polyrotaxane Network

To obtain a structurally defined polyrotaxane network by the reduction of structural diversity derived from the linking of multifunctional wheel components of main-chain-type polyrotaxane, a synthesis of main-chain-type polyrotaxane possessing monofunctional α-cyclodextrin (modified α-CD) was planned (Figure 15.1 (I)) [5]. Prior to the preparation of the networked polyrotaxane, the preparation of a main-chain-type polyrotaxane bearing permethylated α-cyclodextrin (PMα-CD) as

Supramolecular Polymer Chemistry, First Edition. Edited by Akira Harada.
© 2012 Wiley-VCH Verlag GmbH & Co. KGaA. Published 2012 by Wiley-VCH Verlag GmbH & Co. KGaA.

Method (I): Linking of Wheels of Main Chain-Type Polyrotaxane

Method (II): Linking of Macrocyclic Unit of Polymacrocycle with Axle Molecule

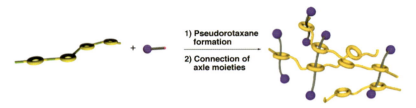

Figure 15.1 Typical synthetic strategies of polyrotaxane networks.

a non-reactive wheel component was carried out, because no successful synthesis of such a type of polyrotaxane with modified α-CD has been reported (Schemes 15.1 and 15.2). No amine-terminated polymer as an axle was required but OH-terminated polymer could be used in this synthetic process, since the end-capping of the axle terminal of pseudopolyrotaxane is not disturbed due to the absence of OH groups on the wheel component.

Scheme 15.1 Synthesis of pseudopolyrotaxane consisting of PMα-CD.

The synthesis of pseudopolyrotaxane as a precursor of polyrotaxane was carried out using PMα-CD and polytetrahydrofuran (PTHF, $M_n = 1400$), according to the Harada method (Figure 15.2) [6, 7]. Initial mixing of the two materials by sonication in water afforded 100% yield of pseudopolyrotaxane, which had 107% coverage ratio (θ represents the extent of the area of axle component covered with wheel component).

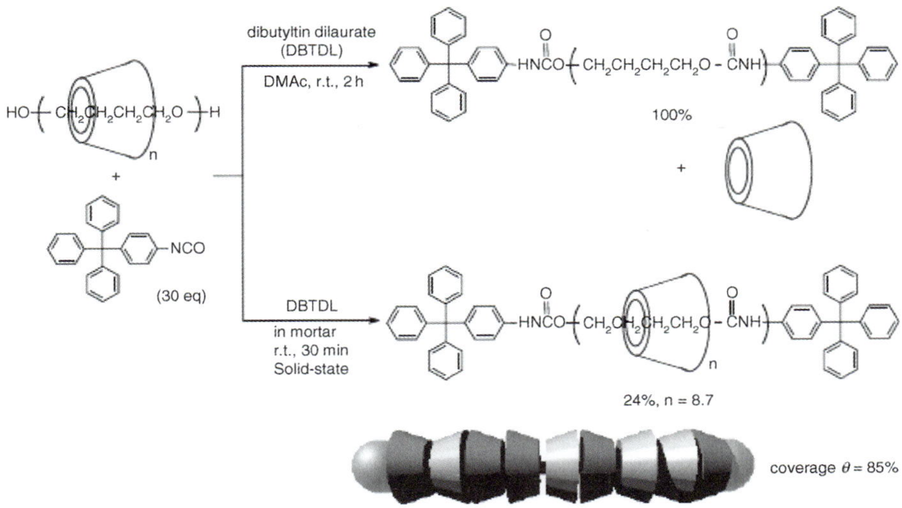

15.2 Linking of Wheels of Main-Chain-Type Polyrotaxane – Structurally Defined Polyrotaxane Network

Scheme 15.2 End-capping reaction of pseudopolyrotaxane in solution and in the solid state.

The fixing of the pseudopolyrotaxane to the polyrotaxane by end-capping was quite difficult to achieve, in contrast to that in the case of unmodified α-CD-based polyrotaxane. The acid-catalyzed reaction of pseudopolyrotaxane with 30 equivalents of a bulky isocyanate (tritylphenyl isocyanate) gave no polyrotaxane in any organic solvent, but only dumbbell-shaped polymer formed by the complete release of the wheel component followed by the terminal acylation (Scheme 15.2, top).

Successful synthesis of the designed polyrotaxane was accomplished by the solvent-free reaction (solid-state synthesis) achieved by grinding a mixture of the pseudopolyrotaxane, a bulky isocyanate, and an acid catalyst in a mortar without solvent at room temperature. A mixture of pseudopolyrotaxane, 30 equivalents of tritylphenyl isocyanate, and a catalytic amount of dibutyltin dilaurate (DBTDL) was thoroughly ground in an agate mortar at room temperature. PMαCD-containing polyrotaxane was isolated in 24% yield as the Et_2O- and MeOH-insoluble part from the mixture after 30 min grinding. The ^1H NMR spectrum of the polyrotaxane in $CDCl_3$ suggests that the two terminal hydroxyl groups were completely converted to urethane groups and that the polyrotaxane had 8.7 wheels per axle on average ($\theta = 85\%$). On extending the grinding time to 90 min, neither θ nor the yield increased.

Figure 15.2 Synthesis of polyrotaxane network by linking of axle terminal of polypseudorotaxane.

On the basis of the above results using permethylated α-CD wheel, synthesis of polyrotaxane with monofunctional α-CD wheel and its derivation to polyrotaxane network were studied. Monoacetoxy permethylated α-CD was prepared according to the reported method (with slight modification [8]) from α-CD via the initial monotritylation, permethylation, deprotection, and acetylation. Pseudopolyrotaxane was obtained similarly from PTHF ($M_n = 2000$) and the modified α-CD (Scheme 15.3). The solid-state end-capping reaction of the pseudopolyrotaxane with tritylphenyl isocyanate proceeded smoothly in the presence of DBTDL to afford the corresponding polyrotaxane in 43% yield. It turned out that the coverage ratio of the resulting polyrotaxane was 56%, suggesting that the polyrotaxane had 8.2 wheels per axle on average, on the basis of ^{1}H NMR analysis.

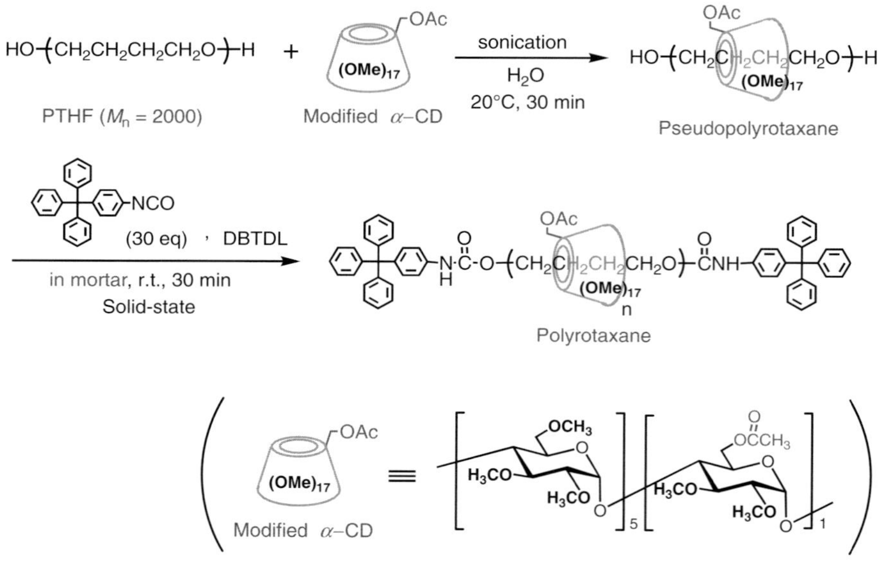

Scheme 15.3 Synthesis of monofunctional α-CD-containing main-chain-type polyrotaxane.

Scheme 15.4 features the synthetic scheme of the polyrotaxane network utilizing the structurally defined main-chain-type polyrotaxane prepared above. Polyrotaxane with acetoxy-functionalized PMα-CD wheel was deprotected by alkaline methanolysis to quantitatively afford the main-chain-type polyrotaxane having one hydroxyl functionality per wheel, which it was hoped would give the structurally defined polyrotaxane network by controlled linking of two wheels. The resulting polyrotaxane was treated with diphenylmethane diisocyanate ([NCO]/[OH] = 0.5) in the presence of DBTDL as a catalyst in CHCl$_3$. Gelation immediately took place to quantitatively afford an insoluble gelled material; that is, the polymer cross-linked via the interlocked bonding. The gel showed remarkable swelling properties despite the fairly

15.2 Linking of Wheels of Main-Chain-Type Polyrotaxane – Structurally Defined Polyrotaxane Network | 335

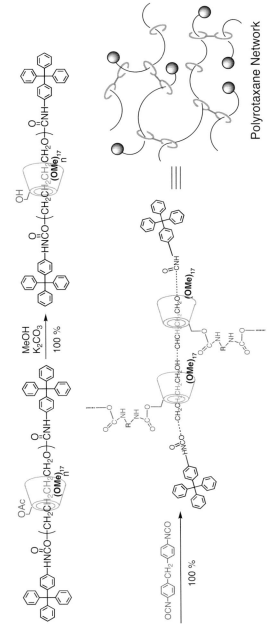

Scheme 15.4 Cross-linking reaction of polyrotaxane at the wheel components to furnish a well-defined polyrotaxane network.

high degree of cross-linking: the degree of gelation produced by typical organic solvents was 1.80 g/g for DMF and 19.3 g/g for $CHCl_3$.

15.3
Linking of Macrocyclic Units of Polymacrocycle with Axle Unit to Directly Yield a Polyrotaxane Network

In the course of development of a novel cross-linking method, we have recently been intrigued by the potential usefulness of macrocyclic moieties in the polymer as the cross-linking points, according to the second strategic approach to the polyrotaxane network (Figure 15.1 (II)). If the sparse macrocyclic moieties in the polymer chain were successfully available as cross-link points, an attachment of topological characteristics to the desired polymer might be possible. According to this idea, we examined the practical synthetic approach of polyrotaxane networks possessing the crown ether or cyclodextrin moiety as the macrocyclic parts placed sparsely in the polymer chain, which actually takes part into the cross-linking, along with their properties.

15.3.1
Polyrotaxane Networks Having Crown Ethers as the Wheel at the Cross-link Points (I)

Firstly, the crown ether moiety was chosen as a wheel component at the cross-link point based on our rotaxane chemistry utilizing a *sec*-ammonium salt as the axle component and crown ether wheel. Figure 15.2 illustrates the strategy to achieve the polyrotaxane network via the intermolecular connection of axle terminals of intermediary polypseudorotaxane [9]. Treatment of the polymacrocycle with an axle component bearing a bulky end-capping group at the terminal end afforded the polypseudorotaxane, in which the axle components are threaded to the macrocyclic cavities in the main chain. Subsequent connection of the axle terminals of the polypseudorotaxane with a bifunctional linear linker gave the corresponding polyrotaxane network.

The actual synthesis of the polypseudorotaxane was carried out using a main-chain-type poly(crown ether) and *sec*-ammonium hexafluorophosphate with a bulky end-capping group and hydroxyl group at the both termini (Scheme 15.5). After considerable experimental work, we found that controlled feed of the axle component enabled us to control the rotaxanated ratio in the polypseudorotaxane. In a typical synthesis of polyrotaxane network, the polypseudorotaxane prepared *in situ* (79% of the rotaxanated ratio) underwent the reaction with 4,4'-methylenebis(phenyl isocyanate) in the presence of DBTDL at 25 °C for 16 h in $CHCl_3$ to afford the corresponding polyrotaxane network. The gel obtained was washed three times with $CHCl_3$ and MeOH to give the polyrotaxane network as a colorless solid in 96% yield. The swelling behavior was investigated in several typical organic solvents for 3 d at 30 °C. The cross-linked polymer was readily swollen in DMSO, the degree of swelling reaching 9800% to give a transparent gel, while that in MeOH was 420%. The high degree of

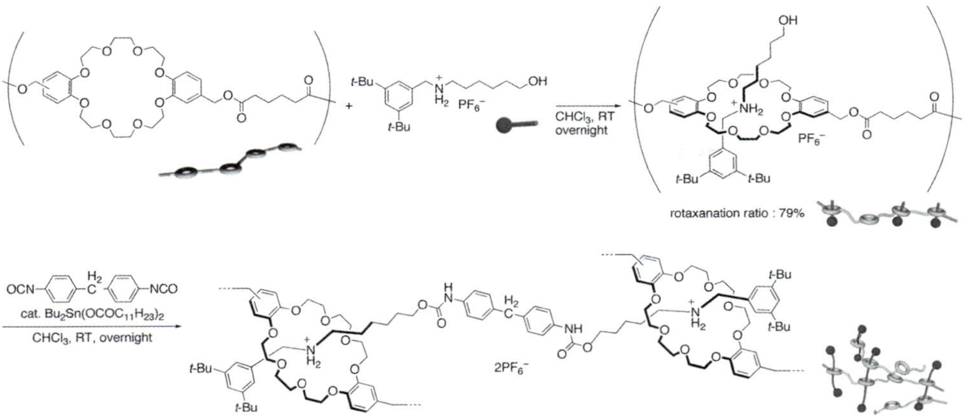

15.3 Linking of Macrocyclic Units of Polymacrocycle with Axle Unit to Directly Yield

Scheme 15.5 Synthesis of polyrotaxane network by linking the axle terminal on the polypseudorotaxane.

swelling in DMSO could be attributed to both the ionic structure and the low degree of cross-linking. Thus, this synthetic method using polypseudorotaxane afforded the polyrotaxane network simply and in high yield.

15.3.2
Polyrotaxane Network Having Crown Ethers as the Wheel at the Cross-link Points (II)

Following the development of a more effective cross-linking protocol, we became intrigued by the synthesis of polyrotaxane network exploiting dynamic covalent chemistry (Figure 15.3) [10]. This protocol was characterized by thermodynamically reversible cross-linking, affording recyclable materials [11]. Since the reaction that occurred during the cross-linking was only that of the axle component in this system, no main-chain polymer apparently takes part in the cross-linking reaction, that is, no main-chain polymer is injured after the cross-linking. From another viewpoint, there is no apparent chemical bond formation or breaking during the cross-linking, because both the wheel and axle components do not change their structures before and after the cross-linking but only change their position, that is, the topology. Thus, this cross-linking protocol is unusual enough to be noteworthy from both purely scientific and a practical points of view.

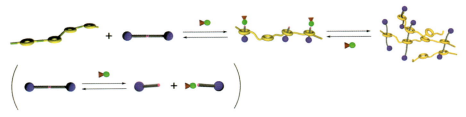

Figure 15.3 Synthetic strategy of polyrotaxane network using dynamic covalent chemistry.

338 | *15 Polyrotaxane Network as a Topologically Cross-Linked Polymer: Synthesis and Properties*

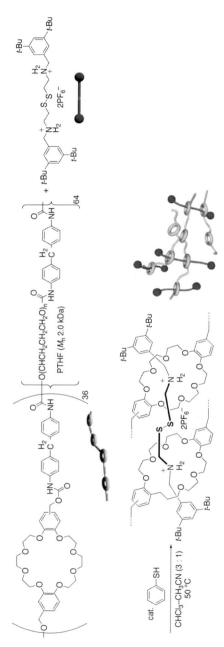

Scheme 15.6 Synthesis of polyrotaxane network using reversible cleavage of disulfide linkage.

15.3 Linking of Macrocyclic Units of Polymacrocycle with Axle Unit to Directly Yield

The formation of a polyrotaxane network using a thiol–disulfide interchange reaction, which is one of the important reactions in dynamic covalent chemistry, is illustrated in Scheme 15.6 [12]. Treatment of a bifunctional *sec*-ammonium salt on the axle component bearing a central disulfide linkage and two bulky end-caps with poly(crown ether) as the main-chain polymer afforded the polyrotaxane network as a thermally equilibrated mixture in a less polar solvent in the presence of a catalytic amount of benzenethiol. The reaction mixture gradually changed to a transparent insoluble material after about 3 d, because the crown ether units at the cross-link points are placed far from each other (the calculated mean distance between the crown ether moieties is about 50 nm.). The resulting cross-linked polymers were repeatedly washed with $CHCl_3$ and dried under reduced pressure at room temperature overnight to yield a translucent hard polymer in a quantitative yield.

The chemical recyclability of the resulting polyrotaxane network was demonstrated as shown in Scheme 15.7. Exposure of the polyrotaxane network to a small amount of an alkane thiol having a half structure of the disulfide cross-linker in DMF at 60 °C enabled the smooth decross-linking to afford a homogeneous mixture, which was subjected to precipitation with MeOH to give the main-chain polymer as a white solid in quantitative yield. After filtering off the precipitate, the filtrate was precipitated with H_2O to give a mixture of dumbbell-shaped ammonium salt and the thiol catalyst in high yield, strongly suggesting the effective recyclability of the polyrotaxane network based on the reversible cleavage of disulfide linkage.

15.3.3
Polyrotaxane Network Having Cyclodextrins as Cross-link Points: Effective Use of Oligocyclodextrin

According to the concept of exploiting macrocyclic cavities as cross-link points, oligocyclodextrin instead of poly(crown ether) was utilized for the synthesis of polyrotaxane network. Figure 15.4 shows a new cross-linking protocol using oligocyclodextrin and macromonomer having a bulky end group. Treatment of the oligocyclodextrin with the macromonomer as the axle component can afford a quasi-stable polymer network formation, that is, a pseudopolyrotaxane network, by the threading of two or more cyclodextrin moieties. Subsequent radical polymerization of the axle terminals of the network readily gave the polyrotaxane network [13].

The oligocyclodextrin was prepared by the polycondensation of α-cyclodextrin with poly(propylene glycol)-tolylene-2,4-diisocyanate (M_n 1000) in DMF in the presence of DBTDL. Characterization of the polymer obtained was done after peracetylation of the wheel components; its molecular weight was $M_n = 7000$, while the molecular weight distribution (M_w/M_n) was 2.5. Meanwhile, the macromonomer as the axle component was prepared by the addition of poly(ethylene glycol)monomethacrylate (M_n 360) to 3,5-dimethylphenylisocyanate in CH_2Cl_2 in the presence of DBTDL. A mixture of the oligocyclodextrin and the macromonomer in 0.1 M NaOH was subjected to sonication at room temperature for 30 min to give a gel, indicating the formation of polypseudorotaxane, originating from the pseudo-cross-linking reaction based on multiple

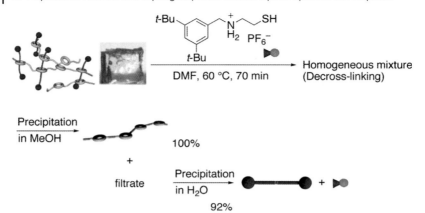

Scheme 15.7 Chemical recycle of polyrotaxane network based on dynamic covalent chemistry.

Figure 15.4 Synthetic strategy of polyrotaxane network exploiting the initial threading complexation and subsequent radical polymerization at the axle terminal.

penetrations of the macromonomer to cyclodextrin moieties (Scheme 15.8). The radical polymerization of the vinylic moieties of the resulting inclusion complexes in the presence of 2-hydroxy-4'-(2-hydroxyethoxy)-2- methylpropiophenone as a photoinitiator proceeded smoothly under UV irradiation at room temperature for 3 min to give the corresponding polyrotaxane network.

Through the radical polymerization behavior of the polypseudorotaxane, we envisioned that the use of the inclusion complex as a supramolecular cross-linker in the presence of vinyl comonomers could lead to the corresponding polyrotaxane network. The supramolecular cross-linker features (i) no requirement for pre-synthesis of polyrotaxane for the construction of polyrotaxane network and (ii) the structural versatility of vinyl comonomers to give the corresponding polyrotaxane network with unique properties originating from the vinyl comonomer structure and movable cross-linking points. The copolymerization of the polypseudorotaxane with vinyl comonomers such as N,N-dimethylacrylamide was also performed under a similar protocol to give the corresponding copolymer polyrotaxane network.

The swelling behaviors of polyrotaxane networks are summarized in Figure 15.1. While the cross-linked polymer (I) was well swollen in $CHCl_3$, the degree of swelling reaching 700%, that in H_2O or MeOH was about 200%, probably depending on the affinity of each component for organic solvents. In contrast to the polyrotaxane network (I) without any comonomer unit, the degree of swelling in H_2O of the

15.3 Linking of Macrocyclic Units of Polymacrocycle with Axle Unit to Directly Yield | 341

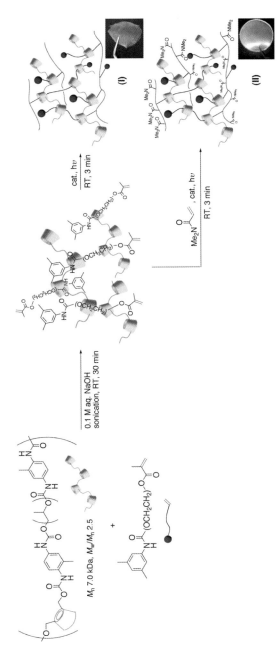

Scheme 15.8 Synthesis of polyrotaxane networks according to the synthetic protocol of Figure 15.4.

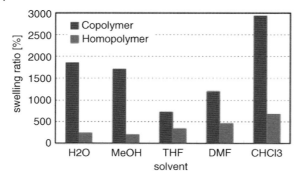

Figure 15.5 Swelling ratio of polyrotaxane networks in typical organic solvents: swelling ratio (%) = $(W_{swell} - W_{dry})/W_{dry} \times 100$.

polyrotaxane network (II) reached 1800%, depending on the properties of the polyacrylamide moiety. The results suggest that the use of various comonomers can affect the versatile hybridized polymers, showing that their properties are based on not only the comonomer unit but also on the rotaxane cross-link point.

15.4
Linking of Wheels of Polyrotaxane Cross-linker to Afford Polyrotaxane Network: Design of the Cross-linker

For a simpler and easier construction of polyrotaxane network, we designed a polyrotaxane cross-linker. Figure 15.5 shows the concept of the polyrotaxane cross-linker having a controlled movable range of the wheel components. The simplest structure of polyrotaxane cross-linker is [3]rotaxane, capable of showing a topologically cross-linking nature. The chemical structure of the cross-linker directly affects the properties of the corresponding polyrotaxane network. On the basis of these ideas, we envisioned that the polyrotaxane network consisting of [3]polyrotaxane structure providing movable cross-linking points could have superior properties, such as higher swelling, elasticity, and stress-relieving ability, to a structure including cross-link points with limited mobility (Figure 15.6). Thus, novel cross-linkers, [3]polyrotaxanes bearing two ditopic wheel moieties with well-defined topological characteristics, were designed and used for the synthesis of polyrotaxane networks with controlled properties [14].

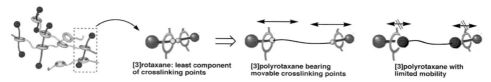

Figure 15.6 Design concept of polyrotaxane cross-linker with and without mobility of the wheel components.

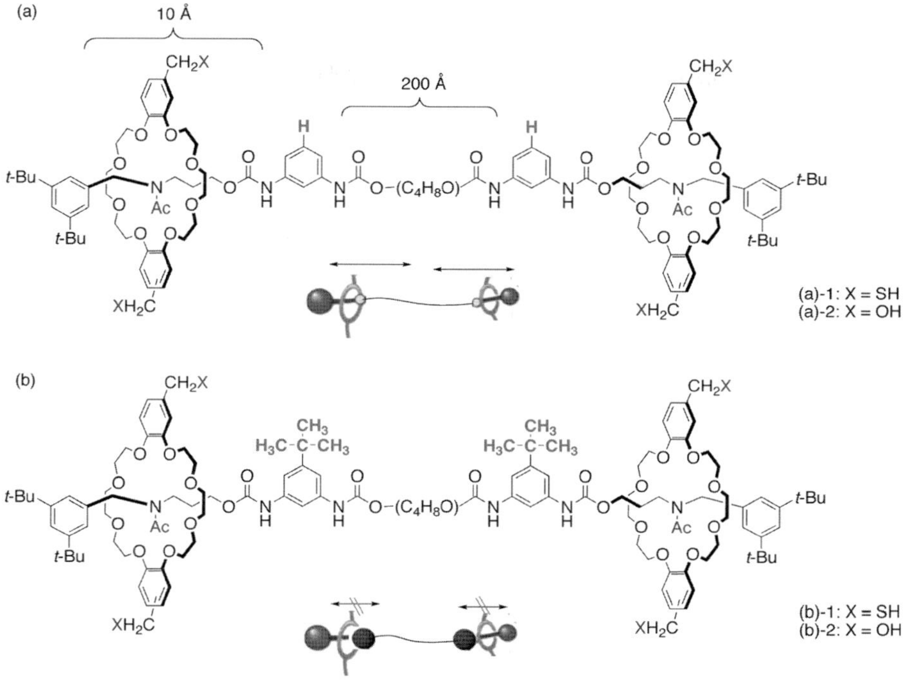

15.4 Linking of Wheels of Polyrotaxane Cross-linker to Afford Polyrotaxane Network

Figure 15.7 Structures of [3]polyrotaxane cross-linkers: (a) [3]polyrotaxane having wheels with high mobility and (b) with limited mobility.

Figure 15.7 features the detailed structures of the two polyrotaxane cross-linkers. Structure (a) consists of two wheel components each of which has ditopic hydroxyl groups as the cross-link points and polymeric dumbbell-shaped neutral axle component on which the wheel components can move freely. The movement distance of the wheel components was calculated as more than 200 Å. On the other hand, structure (b) contains two *tert*-butyl groups as the steric barriers to prevent the free movement of the wheel components in addition to the framework of structure (a). The wheel components of structure (b), therefore, have much less mobility than those of structure (a).

The synthesis of polyrotaxane networks exploiting the above-mentioned [3]polyrotaxane cross-linkers (a)-2 and (b)-2 (Figure 15.7) was carried out by the polyaddition between them and PTHF-*m*-phenylene diisocyanate terminated in the presence of DBTDL in CHCl$_3$ at room temperature for 1 d (Scheme 15.9). The cross-linking reactions smoothly proceeded to give the gelled products as the corresponding polyrotaxane networks. The properties of the gels were compared to examine the effect of the wheel mobility. The polyrotaxane networks showed almost similar T_gs: T_g1. Values for the PTHF chain were −5.0 °C {(a)-2} and −3.4 °C {(b)-2}, while T_g2 due to the urethane chain was 56 °C {(a)-2} and 50 °C {(b)-2}. While the degree of swelling in CHCl$_3$ of polyrotaxane network derived from (a)-2 reached 2500%, that of

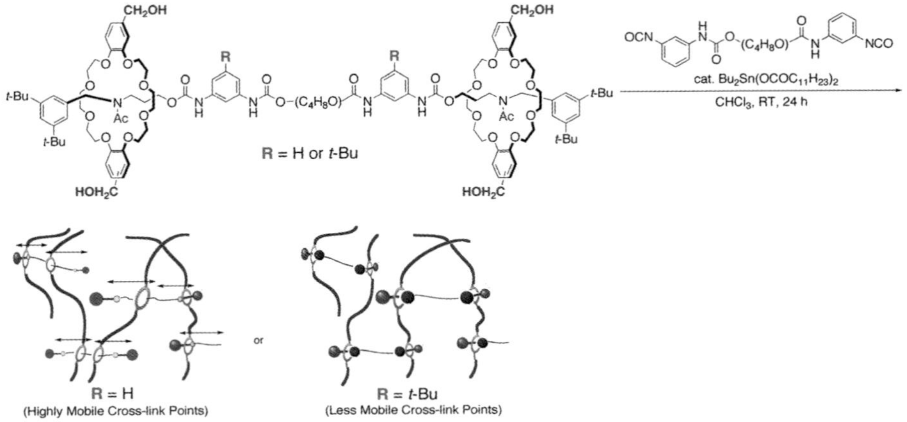

Scheme 15.9 Synthesis of polyrotaxane networks with and without mobility of the cross-link point.

Table 15.1 Effects of movable range of wheel components of [3]polyrotaxanes on swelling behavior with organic solvents: swelling ratio (%) = $(W_{swell} - W_{dry})/W_{dry} \times 100$.

Swelling ratio	DMF	DMSO	CHCl$_3$	CH$_3$CN	MeOH
Structure (a)[a]	1400	550	2500	130	140
Structure (b)[a]	530	260	1100	140	140

a) See Figure 15.7.

(b)-2 was 1100% (Table 15.1). Mechanical properties in the dry state were as follows: the polyrotaxane network from (a)-2 displayed 800% elongation, whereas that of structure (b)-2 was 400%, while the thermal properties of the two polymers were similar. These results strongly supported the view that the difference in mobility of wheel component of the [3]polyrotaxane cross-linkers was due to the difference in their chemical, physical, and mechanical properties. The shapes of the stress-strain curves were very similar to each other, except in the high-extension region and were consistent with the network structure. The stress-strain curve of polyrotaxane network derived from (a)-2 at the high extension region seems to indicate that the degree of movable range of cross-link points is directly related to the stressrelieving ability.

15.5
Conclusion

The present review has demonstrated some new cross-linking methods exploiting poly(oligo)macrocycles based on crown ether or cyclodextrin which have recently been developed by Takata *et al.* It has also revealed that the macrocyclic cavities sparsely placed in the polymer chain take part into the cross-linking to afford an attachment of topological characteristics to a desired polymer. In addition, the design of the chemical structure of the cross-link points on the polyrotaxane network enables control of topological characteristics such as swelling behavior, elasticity, stress relieving ability, and so on. Further studies directed toward the application of the present synthetic methods and polyrotaxane networks with controlled properties are currently in progress.

References

1 de Gennes, P.G., (1999) *Physica A.*, **271**, 231.
2 (a) For selected reviews concerning polyrotaxane and polyrotaxane network, see: Gibson, H.W., and Marand, H., (1993) *Adv. Mater.*, **5**, 11; (b) Sauvage, J.-P., and Dietrich-Buchecker, C., (1999) *Molecular Catenanes, Rotaxanes, and Knots*, Wiley-VCH, New York, Weinheim, (c) Huang, F., and Gibson, H.W., (2005) Polypseudorotaxanes and Polyrotaxanes. *Progr. Polym. Sci.*, **30**, 982; (d) Huang, F., Pederson, A.M., and Gibson, H.W., (2007) *Physical Properties of Polymers Handbook* (ed. J.E. Mark), Springer-Verlag, New York, p. 693; (e) Takata, T., Kihara, N., and Furusho, Y., (2004) *Adv. Polym. Sci.*, **171**, 1; (f) Takata, T., (2006) *Polym. J.*, **38**, 1.
3 (a) Okumura, Y., and Ito, K., (2001) *Adv. Mater.*, **13**, 485; (b) Shimomura, T., Arai, T., Abe, T., and Ito, K., (2002) *J. Chem. Phys.*, **116**, 1753; (c) Karino, T., Okumura, Y., Ito, K., and Shibayama, M., (2004) *Macromolecules*, **37**, 6177; (d) Araki, J., Zhao, C., and Ito, K., (2005) *Macromolecules*, **38**, 7524; (e) Kidowaki, M., Kataoka, T., and Ito, K., (2006) *Drug Delivery System*, **21**, 592; (f) Ito, K., (2007) *Polym. J.*, **39**, 489; (g) Watanabe, J., Ooya, T., and Yui, N., (1999) *J. Biomater. Sci., Polym. Ed.*, **10**, 1275; (h) Watanabe, J., Ooya, T., Park, K.D., Kim, Y.H., and Yui, N., (2000) *J. Biomater. Sci., Polym. Ed.*, **11**, 1333; (i) Ochi, T., Watanabe, J., Ooya, T., and Yui F N., (2001) *Biomacromolecules*, **2**, 204.
4 (a) For related reports concerning synthesis of polyrotaxane network via statistic protocol, see: Delaviz, Y., and Gibson, H.W., (1992) *Macromolecules*, **25**, 4859; (b) Gong, C., and Gibson, H.W., (1998) *Macromol. Chem. Phys.*, **199**, 1801; (c) Gibson, H.W., Nagveker, D.S., Powell, J., Gong, C., and Bryant, W.S., (1997) *Tetrahedron*, **53**, 15197; (d) Gong, C., and Gibson, H.W., (1997) *J. Am. Chem. Soc.*, **119**, 5862; (e) Gong, C., and Gibson, H.W., (1997) *J. Am. Chem. Soc.*, **119**, 8585; (f) Gibson, H.W., Nagvekar, D.S., Yamaguchi, N., Bhattarcharjee, S., Wang, H., Vergne, M., and Hercules, D.M., (2004) *Macromolecules*, **37**, 7514; (g) Zada, A., Avny, Y., and Zilkha, A., (1999) *Eur. Polym. J.*, **35**, 1159; (h) Zada, A., Avny, Y., and Zilkha, A., (2000) *Eur. Polym. J.*, **36**, 351; (i) Zada, A., Avny, Y., and Zilkha, A., (2000) *Eur. Polym. J.*, **36**, 359; (j) Zilkha, A., (2001) *Eur. Polym. J.*, **37**, 2145; (k) Oike, H., Mouri, T., and Tezuka, Y., (2001) *Macromolecules*, **34**, 6229; (l) Kubo, M., Hibino, T., Tamura, M., Uno, T., and Itoh, T., (2002) *Macromolecules*, **35**, 5816; (m) Kubo, M., Kato, N., Uno, T., and Itoh, T., (2004) *Macromolecules*, **37**, 2762; (n) Kubo, M., Matsuura, T., Morimoto, H., Uno, T., and Itoh, T., (2005) *J. Polym. Sci.: Part A: Polym. Chem.*, **43**, 5032.
5 (a) Kihara, N., Hinoue, K., and Takata, T., (2005) *Macromolecules*, **38**, 223; (b) Liu, R., Harada, A., and Takata, T., (2007) *Polym. J.*, **39**, 23; (c) Arai, T., and Takata, T., (2007) *Chem. Lett.*, **36**, 418; (d) Takata, T., Liu, R., Maeda, T., Kihara, N., and Harada, A., (2007) *J. Polym. Sci., Part A: Polym. Chem.*, **45**, 1571.
6 Okada, M., Kamachi, M., and Harada F A., (1999) *Macromolecules*, **32**, 7202.
7 For a related study concerning the efficient pseudopolyrotaxane formation of PMα-CD in hydrocarbon solvent

through a heterogeneous reaction, see: Nakazono, K., Takashima, T., Arai, T., Koyama, Y., and Takata, T., (2010) *Macromolecules*, **43**, 691.

8. (a) Melton, L.D., and Slessor, K.N., (1971) *Carbohydr. Res.*, **18**, 29; (b) Tanaka, M., Kawaguchi, Y., Niinae, T., and Shono, T., (1984) *J. Chromatogr.*, **314**, 193; (c) Kaneda, T., Fujimoto, T., Goto, J., Asano, K., Yasufuku, Y., Jung, J.H., Hosono, C., and Sakata, Y., (2002) *Chem. Lett.*, **31**, 514.

9. (a) Kohsaka, Y., Konishi, G., and Takata, T., (2007) *Polym. J.*, **39**, 861; (b) Takata, T., Kohsaka, Y., and Konishi, G., (2007) *Chem. Lett.*, **36**, 292; (c) Bilig, T., Koyama, Y., and Takata, T., (2008) *Chem. Lett.*, **37**, 468; (d) Sato, T., and Takata, T., (2008) *Macromolecules*, **46**, 2549.

10. (a) For related reviews, see: Rowan, S.J., Cantrill, S.J., Cousin, G.R.L., Sanders, J.K.M., and Stoddart, J.F., (2002) *Angew. Chem. Int. Ed.*, **41**, 898; (b) Lehn, J.-M., (2005) *Prog. Polym. Sci.*, **30**, 814; (c) Takata, T., and Ohtsuka, H., (2006) *Yuki Gosei Kagaku Kyokaishi*, **64**, 4.

11. For a selected review, see: Takata, T., and Koyama, Y., (2008) *Kobunshi*, **57**, 346.

12. (a) Furusho, Y., Oku, T., Hasegawa, T., Tsuboi, A., Kihara, N., and Takata, T., (2000) *Chem. Lett.*, **1**, 18; (b) Takata, T., and Kihara, N., (2000) *Rev. Heteroatom Chem.*, **22**, 198; (c) Furusho, Y., Oku, T., Hasegawa, T., Tsuboi, A., Kihara, N., and Takata, T., (2003) *Chem. Eur. J.*, **9**, 2895; (d) Oku, T., Furusho, Y., and Takata, T., (2004) *Angew. Chem. Int. Ed.*, **43**, 966; (e) Bilig, T., Oku, T., Furusho, Y., Koyama, Y., Asai, S., and Takata, T., (2008) *Macromolecules*, **41**, 8496; (f) Yoshii, T., Kohsaka, Y., Moriyama, T., Suzuki, T., Koyama, Y., and Takata, T., (2011) *Supramolecular Chem.*, **23**, 65.

13. Arai, T., and Takata, T., manuscript in preparation.

14. Shioya, M. and Takata, T., (2007) *Network Polymer (Japanese)*, **28**, 2.

16
From Chemical Topology to Molecular Machines

Jean-Pierre Sauvage

16.1
Introduction

Although the field of *catenanes* and *rotaxanes* is long established [1], it is only relatively recently that it has become so remarkably active, mostly in relation to the novel properties that these compounds can exhibit (electron transfer, controlled motions, mechanical properties, and so forth). In addition, *catenanes* represent attractive synthetic challenges in molecular chemistry. The creation of such complex functional molecules as well as related compounds of the *rotaxane* family demonstrates that synthetic chemistry is now powerful enough to tackle problems whose complexity is sometimes reminiscent of biology, although the elaboration of molecular ensembles displaying properties as complex as biological assemblies is still a long-term challenge. This field has interesting connections to the field of controlled dynamic molecular systems, namely 'molecular machines,' which are molecular systems able to perform large-amplitude motions under the action of an external signal. In the following review article we will mostly focus on the results obtained in Strasbourg in the fields of molecular topology and molecular machines, although highly creative work has also been carried out in several other research teams.

16.2
Copper(I)-Templated Synthesis of Catenanes: the 'Entwining' Approach and the 'Gathering and Threading' Strategy

The most efficient strategies for making catenanes are based on **template** effects. The first templated synthesis [2] relied on copper(I). The use of Cu(I) as a template allows two organic fragments to be entangled around the metal center before incorporating them in the desired catenane backbone. The strategies used are represented in Figure 16.1. The first strategy (**A**) is very straightforward but will tend to lead to symmetrical catenanes consisting of two identical interlocking rings. The second approach (**B**) necessitates the preliminary synthesis of a coordinating ring, but in this case catenanes consisting of two different rings can be produced using this stepwise

Supramolecular Polymer Chemistry, First Edition. Edited by Akira Harada.
© 2012 Wiley-VCH Verlag GmbH & Co. KGaA. Published 2012 by Wiley-VCH Verlag GmbH & Co. KGaA.

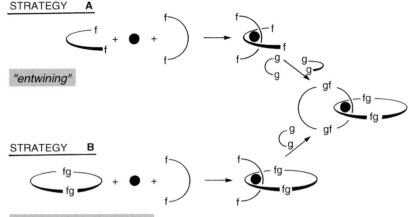

Figure 16.1 General transition metal-templated strategy; the black dot represents a transition metal (copper(I)); the arcs of a circle and the rings contain coordinating fragments. The success of the approach relies on the formation of an 'entanglement' before the cyclization reaction(s) is(are) carried out [2].

procedure. Figures 16.2 and 16.3 show the actual molecules and the chemical reactions leading to the desired catenanes. It should be noted that the high yield (27%) immediately obtained for the first copper(I)-complexed catenane allowed us to prepare this molecule on the scale of several hundred milligrams. The procedure has since been gradually improved, culminating in a 92% yield for the double cyclization reaction using the ring-closing metathesis methodology in collaboration with Grubbs and coworkers (Figure 16.4) [3].

Organic templates assembled via formation of aromatic acceptor–donor complexes and/or hydrogen bonds have also been very successful. Hydrophobic interactions have also been exploited in a very successful manner. Today, numerous

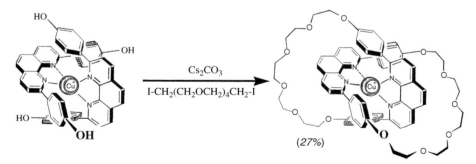

Figure 16.2 The first templated synthesis of a [2]catenane. The copper(I) complex obtained ('catenate') is extremely stable due to the interlocking nature of the ligands.

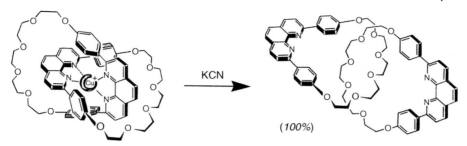

Figure 16.3 Removal of the copper(I) center and liberation of the free catenane ligand ('catenand').

template strategies are available which have led to the preparation of a myriad of catenanes and rotaxanes incorporating various organic or inorganic fragments and displaying a multitude of chemical and physical functions [4].

16.3
Molecular Knots

Knots are fascinating nontrivial topological entities. They are not only present in art and history but in many scientific fields as well, ranging from mathematics to biology. By targeting this tantalizing structure, chemists have contributed to the promotion of their beauty.

In mathematics, knots and links occupy a special position. They have been the object of active thinking for more than a century [5]. Any interested reader should have a look at a relatively recent and accessible small book entitled *The Knot Book* [6]. In Figure 16.5 are represented the first ten prime knots. These knots are single-knotted loops, in contrast to 'links' (or catenanes, in chemistry), which are sets of knotted or unknotted loops, all interlocked together. The discovery that DNA forms catenanes and knots, some of them of extreme complexity, has initiated a new field of research that has been called 'biochemical topology' [7]. Besides naturally occurring DNA, catenanes, and knots, a fascinating family of related molecules have been synthesized and described by Seeman and coworkers [8]. The elegant approach of this group utilizes synthetic

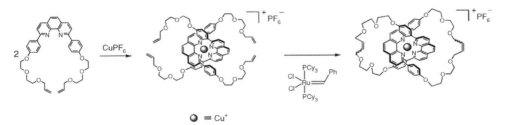

Figure 16.4 High-yield synthesis of a [2]catenane using the methodology proposed by Grubbs and his group. The overall yield is close to quantitative [3].

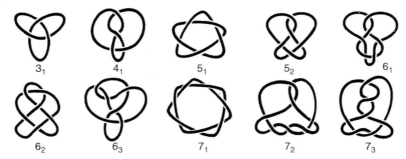

Figure 16.5 The ten first prime knots. Even the simple 4-crossing knot 4_1 still remains to be made at the molecular level. 3_1, 5_1 and 7_1 are called 'knots of the torus' (they can drawn on the surface of a torus without crossing points). They are topologically chiral, in contrast to 4_1 [6].

single-stranded DNA fragments, which are combined and knotted by topoisomerases. Interestingly, DNA is not the only biological molecular system to have the privilege of forming catenanes and knots. Liang and Mislow examined X-ray structures of a large number of proteins and, to the surprise of many molecular chemists and biochemists, they found catenanes and even trefoil knots [9, 10].

The trefoil knot occupies a special position among the knots: it is of course the simplest nontrivial knot, and it has been used in art many times by many civilizations.

The success encountered in the synthesis of various catenanes following the strategy depicted in Figure 16.1 led to a molecular trefoil knot synthesis by extending the initial synthetic concept from one to two copper(I) ions [11]. As shown in Figure 16.6, two bis-chelating molecular threads (**A**) can be interlaced on two transition metal centers, leading to a double helix (**B**). After cyclization to **C** and demetalation, a knotted system (**D**) should be obtained. An important prerequisite for the success of this approach is the formation of a helical binuclear complex (**B**).

Using 1,10-phenanthroline derivatives, this approach turned out to be successful, although the yield of the reaction was, at first, far from satisfactory (3%) [11]. However, it was gradually improved by modifying key structural features of the precursors and the final double cyclization reaction [12]. Once again, the ring-closing metathesis of olefins turned out to be spectacularly efficient, leading to a yield of 74% for the last

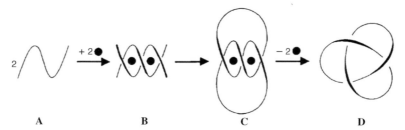

Figure 16.6 The strategy leading to a trefoil knot is represented on the bottom line. It involves two metal centers and two coordinating organic threads [11]. On the upper line, the synthesis principle of a [2]catenane is also given, to show the analogy between both strategies.

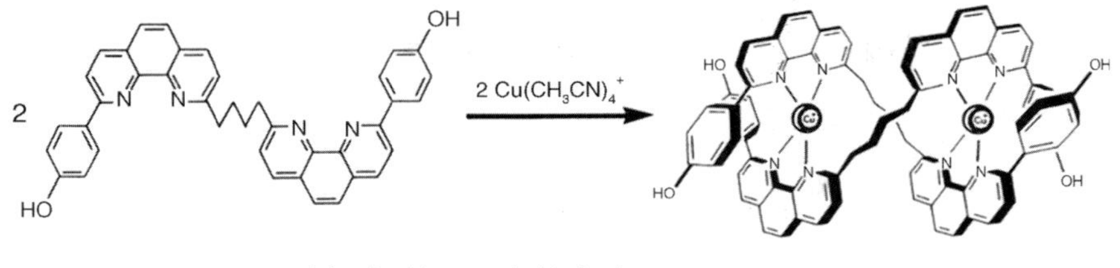

16.3 Molecular Knots

[Figure showing formation reaction with 2 Cu(CH₃CN)₄⁺]

Figure 16.7 Formation of the double-stranded helical precursor.

double cyclization step [13]. The sequence of reactions originally developed and affording the first molecular trefoil knot (1989) is indicated in Figures 16.7 and 16.8.

The first copper(I)-complexed trefoil knot could be crystallized, and the crystals obtained were subjected to X-ray diffraction [14]. The structure obtained is shown in Figure 16.9. Remarkably, the compound seems to undergo spontaneous resolution, since only one enantiomer is found in a given crystal. This property seems to be fairly general within the series of copper(I)-complexed knots made in Strasbourg. Among the few X-ray structures available for these complexes, only one shows that the compound crystallized as a racemate and not as a conglomerate of enantiomerically pure crystals. By demetalation, a chiral knotted ligand is obtained, as shown on Figure 16.10. A particularly noteworthy synthetic improvement has been to use meta-phenylene as linker between the chelates, as depicted on Figure 16.11.

Interestingly, the work of Vögtle and coworkers on the templated synthesis of [2] catenanes based on hydrogen bonds led to a very efficient preparation of a molecular knot [15]. This impressive synthetic achievement was probably not totally predicted

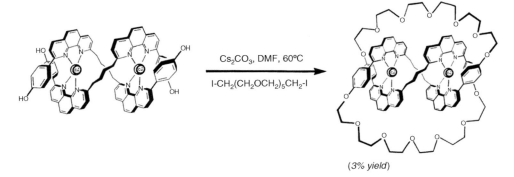

(3% yield)

Figure 16.8 The dicopper-trefoil knot is obtained via a double cyclization reaction from the double-stranded helical precursor [11]. Since the isolation and the purification of this precursor seemed to be practically impossible due to the complexity of the mixture obtained by mixing the organic coordinating threads and the corresponding stoichiometric amounts of copper(I), the ring-forming reaction was performed on the crude mixture containing only a relatively small proportion of the helical complex (~15%). Once the topology is 'frozen,' that is, the ring has been closed, the compound is stable and can be isolated pure by chromatography. Several columns were required, but, in spite of the very low preparative yield, amounts in the range of 20–70 milligrams could be obtained per reaction.

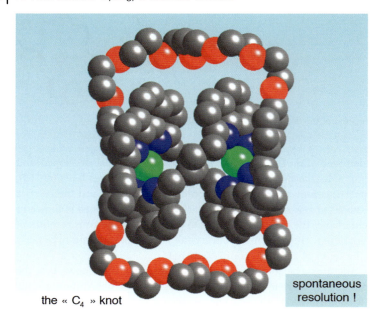

the « C$_4$ » knot — spontaneous resolution !

Figure 16.9 X-ray structure of the original trefoil knot prepared in 1989 and crystallized in 1990 [14]. The knotted macrocycle is an 84-membered ring. The copper(I) atoms are green; the nitrogen and oxygen atoms are blue and red respectively. The linker between the two 1,10-phenanthroline nuclei of each strand is a -(CH$_2$)$_4$- fragment. This linker is not very well adapted to the formation of a double-stranded helix. In contrast, the use of a 1,3-phenylene spacer leads to quantitative formation of the double helix and, subsequently, to the formation of the trefoil knot in a very acceptable yield, as shown in Figure 16.11.

by the authors themselves but the result is that another novel and truly preparative method is now available.

Where will the chemistry of synthetic knots lead us? Today, it is of course difficult to know whether practical applications will be found, although one can easily imagine that polymers containing knotted fragments could be interesting organic materials or that knotted compounds able to interact in a specific way with DNA could display new biological properties. The field still remains fascinating from a purely fundamental viewpoint. The challenge of making nontrivial prime knots beyond the trefoil knot is certainly worth considering, although when one looks at the beautiful but very complex knots of Figure 16.5, one can foresee great chemical difficulties. Obviously, chirality is an essential property in molecular chemistry, and knots are exciting systems in this regard [15, 16]. With a touch of fantasy, it could be conceived that some of the chemical processes for which chirality is crucial (enantioselection of substrates, asymmetric induction and catalysis, cholesteric phases and ferroelectric liquid crystals, molecular materials for nonlinear optics, and so forth) could one day use enantiomerically pure knots. The future of molecular knots will, to a large extent, be determined by their accessibility, and, even though the transition metal-templated strategy and the hydrogen bond approach [15] represent interesting synthetic

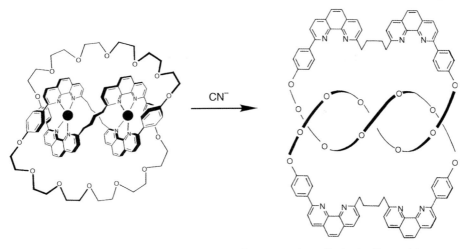

Figure 16.10 The complexed knot can be demetalated quantitatively to afford a chiral knotted ring, ready to undergo further coordination chemistry with a variety of cationic metal centers, as shown by subsequent studies carried out in our group.

achievements, there is still a long way to go before molecular knots can be made on a preparative scale compatible with industrial applications.

16.4
Molecular Machines Based on Catenanes and Rotaxanes

A particularly promising area is that of synthetic molecular machines and motors [17]. In recent years, several spectacular examples of molecular machines leading to real

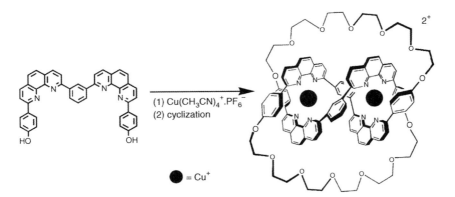

Figure 16.11 The use of the meta-phenylene connector between the two 1,10-phenanthroline nuclei of the same thread results in much better yields for the double-stranded precursor and for the trefoil knot (30%) [12].

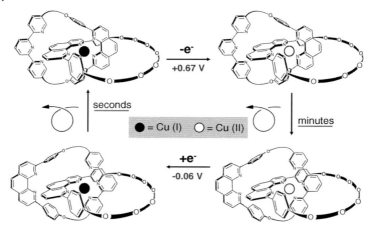

Figure 16.12 The prototypical bistable copper-complexed catenane [18a]. The compound undergoes a complete metamorphosis by oxidizing Cu(I) or reducing Cu(II). The process is quantitative but slow.

devices have been proposed, based either on interlocking systems or on noninterlocking molecules [18]. In parallel with this, more and more sophisticated molecular machines have been reported, frequently based on multicomponent rotaxanes. Particularly noteworthy are the muscle-like compounds reported by two groups [19, 20] as well as a molecular elevator [21], illustrating the complexity that dynamic threaded systems can reach.

One of the prototypical systems is a bistable catenane whose motions are triggered by an electrochemical signal. The compound and its various forms are represented in Figure 16.12 [18a]. Copper is particularly well adapted to the design of molecular machines since its two oxidation states have distinct stereo-electronic requirements: whereas copper(I) is fully satisfied in a 4-coordinate (tetrahedral) geometry, copper(II) requires more ligands in its coordination sphere. A 5-coordinate situation is more adapted to the divalent state, as illustrated on Figure 16.12, Cu(II) being coordinated to both a 1,10-phenanthroline ligand and a 2,2′,2″,6″-terpyridine.

In the course of the last 14 years, the response times of the various molecular machines made in Strasbourg have been considerably shortened. The fastest system is a rotaxane able to undergo a 'pirouetting' motion under the action of the same redox signal as that for the catenane (CuII/CuI) and whose axis incorporates a nonsterically hindering chelate of the 2,2′-bipyridine type. Today, the motions take place on a micro- to millisecond timescale [22].

16.5
Two-Dimensional Interlocking Arrays

In recent years, our group has also proposed transition metal-based strategies for making two-dimensional interlocking and threaded arrays [23, 24]. It was found to be possible to prepare large cyclic assemblies containing several copper(I) centers,

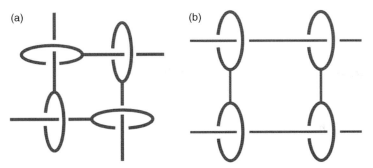

Figure 16.13 Two-dimensional interlocking arrays built via the copper(I)-template strategy. The 'gathering and threading effect' of Cu(I) leads to the quantitative formation of the pseudorotaxane tetramers (A) or (B) from the corresponding organic fragments and stoichiometric amounts of copper(I) [Refs. [23] and [24] respectively].

which open the gate to controlled dynamic two-dimensional systems and membrane-like structures consisting of multiple catenanes and rotaxanes. Two examples are presented in Figure 16.13.

The X-ray structure of the four-copper(I)-center [4]rotaxane of Figure 16.15 was recently solved by the group of Kari Rissanen (Finland) [25]. It is shown in Figure 16.16.

16.6
A [3]rotaxane Acting as an Adjustable Receptor: Toward a Molecular 'Press'

In the course of the last few years, our group has been interested in trying to construct a new molecular machine able to function as a 'press' [26]. The main components of the assembly are represented in Figure 16.14. The principle and the role of the various components are explained in a schematic fashion in Figure 16.17. Essential is the function of two porphyrinic plates, able to glide along the 'rail' which threads the rings to which they are firmly and rigidly attached. Starting from the metal-free [3] rotaxane represented on the bottom line, metalation of the coordination sites belonging to both the rod and the rings will induce a translation motion of the threaded rings: The two porphyrinic plates will come into closer proximity to one another. As a consequence, a substrate trapped between these two plates will be compressed and, possibly, expelled from the rotaxane receptor. For the moment, the molecular system made is still primitive and functions more as an adjustable

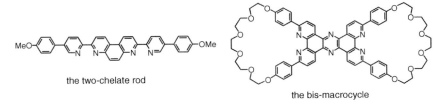

the two-chelate rod

the bis-macrocycle

Figure 16.14 The organic fragments used for preparing a compound similar to (B) of Figure 16.14.

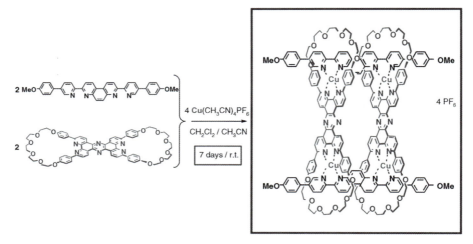

Figure 16.15 The threading reaction, leading to the desired [4]pseudorotaxane in almost quantitative yield [24].

receptor than as a true molecular press. We hope that by better control of the geometrical properties of the compound we will be able to elaborate an efficient molecular press. For space reasons, we cannot discuss the synthetic work carried out in the course of this difficult and ambitious project.

16.7
Conclusion

It is still not sure whether the fields of catenanes, rotaxanes, knots, and molecular machines will lead to applications in the short term, although spectacular results

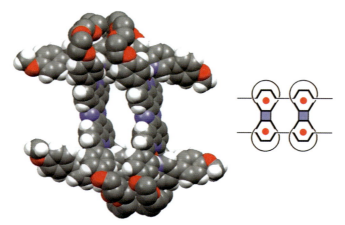

Figure 16.16 X-ray structure of the [2]rotaxane tetramer. The red dots of the Scheme (right) represent the 4 copper(I) atoms [25].

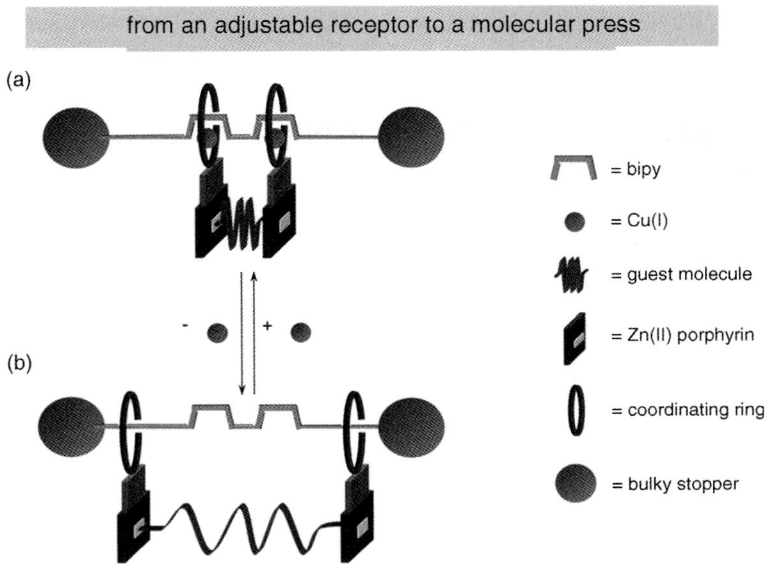

Figure 16.17 Principle of a molecular press whose two plates can be brought into close proximity or moved away from one another.

have been obtained during the last few years in relation to information storage and processing at the molecular level [27]. From a purely scientific viewpoint, the field of molecular machines is particularly challenging and motivating: the fabrication of dynamic molecular systems with precisely designed dynamic properties is still in its infancy and is certain to experience rapid development during the next decades. Besides potential practical applications, the elaboration of topologically nontrivial molecules has led to the discovery of new template principles based on the three-dimensional arrangement of given organic fragments around a transition metal center (usually copper(I)). Such an approach can be considered as a new synthetic method and possesses an obvious characteristic of generality. It has in particular allowed several groups to make and study fascinating compounds displaying highly novel topological properties in conjunction with new chemical and physical features as well as an indisputable esthetic attractiveness.

Acknowledgments

I thank all the motivated and talented researchers who have contributed to the work discussed in the present chapter. I would like to pay special homage to Dr C.O. Dietrich-Buchecker, who initiated the field in our group and whose exceptional skill, perseverance, and creativity have been at the origin of many topologically original compounds made in Strasbourg, such as, in particular, the trefoil knot.

References

1 For early work, see: Schill, G. (1971) *Catenanes, Rotaxanes and Knots*, Academic Press, New York and London.
2 Dietrich-Buchecker, C.O., Sauvage, J.-P., and Kintzinger, J.-P. (1983) *Tetrahedron Lett.*, **24**, 5095–5098; Dietrich-Buchecker, C.O., Sauvage, J.-P., and Kern, J.-M. (1984) *J. Am. Chem. Soc.*, **106**, 3043–3044.
3 Mohr, B., Weck, M., Sauvage, J.-P., and Grubbs, R.H. (1997) *Angew. Chem. Int. Ed.*, **36**, 1308–1310; Beck, M., Mohr, B., Sauvage, J.-P., and Grubbs, R.H. (1999) *J. Org. Chem.*, **64**, 5463.
4 (a) As already mentioned, the field of catenanes and rotaxanes has experienced an extremely rapid growth in the course of the last fifteen years, which makes it impossible to refer to all the contributions published in this area. For a few representative examples, see: Ogino, H. (1981) *J. Am. Chem. Soc.*, **103**, 1303–1304; (b) Dietrich-Buchecker, C.O. and Sauvage, J.-P. (1987) *Chem. Rev.*, **87**, 795–810; (c) Amabilino, D.B. and Stoddart, J.F. (1995) *Chem. Rev.*, **95**, 2725–2828; (d) Sauvage, J.-P. and Dietrich-Buchecker, C. (1999) *Molecular Catenanes, Rotaxanes and Knots*, Wiley-VCH, Weinheim; (e) Johnston, A.G., Leigh, D.A., Pritchard, R.J., and Deegan, M.D. (1995) *Angew. Chem. Int. Ed.*, **34**, 1209–1212; (f) Vögtle, F., Dünnwald, T., and Schmidt, T. (1996) *Acc. Chem. Res.*, **29**, 451–460; (g) Wenz, G., Steinbrunn, M.B., and Landfester, K. (1997) *Tetrahedron*, **53**, 15575; (h) Fujita, M. (1999) *Acc. Chem. Res.*, **32**, 53–61; (i) Hoshino, T., Miyauchi, M., Kawaguchi, Y., Yamaguchi, H., and Harada, A. (2000) *J. Am. Chem. Soc.*, **122**, 9876–9877; (j) Bogdan, A., Vysotsky, M.O., Ikai, T., Okamoto, Y., and Böhmer, V. (2004) *Chem. Eur. J.*, **10**, 3324–3330; (k) Beer, P.D., Sambrook, M.R., and Curiel, D. (2006) *Chem. Commun.*, 2105–2117.
5 Conway, J.H. (1970) *Computational Problems in Abstract Algebra*, Pergamon, New York.
6 Adams, C.C. (1994) *The Knot book*, Freeman, New York.
7 Wasserman, S.A. and Cozzarelli, N.R. (1986) *Science*, **232**, 951–960.
8 Seeman, N.C. (1997) *Acc. Chem. Res.*, **30**, 357–363.
9 Liang, C. and Mislow, K. (1994) *J. Am. Chem. Soc.*, **116**, 11189–11190.
10 Liang, C. and Mislow, K. (1995) *J. Am. Chem. Soc.*, **117**, 4201–4213.
11 Dietrich-Buchecker, C.O. and Sauvage, J.-P. (1989) *Angew. Chem. Int. Ed.*, **28**, 189–192.
12 (a) Dietrich-Buchecker, C.O., Sauvage, J.-P., De Cian, A., and Fischer, J., *Chem. Commun.*, 2231–2232; (b) Dietrich-Buchecker, C.O., Rapenne, G., and Sauvage, J.-P. (1999) *Coord. Chem. Rev.*, **185–186**, 167–176.
13 Dietrich-Buchecker, C.O., Rapenne, G., and Sauvage, J.-P. (1997) *Chem. Commun.*, 2053–2054.
14 Dietrich-Buchecker, C.O., Guilhem, J., Pascard, C., and Sauvage, J.-P. (1990) *Angew. Chem. Int. Ed.*, **29**, 1154–1156.
15 Safarowsky, O., Nieger, M., Fröhlich, R., and Vögtle, F. (2000) *Angew. Chem. Int. Ed.*, **39**, 1616–1618.
16 (a) Rapenne, G., Dietrich-Buchecker, C.O., and Sauvage, J.-P. (1996) *J. Am. Chem. Soc.*, **118**, 10932–10933; (b) Dietrich-Buchecker, C.O., Rapenne, G., Sauvage, J.-P., De Cian, A., and Fischer, J., *Chem. Eur. J*. (1999) **5**, 1432–1439.
17 (a) (2001) *Acc. Chem. Res.*, **34**, 409–522 (Special Issue on Molecular Machines) and references therein; (b) Sauvage, J.-P. (2001) *Structure and Bonding – Molecular Machines and Motors*, Springer, Berlin, Heidelberg, (c) Balzani, V., Venturi, M., and Credi, A. (2008) *Molecular Devices and Machines – Concepts and perspectives for the Nanoworld*, Wiley-VCH, Weinheim, (d) Kay, E.R., Leigh, D.A., and Zerbetto, F. (2007) *Angew. Chem. Int. Ed.*, **46**, 72–191.
18 (a) Livoreil, A., Dietrich-Buchecker, C.O., and Sauvage, J.-P. (1994) *J. Am. Chem. Soc.*, **116**, 9399–9400; (b) Koumura, N., Zijistra, R.W.J., van Delden, R.A., Harada, N., and Feringa, B.L. (1999) *Nature*, **401**, 152–155; (c) Collier, C.P., Mattersteig, G., Wong, E.W., Luo, Y., Beverly, K., Sampaio, J., Raymo, F.M., Stoddart, J.F., and Heath, J.R. (2000) *Science*, **289**, 1172–1175; (d) Leigh, D.A.,

Wong, J.K.Y., Dehez, F., and Zerbetto, F. (2003) *Nature*, **424**, 174–179; (e) Korybut-Daszkiewicz, B., Wieçkowska, A., Bilewicz, R., Domagata, S., and Wozniak, K. (2004) *Angew. Chem. Int. Ed.*, **43**, 1668–1672; (f) Fabbrizzi, L., Foti, F., Patroni, S., Pallavicini, P., and Taglietti, A. (2004) *Angew. Chem. Int. Ed.*, **43**, 5073–5077; (g) Harada, A. (2001) *Acc. Chem. Res.*, **34**, 456–464.

19 (a) Jiménez, M.C., Dietrich-Buchecker, C., and Sauvage, J.-P. (2000) *Angew. Chem. Int. Ed.*, **39**, 3284–3287; (b) Jiménez-Molero, M.C., Dietrich-Buchecker, C., and Sauvage, J.-P. (2003) *Chem. Eur. J.*, **8**, 1456–1466.

20 Liu, Y., Flood, A.H., Bonvallet, P.A., Vignon, S.A., Northrop, B.H., Jeppesen, J.O., Huang, T.J., Brough, B., Baller, M., Magonov, S.N., Solares, S.D., Goddard, W.A., Ho, C.-M., and Stoddart, J.F. (2005) *J. Am. Chem. Soc.*, **127**, 9745–9759.

21 Badjic, J.D., Balzani, V., Credi, A., Serena, S., and Stoddart, J.F. (2004) *Science*, **303**, 1845–1849.

22 Létinois-Halbes, U., Hanss, D., Beierle, J., Collin, J.-P., and Sauvage, J.-P. (2005) *Org. Lett.*, **7**, 5753–5756.

23 Kraus, T., Budesinsky, M., Cvacka, J., and Sauvage, J.-P. (2006) *Angew. Chem. Int. Ed.*, **45**, 258–261.

24 Collin, J.-P., Frey, J., Heitz, V., Sakellariou, E., Sauvage, J.-P., and Tock, C. (2006) *New J. Chem.*, **30**, 1386–1389.

25 Frey, J., Tock, C., Collin, J.-P., Heitz, V., Sauvage, J.-P., and Rissanen, K. (2008) *J. Am. Chem. Soc.*, **130**, 11013–11022.

26 Frey, J., Tock, C., Collin, J.-P., Heitz, V., and Sauvage, J.-P. (2008) *J. Am. Chem. Soc.*, **130**, 4592–4593.

27 Green, J.E., Choi, J.W., Bunimovich, A.B.Y., Johnston-Halperin, E., DeIonno, E., Luo, Y., Sheriff, B.A., Xu, K., Shin, Y.S., Tseng, H.-R., Stoddart, J.F., and Heath, J.R. (2007) *Nature*, **445**, 415–417, and references therein.

Index

a

A-B heteroditopic cavitand monomer
- bimodal self-assembly cycle 79
- dimer, crystal structure 80
absorption coefficient 58, 318
acceptor–donor complexes 348
N-acetylgalactosamine glutamate ester 242
achiral synthetic polymers
- double-stranded helix formation 56
2-acrylamido-2-methylpropanesulfonate (AMPS) 109
ADA–DAD arrays 7
aligner 67
- absorption spectra 54
- molecules 54, 55
- PPE assembly 54, 55
alkene-containing polymers 185
alkenes 185, 186
amido tetraethyl triurea oligomers 17
amine-terminated polymer 332
ammonium cations 244
amorphous/semi-crystalline oligomers 18
amphiphilic polyrotaxanes 223
amplified colorimetric sensing 58
anionic polymers 200
anthracene chromophoric unit
- characteristic structural bands 318
9-anthryl stopper 318
antibody dendrimers
- for antigens, binding properties of 136–139
- topological structures 136
antibody–divalent antigen complexes 134
antibody supramolecules 130
- structural properties of antibody molecules 130
ascorbic acid 60
association constant 112, 313

atomic force microscopy (AFM) 16, 20, 23, 61, 119, 129, 130, 139, 140, 143, 145, 279, 280
atom transfer radical homopolymerization (ATRP) 117
azobenzene 47, 103, 104, 110, 117, 251, 253, 255
trans-azobenzene 251
azobenzene functionalized hydropropyl methylcellulose (Azo-HPMC) 117
azobenzene-modifed α-CD 251
azobisisobutyronitrile (AIBN) 100, 116, 323
2,2′-azobis(isobutyronitrile) (AIBN)
- radical polymerization with 116
4,4′-azodibenzoic acid (ADA) 253
- *trans*-4,4′-azodibenzoic acid (ADA) 114

b

β-cyclodextrin (β-CD) 29
benzenethiol 160, 339
benzene-1,3,5-tricarboxamide (BTA) motif 23
bicycloalkanes
- of C_nH_{2n-2} 296
- isomers of C_5H_8 296
bifunctional metal complexes 239
bimodal self-assembly protocol 80
binaphthylbisbipyridine-Cu(I) complex 259
biodegradable polyrotaxane 206
biological degradation 211
biological supramolecular polymers 231
biological systems 195
biomedical-oriented polyrotaxanes 204
biomedical polyrotaxanes 196
bis(5-carboxy-1,3-phenylene) crown ether 161
bis(2,4-dinitrophenyl)-PEG (PEG-DNB2) 209
2,6-bis(1′-methylbenzimidazolyl)-4-pyridyl groups 239, 259

Supramolecular Polymer Chemistry, First Edition. Edited by Akira Harada.
© 2012 Wiley-VCH Verlag GmbH & Co. KGaA. Published 2012 by Wiley-VCH Verlag GmbH & Co. KGaA.

bis(*m*-phenylene)-32-crown-10 peripherymoieties 241
bis-N^6-anisoyladenocine-modified pTHF 237
2,6-bis(2-oxazolyl)pyridine (PYBOX) 55, 56
bisphenol A 238
N,*N*′-bis(salicylidene)-1,n-alkanediamines 241
bistable copper-complexed catenane 354
bisvelcrand
– A-A:A-A homopolimerization 77
– chloroform solution 78
– ^{1}H NMR spectrum 78
– tendency 78
block copolymers 3
– phase separation in 18
Borromean rings 305
branched supramolecular polymers 42
β-sheets 241
bulk heterojunction 60
4-tert-butylpyridine (tbpy) 185
6-*O*-(*tert*-butylthio)-β-CD molecules
– crystal structure 30–32
– X-ray crystallographic study 31

c

capping reagents 225
carbon nanotubes (CNTs) 60, 61
catalyst (Mn1) 185
[2]catenane
– high-yield synthesis 349
– templated synthesis 348
catenane ligand 349
catenanes 57, 127, 129, 144, 145, 347, 351
catenation reaction
– between nicked DNA and circular DNA 142
– with topoisomerase I 141–143
cationic polymers 199
– polyplex formation of 199
cationic telechelic precursor 298
cavitand-based supramolecular polymers 84
cavitands 72–75. *see also* ditopic cavitand monomers
– A-B heteroditopic polymerization mode 84
– classes 76
– heteroditopic monomers 75
– homoditopic monomers 75
– supramolecular polymerization motifs 76
Cayley tree 294
CDCl$_3$
– ^{1}H NMRexperiments in 81, 82
– partial ^{1}H NMR spectra in 83
chain breaking process 298
chemical gel 205

– with fixed network junctions 208
– swelling behavior 205
chemically responsive gel-to-sol transition 246
chemical-responsive systems 246–249
chiral insulated polythiophene
– by SPG wrapping 62
chromophore-modified cyclodextrin 33
3-CiNH-α-cyclodextrin
– supramolecular polymers construction 40
6-CiNH-α-cyclodextrin 38
6-CiO-α-cyclodextrin 38
circular dichroism 246
click process 301, 302
Co(II)/4-dodecyloxyethyl-1,2,4-triazole complexes 237
complex self-assembly processes 4
Con A-induced hemagglutination test 197
conjugated polymers (CPs) 53
– alignment 54
– – via supramolecular bundling approach 55
– π-conjugated polymers 104
π-conjugated systems, in CNT 61
contraction and expansion systems 110
cooperative polymerizations 6
copolymerization reactions 100, 106, 108, 111, 116, 323, 324, 340
copper(I)-complexed catenane 348
copper(I)-complexed knots 351
Cotton bands 41
covalent fixation process 298, 299, 302
cross-linked polymeric materials 205
cross-linked SPs networks 276
– semidilute entangled SPs networks, properties of 283–285
– semidilute unentangled SPs networks, reversibility in 276–283
– sticky reptation model 285, 286
cross-linking protocol 160, 336, 337, 339
crown ethers 151, 315, 336, 337, 339, 344
crown functionalized polymers 179
crystalline [2]pseudorotaxanes formation 308
cucurbit[7]uril (CB7) 116
Curdlan (CUR) 63
– microscopic observation 63
– regioselective bromination- azidation 63
current hydrogen-bonding motifs 10
cyclic polymers
– linking betenoxy group 301
– topological isomerism 297
– trefoil topology 299
– types 299
cyclic supramolecular complexes 128
cyclization reaction 350

α-cyclodextrin (α-CD) 29, 331, 339
– cooperative complexation 115
– single-crystal X-ray diffraction 29
β-cyclodextrin (β-CD)
– dimer, ROESY NMR spectra 44
– guest molecule for 43
– host–guest complexes 103
– monoaldehyde derivative, attachment 101
– single-crystal X-ray diffraction to 29
– tosylation 98
cyclodextrin-based supramolecular polymers
– c2-daisy chain, artificial molecular muscle 45–47
– CD aliphatic tethers 30, 31
– β-CDs aromatic tethers 31–33
– CDs–guest conjugates
– – homo-intramolecular/intermolecular complexes formation 33–43
– intermolecular complexes, by CD and guest dimers formation 44
– modified CDs
– – hetero-supramolecular structures formed by 42, 43
– 3-modified α-CDs
– – with guest molecules 36
– – supramolecular structures formed by 40–42
– 6-modified α-CDs
– – with guest molecules 34
– – supramolecular structures formed by 33–39
– monomer units, chemical structures 37
– in solid state 29–33
– striking property 102
β-cyclodextrin-ethyl cinnamate complex
– trans-azobenzene -modified 39
– crystal structure 39
cyclodextrins (CD) 97
– containing covalent networks 97
– 2,6-dimethyl-β-cyclodextrins 106
– direct cross-linking 97
– inclusion complexes
– – crystal packing structures, classification 30
– monofunctional derivatives 101
– monofunctional monomers 99
– mono 6^I-amino-6^I-desoxy-cyclodextrins 99
– mono-monomer synthesis 99
– polyrotaxane network having 195, 339–342
– as side groups
– – polymers with covalently bonded 97, 98
– – structure–property relationship of polymers containing 102–105
cyclopentadienyl ligands 309

cyclopropane (C_3H_6) 294
cystamine 200
cytocleavable polyrotaxanes 196, 199, 200
– cleavage role in intracellular pDNA release 201
– efficiency 201
– for gene delivery 200
– as gene vectors 200
cytotoxicity 200

d

DAA– ADD arrays 7
DADA hydrogen-bonding motif 8, 9
DADA motif, dimerization constant 10
daisy chains 153–156
– oligomer 38
degree of polymerization (DP) 6, 13, 19, 100
– theoretical dependence 14
degree of substitution (DS) 102
dendritic polymers 196
dendritic rotaxanes 152, 157, 158
dendronized polymers 152, 158
deprotonated 2,2,2-cryptand-captured K^+ 242
Dess–Martin periodinane 100
dethreading reaction
– intermediate 318
– mechanism 190, 191
– of polymers 189
– thermodynamic parameters 190
deuterated DMSO (d-DMSO) 211
dextran 102, 196
dextran-based azobenzene 255
diamines 4,4′-oxydianiline 162
diamino-terminated PEG (PEG-DAT) 209
1,8-diazobicyclo[5.4.0]undec-7-ene (DBU) 88
dibutyltin dilaurate (DBTDLx)
– catalytic amount 333
dicopper-trefoil knot 351
N,N-dicyclohexylcarbodiimide (DCC) 113
dicyclohexylphosphine 240
Diels–Alder reaction 313
differential scanning calorimetry (DSC) 117, 310
diffusion coefficient 41, 82
dimerization constant 8, 14, 18
N,N-dimethylacrylamide 109
dimethylaminoethylcarbamoyl (DMAEC) groups 200
dimethyl sulfoxide (DMSO) 60, 210, 246
dinitrophenyl groups 209
diphenylmethane diisocyanate 334
diphenylphosphine 240
dipole–dipole interactions 4, 151

dissociation/association process 115, 116
disulfide bond 257
3,5-di(*tert*-butyl)phenyl blocking group 188
ditopic cavitand monomers
– cavitands 72–75
– combining solvophobic interactions
–– and hydrogen bonding 82–84
–– and metal–ligand coordination 78–82
– heteroditopic self-assembled via host–guest interactions 84–88
– homoditopic self-assembled via host–guest interactions 88–91
– homoditopic via solvophobic π-π stacking interactions 77, 78
– polymerization with 71, 72
– self-assembly 75–91
– supramolecular polymerization
–– structural monomer classification 75–77
DNA catenanes
– AFM images of 143, 144
– preparation by addition of T4 DNA ligase 144
– topological structures 145
DNA ligase 144, 145
DNA polymerase III 185
DNA supramolecules 139
– catenation reaction, with topoisomerase I 141–143
– DNA catenanes, AFM images of 143, 144
– DNA [n]catenanes preparation 144, 145
– plasmid DNA molecules, imaging 139, 140
– preparation of Nicked DNA 140, 141
2D NOESY NMR analysis 109, 110
dodecamethylene linker 253
dodecavalent hydrogen bonds 10
dodecyl-modified poly(acrylic acid) 253
dopaminergic neurotoxicant 130
double-cyclic polymers 297
double-grafted polymer, complexation 117
double-stranded helical precursor formation 351
double-threaded dimer 43, 46
doxorubicin hydrochloride (DOX) 117
drug delivery system (DDS), application 316
dynamic light scattering (DLS) 71

e

electrospray ionization mass spectroscopy (ESI-MS) 82, 90, 158
electrostatic interactions 51, 63, 127, 195, 233, 234
– hierarchical architecture 64

electrostatic self-assembly process 298, 299, 302
end-capping strategy 316
entwining approach 347–349
enzyme-linked immunosorbent assay (ELISA) 132, 133
2-ethylhexyl-modified 4,4′-bipyridine 256

f

Fab fragments of IgG 128
fatty dimer acids 17
ferrocene-carboxylic acid (FCA) 114, 258
ferrocene-containing crown ether synthesis 308, 314, 324
ferroelectric liquid crystals 352
field effect transistors 58
figure-eight cross-links 207
flash photolysis time-resolved microwave conductivity measurements 66
flexible glycosyl bonds 60
flexible light-emitting diodes 58
flexible polymer molecules 297
fluorescence resonance energy transfer (FRET) procedure 319
fluorescence spectroscopy 109, 119
fluorescent chemosensor 33
force-responsive systems 239–241
four-copper(I)-center [4]rotaxane
– x-ray structure 351

g

gathering and threading strategy 347–349
γ-cyclodextrin (γ-CD) 244
gel-to-sol transition 237
– of Py-βCD/SWNT 119
– using photo-dimerization of a stilbene-carrying organo-gelator 250
glass transition temperature 17
graft copolymers 3, 13
graft density 198
Grubbs catalyst 301

h

H-bonding polymerization 72
helical supramolecular ureidotriazine polymer 15
heptamethylene 251–252
hermaphrodite monomer 36
hexamethylene diisocyanate (HMDI) 216
highly oriented pyrolytic graphite (HOPG) 80
high-molecular-weight polymer chain
– entanglements 16
^{1}H NMR experiments 80
homo-metathesis reaction 322

host–guest assemblies formation   105
host–guest chemistry   53, 60, 61, 153, 156, 158, 164, 176, 179
host–guest complexes   97
– of βCD and adamantyl groups   103
– with the decyl moieties in   226
– free-radical polymerization   108
– in water   108
host–guest interactions   72, 84–91, 87, 103, 119
host-guest organogels concept   52
hybrid polymer gels   286–288
hydrodynamic radii   272
hydrogels   238
– formed from SWNT solubilized with   248
– photo-responsive   250
– using biodegradable CD polyrotaxane   206
hydrogen-bonding motifs   4, 6–10, 7, 8, 15, 18, 23, 24
– association constant   14
– with high dimerization constant   14
– inspired on self-assembly as found in nature   11
– linear   6, 7
– multiple hydrogen bonds arrays   6–8
– preorganization through intramolecular hydrogen bonding   8, 9
– self-assembly   11
– tautomeric equilibria   9, 10
hydrogen bonds   4, 6, 8, 10, 15, 17, 19, 24, 57, 151, 168, 195, 206, 233, 286, 302, 348, 352
– donor   6
hydrophobically modified alkali-soluble emulsion (HASE) polymers   119
hydrophobically modified water-soluble polymers (HMASPs)   114
hydrophobic block copolymer
– preparation   118
– structures   118
hydrophobic compartmentalization   15
hydrophobic–hydrophilic interactions   151
hydroxypropylated α-cyclodextrins   211
hydroxypropylated polyrotaxane (H-PR)   211

i

IgG molecule, AFM images   130
immunoglobulin G (IgG)   128
immunosorbent assay   128
π-π interactions   15
interconversion process   298
interlocking arrays   354
– two-dimensional   354
interlocking systems   354
inter-penetrating polymer network (IPN)   220
interrotaxane π-π stacking interaction   311
iodosylbenzene   185, 186
ion pairing   151
IR-and ^{1}H NMR spectroscopy   111
isodesmic polymerization mechanism   22
isomerism concept   296
N-isopropylacrylamide (NIPAM)
– copolymerization with   115
isothermal titration calorimetry (ITC)   85, 152
– thermodynamic parameters derived from   86
– titration, K_a values for   90

j

Janus[2]rotaxane   47
– water-induced contraction of   247
Janus [2]rotaxanes   246, 247, 251, 252
Janus unit   45

k

kite-kite dimer, diagnostic for   78
Kuhn monomers   274

l

Langmuir-Blodgett techniques   67
lanthanoid metal ions   239
lateral interactions   20, 21
– influence   21
– influence of the strength of   21
– and phase separation   19
– in supramolecular polymers in the solid state   18
– surface adsorption amplifies   72
layer-by-layer methods   67
ligand-conjugated polyrotaxanes   196
– multivalent binding
– – with receptor sites of plasma proteins   197
– multivalent interaction using   196
linear chain formation   5
linear oligo-Janus[2]rotaxanes, photoisomerization behavior   45
linear oscillatory rheology   283
linear polymers   13
linear SPs   41, 270, 272, 273
– in solid state   275, 276
– theory related to properties of   274, 275
liquid crystalline polyrotaxane (LCPR)   216
lower critical solution temperature (LCST)   115, 117, 216
low-molecular-weight model   111

m

macrocycle-containing polymer
– rotaxanative cross-linking   331

macrocyclic compound   185
– thread onto the polymer and glide   192
macroscopic responses   261
MALDI-TOF measurements   80
maltose-conjugated polyrotaxanes   197
[4-(m-aminophenoxy)-phenyl] phenylphosphine oxide (m-BAPPO)   162
manganese(III) porphyrin complex   185
mesoporous silica nanoparticles   243
meso-tetrakis(2-methoxyphenyl) porphyrin   186
metal–ligand bond   279, 282
metal–ligand interactions   80
metallo-supramolecular networks   282
metal-organic framework (MOF)   305
metal-organic-rotaxane framework (MORF)   305, 306
meta-phenylene connector   353
metathesis polymer condensation
– for topologically unique polymers   301
metathesis polymer cyclization (MPC)   301
methylated polyrotaxane (MePR)   216
N,N-methylbutylammonium iodide   88
4,4′-methylenebis(phenyl isocyanate)   236
1-methyl-4-phenyl-1,2,3,6-tetrahydropyridine   130, 130
methyl viologen   130, 132
microgravimetric quartz-crystal-microbalance (QCM)   128
microscopic techniques   66
mobile polyrotaxanes, supramolecular characteristics   196
molecular machines, chemical topology to   347
– based on catenanes and rotaxanes   353–354
– catenanes, copper(I)-templated synthesis   347–349
– molecular knots   347–353
– [3]rotaxane acting as adjustable receptor   355
– two-dimensional interlocking arrays   354
molecular polymerization   4
molecular press principle   357
molecular receptors   72
– classes   72
molecular recognition, by biomacromolecules   127
molecular system   355
monoclonal antibodies   128, 133
mono-6^I-iodo- 6^I-desoxy-β-cyclodextrin (β-CD)   102
monomer-stopper interaction   86
multi-component systems   24
multicyclic polymer topologies

– by electrostatic polymer self-assembly   300
multiple hydrogen-bonded supramolecular polymers
– future perspectives   23–25
– historical background   3, 4
– hydrogen-bonding motifs, general concepts   6–10
– hydrophobic compartmentalization   14, 15
– main-chain supramolecular polymers   10–15
– phase separation/additional lateral interactions in solid state   18, 19
– supramolecular chemistry   4
– supramolecular polymerization mechanisms   4–6, 13, 14
– supramolecular polymers, establishment   10–13
– supramolecular thermoplastic elastomers based on   19–23
– thermoplastic elastomers   16–18
multivalent antigens   128
multivalent linear supramolecular polymers
– polymerization   13
myosin–actin complex   45
myricetin   243

n

nano-fibers   23
– atomic force microscopy image   16
– supramolecular polymers   23
nanoparticles   244
naphthalenediimide-based organogelator (Naph(imide)$_2$)   57
– one-dimensional fiber structure   58
napy 4·4 napy tetramer formation
– partial ^{1}H NMR spectra in CDCl$_3$ 83
N,C,N-pincer metal complexes   270–272
6- NH$_2$CiO-β-cyclodextrin
– crystal structure   40
p-nitrophenol   225
NMR analysis   33, 35, 39, 43, 44, 45, 78, 82, 106, 108, 112, 117, 178, 249, 333, 334
noncovalent interactions
– complementarity   51
– principle   51
noncovalent polymerization mechanism   4
nonlinear polymer topologies
– classification   293, 295
nonviralDNAvectors   202
N/P ratio   200, 201
nucleation mechanism   190
nucleobase hydrogen-bonded supramolecular polymers   20

o

octaamide-porphyrin
- fullerene-binding motivated self-assembly 57
oligobisvelcraplex kite conformers 78
oligo(ethylene glycol) (OEG) 47
oligosaccharides 210
organic fragments 355
organic–inorganic hybrid materials 219, 220, 223
organic–inorganic hybrid slide-ring gel 223
organic substances
- advantages 58
organogel 256
organometallic rotaxanes
- experimental section 324–326
- ferrocene-containing [3]rotaxane
-- photochemical properties 318, 319
-- and side chain polyrotaxane 320–323
- ferrocene-containing [2]rotaxanes synthesis by 312–316
- pseudorotaxanes crystals 307–312
- rotaxane-like complex, dethreacting reaction 316–318
- [3]rotaxanes, strategies and synthesis 320, 321
- side-chain type polyrotaxane, strategies and synthesis 321–323
- structure and dynamic behavior 305–307
ORTEP drawing 309

p

paraquat-type cyclic molecules 305
pBR322 plasmid 144
PDMS-based pseudo-polyrotaxanes 221
PDMS-γ-CD polyrotaxane 222
pDNA polyplex 200, 201
PEG-based polyrotaxane 211
permethylated α-cyclodextrin (PMα-CD) 331
PGSE NMR spectroscopy 47, 82, 87
- measurements 88
phase transitions 231
1,10-phenanthroline derivatives 350
6-O-[(R)-1-phenylethyl]amino-β-CD molecules
- crystal structure 32
photochemical data 319
photodimerization 233
- compounds for 234
photoisomerization 233, 250
- of azobenzene moieties 255
- compounds for 234
- from trans to cis 254
photo-responsive systems 249–255
photovoltaic cells 58

pH-responsive systems 241–245
pH/thermosensitive physical polymer network 103
pincer-ligand complexes 279
plasmid DNA (pDNA) 200, 201
polar solvents 73, 306, 313, 316, 339
- as N,N-dimethylformamide 100
- leads to the complete suppression of 88
poly(acrylic acid) 196, 253
poly(amino acids) 196
poly(aspartic acid) 196
1,4-polybutadiene 225
polybutadiene (PBD) 186, 189, 224
- chain, oxidation of 190, 191
poly(ε-caprolactone) oligomers 19
polycatenanes 162
polycatenenes 162
poly(crown ether) 336, 339
(poly)cycloalkane molecules 293
polydienes 224, 226
poly-ε-caprolactone (PCL) 216
poly(ether ketone) (PEEK) 106
poly(ether sulfone) synthesis, and 1H NMR characterization 106
poly(ethylene-butylene) oligomer 20
polyethylene glycol (PEG) 206, 211
poly(ethyleneimine) (PEI) 199, 244
poly(ethylene oxide) (PEOMA) 117
poly(isobutene-*alt*-maleic anhydride) (PiBMA) 114
poly(L-lysine) 199
polymer-analogous reaction 100, 102
polymer-bonded α-cyclodextrin (α-CD)
- interactions 104
polymer complexations 152
polymeric topological isomers
- through electrostatic polymer self-assembly 300
polymerization mechanism 3, 5, 22, 23, 76
- A-B heteroditopic polymerization mode 84
- cationic 118
- cyclic *vs.* linear 83
- kinetically-controlled simultaneous 220
- of monofunctional cyclodextrin monomers 98–100
- supramolecular 4–6, 13, 14, 19
-- N,C,N-pincer metal complexes and ligands for 271
-- structural monomer classification 75–77
- utilizing the acetylene functional group 170, 171
polymer network structures 206
polymer-porphyrin rotaxane catalytic system 187

polymers
– brushes, structures 118
– characterization 339
– with covalently bonded cyclodextrins as side groups 97–105
– – structure-property relationship 102–105
– definition 3
– electrophoretic mobility 114
– growth mechanism 77
– macroscopic properties 3
– mobility of ligands in 197
– monofunctional cyclodextrin
– – polymer-analogous reaction with 100–102
– monofunctional cyclodextrin monomers
– – synthesis and polymerization 98–100
– properties 269
– side chain polypseudorotaxane (polymer (polyaxis)/cyclodextrin (rotor)) 111–120
– side chain polyrotaxanes 106–111
– structures 112, 115
– – with CDs as side groups 98
– supramolecularly attached cyclodextrins 97
poly(N-methacryloylphenylalanine) 113
poly (ethylene glycol) (PEG) 195
polyphosphazene (PN) 118
poly(propylenimine) dendrimers 241
polypseudorotaxanes 152, 161
– copolymerization 340
– polymer (polyaxis)/cyclodextrin (rotor), side chain 111–120
– – properties 106
– radical polymerization behavior 340
– side chain
– – conversion 110
polyrotaxane network. *See also* polyrotaxanes
– chemical recyclability 339, 340
– construction 342
– cross-linker
– – design 342–344
– – structures 343
– cross-linking reaction 335
– crown ethers 336–339
– cyclicvoltammogram 323
– cyclodextrins 339–342
– end-capping reaction 333
– ^{1}H NMR spectrum 333
– main-chain-type polyrotaxane, linking of wheels 331–336
– monofunctional α-CD-containing main-chain-type, synthesis 334
– polymacrocycle, linking of macrocyclic units 336–342

– polypseudorotaxane synthesis, by linking of axle terminal 333
– side-chain 106–111
– – characteristics 106
– – chemoenzymatic synthesis 108
– – hetero, synthesis 111
– – ^{1}H NMR spectrum 107
– – photosensitive, structures 109
– – properties 106
– – strategies for 320
– – structures 107, 108
– stress-strain curve 344
– swelling behaviors 340
– swelling ratio 341
– synthesis 337, 338, 340, 344
– synthetic strategy 332, 337, 340
– thermal gravimetric analysis (TGA) 323
– as topologically cross-linked polymer synthesis and properties 331
polyrotaxanes 152, 163, 200, 207
– crown ether-based 153
– intramolecular friction, control strategy for 224
– loops as a dynamic interface 197–199
– mobile ligands linked to CD molecules 196
– modification 210
– of PEG and α-CD capped at both ends with adamantane 210
– from PEG-COOH 210
– slide-ring materials using (*See* slide-ring polymeric materials)
– synthesis scheme 221
– threaded onto Kan-Ei-Tsuho coins 203
– used in Asian and Chinese culture 203, 204
[3]polyrotaxanes
– movable range effects 344
poly[2]rotaxanes 152
poly[3]rotaxanes 152, 173–176
polysaccharides 196, 210
polystyrene (PS) 118
– steric effects of 179
polytetrahydrofuran (PTHF) 188, 332
poly(tetrahydrofuran) (pTHF) 237, 240
poly(4-vinylpyridine) (PVP) 239, 282
Por(amide)$_8$
– fluorescence 57
porphyrinatozinc oligomer 54
porphyrin dimers 133
porphyrinic plates 355
porphyrins 57, 133, 137–139, 190
Por(PYBOX)$_2$/PTMI complex
– CD spectrum 56
PPEs juxtaposition 54
pressure-responsive systems 239–241

processable supramolecular/macromolecular composite 66
programmed molecular systems 65
protein (RecA) 145
protonated dendrimers 241
pseudopolyrotaxane
– consisting of α-CD molecules and 203
– solid-state end-capping reaction 334
– synthesis of 206
pseudorotaxanes 152, 165, 176, 181, 305. see also organometallic rotaxanes
– axle, end-functionalization reactions 312
– crystals 307–312
– DB24C8, 306
– – formation 313
– dialkylammoniums formation 313
– dialkylammonium structure 306, 307
– IR data 309
– monomer 322
– polymeric end group 176
– single-crystal-to-single-crystal phase transition reaction 324
– thermal properties 310
PTHF-m-phenylene diisocyanate 343
PTMI, estimated length 56
PT-PT interpolymer interaction 62
p-type/n-type semiconducting molecules
– self-sorting supramolecular fiber formation 59
pulley effect 208, 209
pulse-field gradient spin-echo (PGSE) NMR experiments 47, 71, 78, 114
pyrimidin-4-ol tautomer 9

q

quadruple hydrogen-bonding motifs 8
quartz crystal microbalance (QCM) measurements 65, 128, 199
quasi-elastic light scattering (QELS) 211
quasi-stable polymer network formation 339
quenching pathway 319
quercetin 243
quinoxaline kite velcrand 82
– cyclic vs. linear polymerization 83

r

radical polymerization mechanisms 100. See also polymerization mechanism
recyclable cross-linked polyrotaxane gels 206
redox responsible hydrogel system 113
redox-responsive systems 255–259
redox signal 354
responses 234
– capture and release of chemicals 235

– change in color 236
– chemical reactions 235
– gel-to-sol and sol-to-gel transitions 236
– movement 235
– viscoelastic properties 236
reversed-phase chromatography (RPC) technique 300
rheological measurements 237
rhodamine B 244, 245, 255, 257, 259
ring-chain mechanism 5
[2]rotaxane 320
– neutral, structure and yield 316
– structure and yield 315
– synthesis 314
– – with N-acryloyl group 321
– tetramer, X-ray structure 356
– threading-followed-by-end-capping strategy for 312
[3]rotaxane
– free-radical copolymerization 323
– ^{1}H NMR spectrum 320
– synthesis strategies 320
– synthesis via cross-metathesis reaction 321
– synthesis via homo-metathesis reaction 322
rotaxane-like complex
– dethreacting reaction of 316
– first-order kinetic rate constants 317
– thermodynamic parameters 318
– threading-followed-by-end-capping strategy for 312
rotaxanes 151, 160, 305, 347. see also organometallic rotaxanes
– N-acylation 315
– advantages 313
– aggregation 305
– dethreading reaction
– – first-order kinetic rate constants 317
– – thermodynamic parameters 317
– main chain, based on polymeric crowns 161–165
– side chain, based on pendent crowns 166–173
– synthesis via end-capping strategy 313
– synthesis via metathesis reactions 320
– using cyclodextrin (CD) as the ring molecule 206
– X-ray results 306
Ru-carbene complexes 313

s

scanning electron microscopy (SEM) 78
scanning tunneling microscopy (STM) 72
schizophyllan (SPG) 60

- advantages 61
- chemical modification 64
- denature/renature process 61
- denature/renature process of 61
- galactose-functionalized 64
- polymer backbone 63
- s-SPG 61
- stereochemistry 63
- structure 60
- SWNTs, noncovalent wrapping 62
secondary interactions, attractive and repulsive
- influence on association constant 7
selective nucleophilic ring-opening reaction 298
self-assembled systems 23
self-assembly processes 18
- A-A:B-B alternating copolymer formation by 89
self-assembly strategy 293
self-oscillating gel device 206
self-sorting organogel system
- advantages 60
septum-capped test 59
sergeant-and-soldiers principle 52
silicone-based polyrotaxane 223
single-walled carbon nanotubes (SWNTs) 61
- applications 119
- noncovalent wrapping, by schizophyllan (SPG) 62
- supramolecular hydrogel 119, 246
size exclusion chromatography (SEC) 71
slide-ring gels
- mechanical properties 213–215
- SAXS vs. SANS 212
- scattering studies 211–213
- stress–strain curve of 214
- synthesis scheme 221
- using QELS 213
slide-ring polymeric materials 207, 208
- design of materials
-- from intramolecular dynamics of polyrotaxanes 224–226
- diversification of main chain polymer 216–219
- organic–inorganic hybrid slide-ring materials 219–223
- pulley effect 208, 209
- synthesis 209–211
sliding graft copolymers 215, 216
small-angle neutron scattering 211, 212
sodium 1-adamantane carboxylate 247
sodium hypochlorite 185
sol–gel processing 220
sol–gel transition 205, 211, 216, 238

solid-state phase transition reaction 311
Solomon's knot 305
sonication-responsive systems 239–241
SPG/conjugated polymer complex 62
SPG-Lac/SWNT composite 65
SPG/PT complex 63
spin-spin relaxation time 197
SPR biosensor 131
SPs. See supramolecular polymers (SPs)
SS-introduced PEG-bisamine (SS–PEG) 200
stable free radical polymerization (SFRP) 179
π–π stacking interaction 237
star polymers 179
static light scattering (SLS) 71
- measurements 87
steady shear viscosity 113
stereoisomers 296
3-Sti-α-CDs
- supramolecular self-assembly 42
trans-stilbene 186
stilbene bis (β-CD) dimer 45
trans-stilbene moieties 46
stimuli 231
- chemicals 233
- electromagnetic waves 233, 234
- force 232
- light 233, 234
- pH 233
- pressure 232
- redox 234
- stress 232
- temperature 231, 232
- ultrasound 232
stimuli-responsive monomers 91
stimuli-responsive supramolecular polymer systems 231, 232, 235, 236
stimulus-responsive supramolecular polymers mechanism 47
p-sulfonatocalix[4]arene 255
supra-macromolecular chemistry 51–53
- macromolecular assemblies interactions 63–65
- macromolecules, and macromolecular assemblies interactions 65–67
- macromolecules interactions 60–63
- materials design, with hierarchy 54
- molecular assemblies interactions 58–60
- small molecules, and macromolecules interactions 53–56
- small molecules, and molecular assemblies interactions 56–58
supramolecular approach 59

supramolecular biomaterials, modular approach 22
supramolecular bundling approach
– conjugated polymers (CPs) alignment 55
supramolecular complexes
– formation, schematic illustration 44
– prepared by DNAs 129
– topological structures, by AFM 129, 130
supramolecular copolymers, formation 88
supramolecular dendrimers 136
– antibody dendrimers 136–139
– chemically modified IgG 136
– constructed by IgM 136
supramolecular formation, of antibodies 127–129
– with divalent antigens 131–133
– with multivalent antigens 130, 131
– with porphyrin dimers 133
supramolecular liquid crystalline polymer
– formation by hydrogen bonding 12
supramolecular nano-particles 24
supramolecular polymerization mechanisms. See polymerization mechanism
supramolecular polymers (SPs) 3, 156, 269. See also polymers
– characteristic 29
– characterization 71
– definition 10
– self-assembly 72
– structures 43, 44
supramolecular rubber
– based on hydrogen bonding 17
supramolecular thermoplastic elastomer 20
surface plasmon resonance (SPR) 128
synthetic polymers 3
– it-PMMA/st-PMMA 53
synthetic water-soluble polymers 196

t

tapping mode scanning force microscopy (TM-SFM) 80
telechelic oligomers 18
– hydrogen-bonding motifs to 18
telechelic organic polymers 220
telechelic polyacrylates 301
telechelic poly(ε-caprolactone) 22
telechelic supramolecular poly(ethylene-butylene) polymers
– rheological master curves and tensile testing 21
telechelic supramolecular polymers
– binding constant, influence 19
temperature-responsive systems 236–238

template-driven switching
– from linear to star-branched architectures 87
template effects 347
2,2,6,6-tetramethylpiperidine-1-oxyl radical (TEMPO) 209
thermal gravimetric analysis (TGA) 310
– data 310
thiol–disulfide interchange reaction 339
threading 187–192
– kinetic data for 189
threading equilibrium, to form pseudorotaxane 152
threading reaction 189, 356
time-resolved photoluminescence (PL) 72
topological polymer chemistry 293
– designing unusual polymer rings by 298–302
– future perspectives 302
– nonlinear, systematic classification 293–296
– topological isomerism 296–298
transesterification reaction 225
transition metal templated strategy 348
transmission electron microscopy (TEM) 40, 55, 61, 64, 66, 119
transmittance process 116
trefoil knot 350, 351
– x-ray structure 350
trialkylammonium 180
triblock supramolecular copolymers 13

u

ultra-sensitive chemosensors
– fluorescence signal amplification 52
2-ureido-4[1H]-pyrimidinone (UPy)
– functionalized peptides 22
– – cell adhesion peptides 22
– motif 24
– quadruple hydrogen-bonding DDAA motif 8
– UPy-U nanofibers 21
– UPy-urea model 22
2-ureido-4[1H]-pyrimidinone dimer 9
ureido-pyrimidinone (UPy)
– motif 20
– quadruple hydrogen-bonding motif 4
2-ureido-pyrimidinone motif
– tautomeric equilibria in 9
UV-sensitive azobenzene moieties 103
UV-vis spectra 319
UV-vis wavelength region 61

v

van der Waals interactions 151, 195
vapor pressure osmometry (VPO) analysis 78
velcrand
– bent and linear dimers, metal-directed assembly 81
– dimers, chain and ring formation 81
– mixing 82
– self-assembled nanostructure based 80
1-vinyl-3-butylimidazolium bis(trifluoromethylsulfonyl)imide ([vbim][Tf2N]) 119
viologen-appended polymer 188
viologen dimer–antibody complex 134, 136
viologen dimer molecule 133
viologen (N,N'-dialkyl-4,4'-bipyridinium) trap 188

viscoelastic liquid 17
viscometry 253
viscosity 272, 273

w

water-soluble CDs 206
wide-angle X-ray diffraction (WAXD) 216
Williamson's ether synthesis 308

x

X-ray analysis 43, 79, 85, 90
X-ray crystallography 29, 307
– variable-temperature 310
X-ray diffraction (XRD) 117, 351
X-ray photoelectron spectroscopy (XPS) 199